Connections

Proceedings

of the

36th Annual Conference

of the

American Translators Association

November 8-12, 1995
Nashville, Tennessee

Edited by Peter W. Krawutschke

Information Today, Inc.
Medford, NJ

Published for the American Translators Association
by Information Today, Inc.

Compiled by
Carol Nixon

Manufactured in the
United States of America

ISBN: 1-57387-023-4
Price: $50.00
Order from: Information Today, Inc.
143 Old Marlton Pike
Medford, NJ 08055
USA

Cover Art Direction: Keith Jones
Cover Design: Patrick C. Ryan
The Design Center, Department of Art
Western Michigan University

The papers published in this volume are in the format submitted by the authors. No editorial intervention was undertaken by the Editor unless absolutely necessary and only if the production schedule allowed. The Editor's task was to solicit contributions, to arrange them thematically and sequentially, and to assist in other editorial matters.

PREFACE

The 36th Annual Conference of the American Translators Association will again present a surfeit of information on translation and interpretation while at the same time exhibiting the most advanced equipment, supplies, and services. Some of this information is contained in these Proceedings of our conference.

After reading a number of our past Proceedings, and a few of our recent conference programs as well as after talking with me about the work we do, the young graphic designer who created the cover for these Proceedings concluded that we don't merely translate words but cultures; e.g., even in the global age of the computer, we "localize."

I have found it always profitable to listen when "outsiders" write or talk about us, and it is a positive sign that, as our ATA membership continues to grow, there is a renewed interest in looking at ourselves and how we are "connected" to our national and global communities.

It is my hope, therefore, that future ATA Proceedings continue to document our concern for the social dimension of translation and interpretation, particularly in the medical, legal, and educational environments.

In a time of change, the association and the conference organizer who also edits these Proceedings enjoyed the leadership of Edith Losa, whose sense of fairness and humor bridged many a difficult moment. At the same time, we found in Walter Bacak an executive director who was instrumental in gradually and seamlessly moving the organization of our conferences from the dependency on the good will of corporations or educational institutions to ATA headquarters.

To those who recorded their knowledge and experience on paper for these Proceedings, we and later readers owe our gratitude. By doing so, these authors have created for translation and interpretation a presence on the shelves and in the catalogues of libraries here and abroad. It is this type of dedication to our profession which permeates the remarks made by Harry Obst at last year's conference banquet, and they serve well as an introduction to these Proceedings.

Peter W. Krawutschke
ATA President-Elect

CONTENTS

Introduction

Agencies, Bureaus, and Companies

Freelance Translating

French

General Interest

German

Interpretation

Literary Translation

Portuguese

Public Advocacy, Legal and Regulatory Affairs

Spanish

Terminology

Translation and Computers

Translation Theory

Translator and Interpreter Education

Appendix

Connections

Introduction

TRANSLATORS AND INTERPRETERS IN AMERICA: THE ROAD TO RESPECTABILITY

Harry Obst
Office of Language Services, U.S. Department of State

Everybody wants to be respected, even Rodney Dangerfield. Rodney, the comedian, has two major obstacles to overcome. He does not look like he deserves respect, and he often does not act in a manner which would earn him respect. To a large degree this also holds true of the professions of translation and interpretation in the United States of America.

Here, neither profession is traditionally looked upon as a bona fide profession such as lawyers, architects or ornithologists. Many Americans would not know what an ornithologist is, but the mere sound of the word will convince them that it is a profession. Maybe the fastest road to respectability would be to just call ourselves "translatologists." It certainly has the ring of dignity and big bucks. When you say, "My translatologist charged me 150 bucks an hour," it somehow sounds acceptable. When you say, "My translator charged me 40 bucks an hour," it sounds outrageous.

Even this faint attempt to be humorous indicates that — on the road to respectability — the actual current status of the profession may not matter as much as the general public's perception of the profession. Unfortunately, that perception is bad for a whole number of reasons.

First of all, the profession started off on the wrong foot on this continent. The American natives who dominated this vast land prior to the arrival of the European settlers spoke different languages but rarely relied on written documents. Each tribe had a medicine man and a council of elders but lacked lawyers, written land deeds, and bodies of written codes of criminal, civil, or religious conduct. So there was little need for translation. With hunting, fishing, and fighting other tribes being the preoccupation of many, what interpretation was done consisted mainly of interrogating prisoners, obtaining hunting and travel information, and negotiating terms of surrender.

When the settlers arrived and later the armies to protect them or help expand their territorial holdings, most communication needs, again, were similar and dealt with the native peoples and the Indian tribes who joined or opposed the Spanish, the French, or the British in the decades of wars and skirmishes which preceded the eventual division of the North American continent into Canada, Mexico, and the ever-expanding United States of America.

The few professionally trained translators and interpreters who had arrived with the educated colonists remained confined to the original 13 states and found work in cities like Boston, Philadelphia, New York, and later Washington.

Once the original United States was formed, the next hundred years or so consisted of driving its borders relentlessly westward and into the Southwest while moving most of the surviving Indian tribes off in the same directions. Most of the new breed of settlers arriving in search of gold, silver, furs, timber, and cheap land were not educated people. The West was won by the gun, the saw, and the plough, but not by the pen.

Again, the major communication needs were mostly the same: oral interpretation of information on travel logistics, hunting, farming, and trading, and for the Cavalry, the interrogation of prisoners and negotiating with tribal chiefs who could not produce written land deeds for translation. Written translations were a rare occurrence in the West. Ironically, in our time, more written translations are done each day in California than in New York and Pennsylvania combined.

Yet the traditions of the Old West are still very much with us, and with it the image that language work needs not to be done by professionals. In the federal government, this notion is still prevalent in the agencies which deal with immigration questions, the protection of the borders, illegal drugs, counterfeit money, and firearms. The agents and law enforcement officers working at the front, notably at the border with Mexico, or dealing directly with Haitian or Cuban refugees consider familiarity with the other language to be the key criterion for the function of interpreting and translating, not professional training. Most of the more than 200 "interpreters" for Haitian Creole recently employed by various agencies in connection with the processing of refugees never had any professional instruction in interpreting or translating.

Their work at the front line, however, gives them visibility. The agents who worked with them will go back to Washington, Denver, or Dallas with a certain image of what an interpreter is. They will never buy the notion that interpreting is a profession until they have an equally long exposure to professional interpreters or translators, and most of them will not get that exposure.

So here then is our first big obstacle on the road to respectability: overcoming the negative image of the past plus echoes of that image still emanating from the present. But let us move on to a much larger obstacle.

Professionals are trained at universities or at equivalent schools of higher education. If you want to become a professional translator or interpreter in Europe, you have a wide array of university programs from which to choose. You can go to Florence, Geneva, Paris, Vienna, Heidelberg, Copenhagen, Moscow, or dozens of other reputable programs. In neighboring Canada, you have a number of adequate choices. Not so in the United States of America. When it comes to recognizing the national need for university training programs for professional translators and interpreters, most of academia in the United States is as blind as a bat and firmly embedded in obsolete traditions and unrealistic views. The dozen or so good translation programs which exist in the 50 states and the two or three interpreting programs of any consequence are very much the exception which proves the rule. They cover only a fraction of the national need.

Two years ago, I was told that a certain university had developed an excellent foreign language program, with a very large enrollment. When I contacted the university to check this out, I received an invitation to come and lecture on professional careers in interpretation and translation. During my visit there, I found an excellent mix of languages and a high-quality teaching program for foreign languages supplemented by state-of-the-art technology. Yet not one of the 14 professors I had a working lunch with had a grasp of what translating and interpreting as a working profession was all about. None of them had ever mentioned interpreting to their students as a possible career choice, and only two had mentioned the translation of works of literature as a possible option. They were surprised by my statement that in the marketplace technical translators outnumber literary translators by at least ten to one and that the average income of technical translators exceeds that of their colleagues in the other discipline. The university career center contained no literature on university training programs for translators and interpreters. The number of foreign language students at that university, coupled with its sister university in the same state numbers several thousand, enough to populate its own training program for translators and interpreters, if it had one, without ever having to recruit anywhere else.

At the universities and colleges where such programs exist, they are often hidden in dark corners of the hierarchical structure, poorly funded and poorly staffed. Just ask the incoming president of the ATA. His program in Michigan is desperately struggling for survival and recognition like so many others.

Which planet do American university administrators and chairs of departments of languages and linguistics live on? It cannot be the same planet where the universities of Belgium, Canada, France, Germany, Russia, Switzerland, and dozens of other countries are located who supply their societies with a marvelous crop of high-quality translators and interpreters each year who form that golden bridge across which correct information and billions of dollars in exports flow to other cultures and economies.

The United States does not need any more theoretical linguists. The unemployment lines are full of them and many thousands more fill other jobs in the language field where their university training is more of a hindrance than a help.

In the meantime, the enormous vacuum created by the absence of sufficient training opportunities for young Americans who have the language knowledge and the aptitude to become respected professional translators and interpreters is filled by domestic amateurs and by foreign professionals who are happy to take our corporate dollars back home to Tokyo and Amsterdam. The overwhelming majority of English-Japanese conference interpreters operating in the United States are citizens of Japan or other nations. We are talking here about language professionals many of whom earn six-figure incomes. And that holds true for many professional translators who can handle Japanese, Korean, Russian, or other currently lucrative languages. How much longer can academic decision makers sit on their hands and deny this income to talented young Americans?

Thus academia in the United States is not only a large hurdle on the road to respectability for our profession, it has through its negligence created another hurdle: the tens of thousands of improperly trained translators and interpreters in the United States who in many places overshadow and underbid the true professionals, thus contributing strongly to the negative public perception of the nature of the two professions.

What other obstacles are there to overcome?

An obvious one is the traditional myopia in the American business community on the usefulness of foreign languages, translating and interpreting as tools to create business and open foreign markets. The Dutch, the Swiss, the French, the Germans, and the Japanese, to name but a few, have cleverly used this device to great advantage in vaulting into the top ranks of the world's exporters. It is a fallacy to conclude that because English has become the leading international language of business, we need to pay little attention to this area.

Even today, there are more businessmen and traders in the world who do not speak English than there are those who do. Neglecting to address them in their own language makes no good business sense. Just because something seems plausible or logical on first examination, does not mean that it is.

Let me give you an example. When I had signed on as a salesman with a Washington office machine company many years ago, I was given a territory that was not exactly ideal. The lucrative territories remained with the established salesmen.

I quickly inventoried the few large-sized companies in my territory. One was the Washington headquarters of IBM. I asked our sales manager if anybody had ever called on them. He laughed, "You want to sell business machines to IBM? They are the largest maker of business machines in this country." This seemed plausible and a logical reason not to call on IBM, but to my juvenile mind it was not necessarily relevant. So I made an appointment with the IBM purchasing agent. He was delighted to see me because no office machine salesman had called on him in years. I walked out of my first appointment with a sizeable order for adding machines and printing calculators, neither of which IBM made at the time. Within a few months IBM became my largest account in the entire territory.

Ergo, what seems obvious and logical at first blush, is not necessarily so. The notion that American business can hold its own against tougher and tougher world competition without investing in the use of foreign languages and the professional experts who operate in this field has no practical relevance or merit whatsoever.

Now here is finally some good news. This particular obstacle on our road to respectability should not be around much longer. American business is finally warming up to the idea that the knowledge of foreign languages and business cultures and the utilization of reliable translators and interpreters makes good business sense. We are not out of the tunnel yet. But we can clearly see the light at the end of it.

Once this notion is firmly established in the business world, business leaders will have an incentive to hold the feet of academia's decision makers to the fire when they see their new concept threatened through the lack of training facilities. On the day that corporate grants and scholarships will be as easy to obtain for university programs of translation and interpretation as they are now for MBA programs, academia will take off its blinders and get out of the rocking chair of comfortable but largely useless traditions in the language field. If nothing else seems to be able to move them out of their self-imposed lethargy, believe me, American business will. Virtually all new initiatives of any consequence in this country come out of the private sector.

In the last two decades, our professions have looked to government and academia for solutions, beginning with Jimmy Carter's President's Commission on Foreign Language and International Studies of 1978, the ensuing establishment of miniscule "National Foreign Language Centers" at prestigious universities, and
ending with the language initiatives contained in the recent National Security Education Act.

Some marginal relief for foreign language training has come from all of this, but the dismal state of the professions of translation and interpretation has not improved through these measures. They are no more than lollipops for a sick child in need of strong medicine. And when Senator Paul Simon's proposal for a scholarship grant program for American students who want to become professional translators and interpreters had successfully woven its way through the committees in the Senate and House earlier this year, it died a sudden death in conference committee because of financial anemia and the unwillingness of some in the Department of State to get into the scholarship business.

So let us not fix our eyes so much on government and the American academic institutions because their track record in our field has not been good. Let us take advantage instead of the new curiosity in the business community about what our professions may have to offer. If the promise and the opportunities were not there, do you believe for one moment that AT&T would have jumped into the interpreting and translating business with both feet?

If we can get the corporate community to ride shotgun on our professional wagon, it will clear most of the other boulders from our road to respectability. In this area it should be easy for the ATA, as a strong and respected professional association to become more active. After all so many of you are independent businesswomen and businessmen. The only difference is the annual turnover. You speak the same language already. Call on large companies as I did in my youthful innocence. Get them interested in what our professions can offer them. Organize an occasional symposium for business leaders or whatever it takes to get their attention.

Let us not be like academia and do the same dumb old things over and over again. Agree on new approaches and then stand united behind your new approach and get involved. There has been too much bickering and infighting in some of our professional associations. To outside observers, this makes us look like Rodney Dangerfield at his worst. It makes no sense to be fighting over crumbs on the inside when there are scores of uneaten pies waiting to be uncrated on the outside. Internal division and lack of cooperation between different associations is not moving us down the road to respectability. Nor will we make progress by meeting time and time again just to cry on each other's shoulders about the sad state of affairs. Your approach to your association should be, to paraphrase JFK, "Ask not what your association can do for you. Ask how you can help your association to move the profession forward." If you are not willing to do that, you are part of the problem, not part of the solution. And if you remain part of the problem, you forfeit your moral right to complain.

Another problem our professions face in the United States is that we have never been successfully counted. In a free country there is a lot of strength in factually supported numbers, particularly at election time. Nobody, not even our intelligence agencies, knows the approximate total number of translators and interpreters residing in the United States. All feeble attempts made in the past, for instance, by Princeton University, do not even have an accuracy rate of plus or minus 100 percent. My personal feeling, after 30 years in both professions in this country, tells me that the actual number is probably considerably higher than anybody believes. Time and time again, in my travels I happen upon large groups of translators or interpreters of whom I had never heard before. Yet the facts to support this feeling are nowhere to be found, nor will they be easy to come by. What a feat it would be to get this question put on the census form for the year 2000! But somebody needs to develop a strong initiative to get this accomplished.

Whatever the number, this country has many of the finest translators and interpreters the world has ever seen. Our top professionals are highly respected in other countries. There are people sitting in this room who have enriched the world's libraries with superb translations of original works that require the strongest analytical, linguistic, and creative powers from the translator. In 1987, King Hassan of Morocco sent his best interpreter to Washington for master-level training — not to Geneva or Paris. When a peace conference on the Middle East convened in Madrid in 1991, it called for a mix of languages which had never existed at a major conference before. The then still existing Soviets and the Spanish, who were trying to organize the required language support, finally threw up their hands in despair, stating this just could not be done. Secretary of State James Baker telephoned his chief interpreter in Washington from Madrid and called in the American language cavalry. It was not easy, but the job got done. Baker stated to Stephanie van Reigersberg after the completion of the conference, "You performed miracles!", and he was almost right.

The annual conferences of the American Translators Association, too, are usually very well organized and present a marvelous forum for our professions. This year's meeting was a good example. The attempt to widen the reach of the association to include more professional interpreters and topics on interpretation is also a positive step in the direction of unity among two professions which require the same wide range of linguistic qualifications, intellectual acumen, and knowledge of diverse cultures and divergent fields of human endeavor. The difference is mostly technical, one delivering the product instantly and without recourse to dictionaries and reference aids, the other working with deliberate care and greater accuracy and with the ability to edit out mistakes before they become public.

The interpreter enjoys the ephemeral moment of glory while he or she is allowed on the stage. But after the performance is over, the interpreter and his work are quickly forgotten while the translator's work may be pulled from a library shelf 500 years later. Translators and interpreters are just different practitioners of the same basic profession. They have much in common and little which divides them. Thus it makes great sense for them to work together in harmony.

Translators and interpreters may not yet have the American public's respect. But they certainly deserve it. It is time for all of us to make a greater effort to clear the debris from the road to respectability so we can finally get from here to there. The most promising approach may well be to concentrate the largest effort on the American business community. It furthers the self-interest of American business to work with us and to support us. Self-interest is a powerful force. But first, we must carry the message from this community to the other.

Connections

Agencies, Bureaus and Companies

ISO 9000—CRUCIAL FOR TRANSLATION QUALITY

Linda M. Collins
AT&T Business Translations

Keywords: ISO 9000, Translation Quality, Quality Systems

Abstract: The ISO 9000 requirements makeup a standardized comprehensive quality management system that is commonly used within manufacturing processes. However, the requirements are written in general terminology and can be applied to nearly any type of process. AT&T Business Translations used the ISO 9002 standard to build a formal, documented quality system for their translation business. Several of the ISO elements are discussed with example of how they are applied to a translation process.

1. ISO 9000

What is ISO 9000? How does it apply to the translation industry? What are the benefits to implementing the ISO requirements?

ISO stands for International Organization for Standardization. The ISO 9000 requirements are made up of twenty "elements". These elements makeup a standardized comprehensive quality management system that can be applied to any type of process. The implementation of ISO 9000 quality systems has increased significantly over the last several years. Within the USA primarily manufacturing industries have embraced the standard. However, AT&T Business Translations (BTr) has been certified to the ISO 9002 standard, noting only a few of the elements as not applicable.

ISO 9000 is the consolidation of basic quality system elements developed over time in a variety of industries. By standardizing the elements and requirements of a quality system interrelated industries can share quality terminology and attributes. This common foundation enables customers and suppliers to define quality interfaces more efficiently. The simple statement, "We are ISO certified", allows for an instantaneous shared knowledge of the quality system.

ISO 9000 makes no attempt to judge the quality of a product. Product quality is determined by the product or service provider. The ISO 9000 standards are focused on ensuring a consistent quality process. Thus a "luxury line" and a "budget line" of a similar product can be produced from equally solid but different ISO-certified processes.

It has been an intriguing challenge to develop a formal documented quality system for a translation business around the ISO 9000 standard.

2. TRANSLATION PROCESS

Individuals involved in the translation industry understand that a translation process is made up of many tasks besides the actual conversion of text from one language to another. Initially, requirements for a job must be determined. Then the process involves separate tasks such as analyzing the material to be translated, determining special subject matter terminology, reviewing terminology, translating and editing, formatting the translated text, and publishing the final product. Project management is required to monitor and guide the process. A variety of resources and skills are required to implement this process and ensure a consistent high-quality output.

Companies who specialize in translation services are able to support all of these tasks in a turnkey operation. However, even individuals doing freelance jobs can use the ISO elements to organize and formalize an individualized quality system.

3. ISO 9000 ELEMENTS AND THE TRANSLATION PROCESS

Of the twenty elements within ISO 9000, four do not currently apply to the AT&T Business Translations process. One of these four, design control, applies only to the 9001 standard. Translations typically involves converting material that has been previously designed into another language preserving the design of the source material. On projects where the design of the target language does not necessarily parallel the source design, the customer typically specifies the design of the target language as well. AT&T Business Translations is certified against the 9002 standard which excludes the design element.

The other three elements not covered in the BTr quality system are control of inspection, measuring and test equipment, servicing, and statistical techniques. A translation process simply does not use measuring and test equipment such as calipers and oscilloscopes. Servicing also did not apply since translation support for a product after the translated deliverable is shipped is covered either by the corrective action element or re-enters the process to be updated as a "new" project with reusable material. Statistical techniques was excluded because a formal statistical analysis has not yet been implemented within the BTr process. The sampling techniques however are evolving towards a more formal statistical technique and it is likely that this element will be included as the quality system matures. Evidence that selected elements do not apply must be demonstrated to the ISO registrar. A company is not allowed to pick and choose the elements they wish to comply to and simply ignore the rest.

Several of the remaining sixteen elements have been chosen to illustrate how they have been applied to the translation process.

3.1 Management Responsibility

ISO requires that managers take responsibility for documenting a quality policy, defining an organizational structure, and identifying the resources required to support the translation process. A manager must be appointed and given the authority to establish, implement, and maintain the quality system. At defined intervals managers must review the quality system to determine if it is current and effective.

The requirements under this element lay the groundwork for the entire system. It requires the management team to discuss, tighten up, reach consensus, and formally commit to a documented quality system. There are many ways to implement the details of a translation process and as a company grows it is easy for different groups to diverge down different paths. Documenting one quality system helps set a common system for everyone to follow. A full-time quality manager position was created where previously no one had official and dedicated responsibility for quality initiatives. While sustaining a quality system requires the dedication and commitment from every manager, dedicated quality positions provide the lead and coordinate initiatives across the business.

3.2 Contract Review

The contract review requirements specify that contracts must define and document job/project requirements. There must be evidence of agreement on requirements, and changes to requirements, between the customer and the supplier.

For a translation project this typically means that the language, terminology, format, type of source, delivery schedule, and cost are all clearly defined and agreed upon by Business Translations and the customer. While this may seem like common sense, it was not uncommon in "pre-ISO days" for most projects to have been agreed to in general terms via phone calls and meetings with little to no written evidence of agreements reached and no signatures as evidence of consensus on requirements. The ISO requirements serve as a vital motivator to constantly refine our contract procedures and adhere to a contract process that adequately documents job requirements. The customer benefits from the need to work out many project details prior to the start of a project which leaves less room for complications, surprises, and delays during the translation process.

3.3 Purchasing

The requirements under the purchasing cover any type of material or service that is necessary to execute a process. In general all purchased product or service must demonstrate the ability to meet the quality standards of the purchasing organization

A translation process purchases services rather than materials. Typically these services include subcontracted translators, or desktop publishing and printing. However, it can also cover services provided by organizations within your own company but outside the direct influence of the quality system. For example, BTr is located in a building with other AT&T organizations that share common resources such as the shipping department, and a graphics department. The criteria for selecting and evaluating those services must be specified and records kept of ongoing evaluations. For internal organizations or suppliers without competitors there must still be a mechanism to ensure they will meet the expected quality level.

While the requirements for selecting individual translators were fairly well defined before ISO, other subcontracted service providers were selected on a more casual basis. Ongoing evaluations were rarely done on a formal basis. ISO has helped BTr refine how subcontracted service providers are selected, but more importantly, how their actual performance is evaluated on an ongoing basis. Even the internal service providers are evaluated twice a year and given a "report card". Subcontractors who do not perform according to the specified criteria are either place on probation or removed from the approved vendor list entirely. ISO requires that your business assume full responsibility for the work of all subcontractors.

3.4 Process Control

Process control requirements state that a process must be carried out under controlled conditions. Controlled conditions means that procedures are documented, changes to procedures are reviewed, updates to procedures are consistently available and obsolete versions of procedures are immediately destroyed. These requirements are intended to ensure that the process is consistently executed over time and a variety of individuals. A well documented system adapts easily to high growth or high turnover periods since the tasks are clearly defined and documented.

Within the translation process controlled conditions are ensured by providing written instructions for every task. An initial reaction may very well be that one can not tell a translator how to translate. It is a highly developed skill that professionals spend years perfecting and enhancing. This is certainly true. However, all the tasks and procedures surrounding the actual translation can be more precisely specified. Procedures have been developed for project management, directory and file structures, preconditioning of material prior to a machine translation, marking of files to indicate process status, and transmittal of material between locations and subcontractors. Forms and checklists used within the process are also controlled and referenced from work instructions. Even within the actual translations task there are instructions for proper glossaries to use, and quality criteria.

This element proved to be the most time consuming element to implement due to the fact that few procedures had been formalized, documented, reviewed and controlled. It also requires an extraordinary effort to keep the procedures up to date as our process grows and evolves to keep pace with a wide variety of materials requiring translation. However, the procedures prove invaluable when training new employees or expanding the skill set of existing staff.

3.5 Inspection and Testing

There is a wide variety of requirements for inspection and testing from receipt of incoming product to final inspections. The methods of inspections and testing for each phase of the process must be documented as part of the quality system.

The emphasis for translation is on the inspection of translated material and the final production format, rather than testing. Inspections have been integrated into the translation process as it moves through the individual tasks. The documented procedures state expected inputs and outputs for each task. Actions to take if the material received into a task does not meet expected standards are specified. The translation material being processed is not to leave a task until nonconformities from a previous task have been corrected and the current task is completed as specified. Records are made of significant deviations in quality from a previous task in order to provide feedback and document the need for potential process improvements. For example, if sections of untranslated text are found during final formatting of the files, the text is sent back to the translation task. This nonconformance is described in what we call a "history file". Evidence of successfully completed inspections are recorded on project tracking forms.

3.6 Corrective and Preventive Action

Corrective and preventive action requirements specify that there must be procedures for handling nonconformities or errors in the process or product. These problems may be identified internally as jobs move through the process, or by the customer as feedback, or by various audit activities. The corrective and preventive action procedures must include steps to identify and correct the root cause of the nonconformity rather than just fix the error and move on.

The basic elements of an effective corrective action process remain constant for a translation process. The challenge is to ensure that problems are traced back to a solid root cause and not simply corrected and forgotten. An error in terminology must be corrected in referenced glossaries not simply corrected in the text. Analysis may determine that further investigation is needed to determine why the glossary was originally in error. Errors in document formatting may be traced back to inadequate training or experience of a desk top publisher. While true root cause analysis may seem to take a great deal of time it is required to ensure that the problems will not reoccur on future projects. Delivering a consistent quality translation does not mean delivering the same errors time and time again.

In order to meet the requirements of preventive action, the translation process must contain methods to identify potential causes of nonconformities. Implementation of the previously mentioned history files is one mechanism. Minor problems that may appear insignificant when viewed in isolation can raise in importance when tracked over a period of time. For example, errors made by a translator and corrected by an editor may not seem particularly problematic. But if tracked over time it can become apparent that the translator needs further training in proper punctuation or other translation techniques.

Other preventive action methods may include quality improvement teams. These are small teams of people within a specific area of the translation process such as sales and operations interfaces, text processing, and translators who monitor the effectiveness of current procedures and discuss process improvement ideas.

3.7 Internal Audits

Since a quality system is only as good as its actual implementation, ISO 9000 requires formal procedures for monitoring whether the documented quality system is being followed.

Audit activities are a regular part of many industries, however they are quite rare for a translation industry. The BTr quality system requires that the process be fully audited against all applicable ISO elements twice a year. Trained auditors that are not a part of the translation process conduct these audits. Any part of the translation process that is not being implemented according to the documented quality system and translation procedures is written up and requires corrective action. Areas of the quality system that have had previous problems are checked with special attention. Internal auditors have proven a useful source for independent objective feedback on the status of the quality system, advice on implementing new ideas, and suggestions for addressing subtle quality concerns. Corrective actions written as the result of internal audits are investigated by the external ISO 9000 auditors for timely resolution and effective solutions. Therefore, internal audits provide incentive to sustain the documented quality system.

4. CONCLUSION

Many of the ISO 9000 requirements seem like common sense -- and indeed they are. However, almost every business has experienced the difficulty of sustaining quality systems when the pressure of meeting tight schedules strains process resources. ISO 9000 provides the framework to design a well-disciplined translation process that can provide customers with a consistent translation product. ISO 9000 also ensures that our translation process will continue to evolve and improve. Continuous reviews and audits of the quality system reinforce the common-sense procedures required to meet the quality expectations of our customers.

COPING WITH PMS (PROJECT MANAGEMENT SYNDROME): SURVIVAL TECHNIQUES FOR MONTH TO MONTH RELIEF

Krista M. Conley, Lucy Allan, and Kirk Anderson
Cambridge Translation Resources

Keywords: Project Management, Quality Translation.

Abstract: Project management can be stressful, tense and highly emotional, but it doesn't have to be. As the translation industry matures, agencies and translators need to improve and deepen their commitment to professional project management. Keeping in mind the two critical components of translation delivery, quality *and* speed, this session will discuss three difficult project management situations and invite the audience to develop solutions. The goal is to understand the agency's needs and obligations, as well as the translators' point of view, and to find meaningful and practical solutions to reduce conflict. At the conclusion of the discussion period, the panelists will invite input from participants to outline ideal project management models.

1. INTRODUCTION

PMS can be a drag. No matter what those letters represent. Managing translation projects can create enormous stress on all parties involved: project managers, translators and clients, producing all kinds of nasty symptoms which we'll outline below. Much of this stress can be minimized or even eliminated, if you take the right medicine.

2. THE PERPETUAL CYCLE: QUALITY AND SPEED

You know the drill. Most clients want perfect translation delivered yesterday. But what make the translation perfect—and who is the ultimate judge of that, and doesn't speed necessarily sacrifice quality? Certainly, there are limits to how quickly a quality translation can be produced. To guarantee the quality of every translation job, quality control measures must be followed for every step of the project. To most clients, speed is part of the service. To most translators, speed is quality's enemy. For the people in the middle, the project managers, finding an acceptable middle ground requires preparedness and a sound quality control plan.

At this point, the discussants will ask the audience to give examples of urgent translation requests from their own experience.

3. PROJECT MANAGEMENT SYNDROME: SYMPTOMS

Headaches, cramps, bloating, rapid mood swings. All these painful physical manifestations can be averted if you are prepared. What's the trick? Articulating expectations at the start of each and every project. Everyone involved in a translation project must understand and agree to their responsibilities at the start of the project. Clear guidelines and a project schedule are essential components for a pain-free job.

Once you understand your responsibilities and your delivery schedule, you must also understand flexibility. Projects do evolve, and creating an environment that is responsive to change is a critical component for the emotional and physical health of both project managers and translators.

4. THREE EXAMPLES OF PMS

4.1 Example 1 - *The Headache.* Headaches afflict all of us from time-to-time. There are various immediate remedies, but if you find yourself experiencing bad headaches regularly, you may need to look more deeply into the cause, and treat the root problem!

For project managers the most difficult headache to treat is the persistent headache of a disorganized client: the client who calls you with each individual question at half hour intervals throughout the day, who sends you the project in incomplete sections, who refuses to understand your quality process, and who never seems to read your faxes! A disorganized client can be the most debilitating factor of PMS; it can leave you frustrated and feeling low.

The most effective remedy to avert headaches is Client Education. The more you inform a client about the translation process before the start of a project, the less likely it is that a headache will strike. As every translator, and every project manager knows, there are multiple issues to bear in mind, and many questions that need to be addressed with the client *before* a translation project gets underway. Each piece of information that an agency can give the client before the start a project acts as an Advil/Tylenol—be careful not to administer more than the recommended dose, but client education sheets can give effective relief for your PMS headaches—*fast*!

The discussants will then ask the audience to give examples of project information they feel is vital for completing their assignments.

4.2 Example 2 - *The Mood Swing.* Do you ever get the feeling that your translation project is getting bloated out of proportion? No matter what you do to control the project, some new element seems to crop up to stretch the limits of your multi-tasking capabilities? Translation projects all evolve and some larger projects even move in cycles, going into hibernation for a while, only to awake suddenly, hungry for new deadline demands. The project manager must be comfortable with change, and be prepared for the mood swings. If they are not, clients won't be satisfied, projects won't be completed correctly, and sooner or later, your business will grind to a halt.

A common problem is that project managers often consider change as a problem. This leaves them prone to irritability, and on occasion, tearful outbursts. There is no doubt that change creates more work for the project manager, but to keep pace with change, you need to stick to your quality control system and keep track of your additional costs. Be sure to communicate with your client that change costs money—and be prepared to itemize the costs to help them understand the process. Maybe if they understand that changes mean money, they will be better prepared to give you final instructions *before* you start the project.

The key is to regard change as an adventure, a challenge, and use the energy of the change to give momentum to your project. Mood swings are a natural, and inevitable part of your job as a project manager. They also occur with relentless regularity—you *know* it's going to happen again next month, so be prepared!

The discussants will then ask the audience to give examples of shifting project demands that can frustrate or confuse them.

4.3 Example 3 - *The Cramp.* Picture the scene: you are sitting at your desk, sun streaming through the window, savoring your first coffee of the day, feeling content in the knowledge that a large and complex project is going smoothly, when suddenly it hits you... the cramp! Out of the blue, you receive a phone call from a translator who is unable to deliver on time, leaving you in great discomfort, your good mood evaporating.

The problem with the cramp of late delivery is that, on a large multi-service project, it usually sets off a series of subsequent cramps—editors/typesetters working on the project further down the line have to be informed, and their time often cut so the project manager can fulfill the client's deadline. Delivering late—and hearing all sorts of original excuses—really cramps the project manager's style. They can't give excuses to their client; they must find a way to complete the project on time and on-budget.

Of course people get sick, and emergency situations arise, but sometimes the problem is based in the fact that translators are unable to "just say no." Taking on too much work is, of course, financially tempting, and less-experienced translators don't want to turn down new business from a new source. However, you need to be realistic—know what your real turnover is (including proofreading and checking time), and don't overstretch yourself. The unpleasant cramping feeling of an unnecessary late delivery is far more damaging to your relationship with a translation agency, than turning down work. In fact, we generally view it as a *good* sign that someone is so much in demand that they have to turn projects away!

The discussants will then ask the audience to give examples of troublesome delivery schedules or time crunches with multiple language projects that have affected their personal project management.

5. THE MOTRIN/ADVIL OF THE TRANSLATION INDUSTRY: PROJECT MANAGEMENT MODELS

Working together with the audience, CTR staff will invite opinions to create a quality project management model.

TRANSLATOR SELF-TRAINING: THE VIEW FROM THE TRANSLATION AGENCY

Mordecai Schreiber
Schreiber Translations, Inc.

Keywords: Education, Training, Translation Career, Freelancing, Dictionaries, Computers, Translation Certification, Accreditation

Abstract: Of all the academic disciplines and all the professions in the United States requiring at least college-level education, translation is by far one of the most neglected and underdeveloped. While the craft of translation is critical to world economy, world culture and international relations, and while the U.S. is the leader of the post-Cold War world, little has been done by government agencies, academic bodies or private entities to cultivate this vital profession. This state of affairs prompted Schreiber Translations, Inc. (STI) in Rockville, Maryland to initiate a correspondence course for translators in 1992. In early 1995, STI started a new division called Schreiber Publishing, with the express purpose of providing self-training and reference materials for translators. The new line of publications is designed to serve a dual purpose: help aspiring translators break into the field of professional technical translation, and provide working translators with a set of references designed to enhance their translation work. This paper discusses the first version of the ***Translator Self-Training Program,*** namely, Spanish/English, and the specific problems of translating English into Spanish in the United States. It also alludes to the other current versions, namely, French, German, Russian and Japanese, and the ones planned for next year, viz., Chinese, Italian, Portuguese, Arabic and Hebrew. In addition, the paper discusses practical issues of particular interest to freelance translators, such as translator-agency and translator-client relations.

1. TRANSLATION IN THE POST-COLD WAR WORLD

Two years ago at the 34th Annual Conference of the American Translators Association in Philadelphia, I gave a paper on translator/interpreter training and continuing education. As someone who spends practically every waking moment on the forefront of the translation scene of both U.S. Government agencies and public or private companies, firms and organizations, during the past two years I have witnessed a radical change and transformation of the translator's world. Global realities are creating before our very eyes a world where languages are interacting more closely, more frequently, in more critical areas of business, science, technology and international affairs than ever before in recorded history, and, in the midst of it all, the role of the translator is becoming more critical, more complex, and, yes, more challenging, exciting and profitable than ever before.

The upshot of all of this is twofold: first, we need more translators. And second, in order to become more competitive, we the practitioners of the translation craft need to continue sharpening our skills and learning new ones, familiarize ourselves with the latest computerware, both hard and soft, and gain a better understanding of the realities of our craft in the global village.

It is no secret to those of us who have had anything to do with translation for any length of time, that ours is one of the most neglected, misunderstood, and looked down upon occupations of all the ones that require at least college-level education. This is true in many parts of the world, but it is particularly true here in the United States. Even though we live in the richest, most powerful country in the world, even though our country is the one superpower left in the world, playing a critical role in nearly every corner of the globe, translation has never been accorded a place of importance in the curricula of American academic institutions, in the operations of American government agencies, or in the overall corporate plan of American business. The arrogant assumption still prevalent in wide segments of American society is that others better speak our language, because we are the big kid on the block. Meanwhile, other major players in the global marketplace, such as the Japanese, the Germans, the French, the Chinese, the Russians and others, have a great deal to say in their own language which we, as Americans, must listen to. Without translation and interpretation much of this vital information is lost, and, as anyone who is up on what's happening in the world right now knows, information is becoming by far the most precious commodity of our time.

And so we are faced with an anomaly: on the one hand, translation is one of the most needed and critical professions of our time. On the other hand, it is one of the least developed. In fact, it appears that in the minds of many people today we translators are a species threatened with imminent extinction, since computers will soon replace all of us. All you will have to do is feed a text into a computer and it will instantly produce a perfect rendition in another human language. In a recent article in the ATA Chronicle, the author, who represents a company which produces translation software, makes it clear that MT, or machine translation, does not pose any danger to HT, or human translation, since, to quote, "it does not compete against Human Translation, but rather against non-Translation." In other words, machine translation can produce a raw draft that can more or less convey the general sense of the original document, but certainly not a full and accurate translation. And yet, wishful thinking in the minds of many continues to discount the importance of human translation by viewing the whole affair as only a matter of a few short years before the miracles of electronics, artificial intelligence, and virtual reality will be able to completely replicate the human mind, and, who know, perhaps even the human heart. This miraculous technology may replace not only translators, but also poets, politicians, in short, anyone whose business it is to communicate with the rest of us through words.

As for me, I still believe that no matter how sophisticated and far-reaching human-made technology may become, human language, which represents the processes of the human mind and heart, will require human beings to articulate it and convey it accurately and effectively, whether in creative writing, in generating and communicating ideas, in music and song, and, yes, in translating from one language and culture into other languages and cultures.

2. TRANSLATOR SELF-TRAINING

How does one become an accomplished translator? During the past twelve years I've worked with practically hundreds of translators in over 50 different languages. I've learned a great deal about the inherent difficulties of the translation craft. First, translation requires a variety of different skills which come into play in a harmonious and effective way in one single individual, namely, a flair for languages, good writing skills, intellectual curiosity, good work habits, self discipline, and professional integrity. Second, translation, it seems to me, is more of a natural talent than an acquired skill. Some people's brains can operate in more than one language at the same time, and they are able to repeat something which was said or written in one language accurately and effectively in another language. It is not necessarily a mark of higher intelligence, only the way some people's mind works.

Still, a natural aptitude for translation is not enough. One needs to develop one's talent. As in any other discipline, a good place to start is by taking courses in translation theory and practice in an academic institution. Unfortunately, only a handful of colleges and universities in the United States offer such courses. The other way, which the great majority of practicing translators pursue, is learning on the job. Most of us have come to translation from other fields and occupations, and have learned by doing. We have made our mistakes, learned by trial and error, have had our hard knocks, and those who persevered and went on eventually developed certain techniques, acquired a body of knowledge, learned ways to negotiate the business end of translation, and became either freelancers, career translators, or, in my case, translation administrators.

But the long and the short if it is that translation continues to be to a large extent an occupation for which little if any training is available, for which reference tools on how-to and where-to-find are practically non-existent, and even the vast and expensive body of dictionaries offered by various sources is far from ideal for the needs of the translator.

With all of this in mind, three years ago my company, Schreiber Translations, Inc. in Rockville, Maryland, came up with a new option for translators. We initiated a correspondence course for translators in ten major languages. This program was the subject of my paper at the ATA conference in Philadelphia two years ago. The course aroused interest throughout the country, both on campuses and in the general public. We soon found ourselves besieged with students in all major languages, and we started to train them through the mail and over the phone. The problem was we were not sufficiently funded and staffed for what in effect soon became a tutoring program. Consequently, we gradually changed our approach from tutoring by correspondence to a self-training program, whereby the student is provided with the text to be translated along with instructions and a sample translation with which to compare his or her own version. I would like to explain the nature of this program, and why we at STI believe that it has a great potential for advancing the cause of translation.

The Advantages of a Self-Training Program

Since most translators have to learn how to translate "on the job," without having the benefit of the experience of others, the process is usually very slow, full of basic mistakes, entailing a great deal of wasted of time and effort. This often discourages beginners, who simply find the process to be more trouble than it is worth, and give up. The self-training program consists of three elements, designed to help the translation student to overcome some of the key problems: A book titled *Guide for Translators*, which provides basic information on translation techniques, how to pursue translation careers and freelance work, key dictionaries in ten major languages, academic programs for translation and interpretation, and a directory of translation companies. The second element is a translator workbook, which includes tips on effective translation and annotated exercises in such major areas of technical translation as social sciences, law, medicine, business and economics. The exercises are presented in an ascending level of difficulty, and enable the translation student to assess his or her level of expertise. The third element is a quarterly newsletter which is mailed to program participants, with updates on the various areas covered by the program.

This program offers several advantages: First, it introduces the beginner to the field of translation, pointing out both the benefits and the drawbacks, and helping the reader decide whether or not to pursue this field. Second, it shortcuts through the many pitfalls awaiting the beginner, compressing two or three years of trial and error into a few days of reading and self-training. Third, it saves a translator of, say, Spanish, the trouble of searching for specialized dictionaries in such a field as, for instance, computers, and provides information on (a) which dictionaries to select, and (b) where to find them. Fourth, it introduces the student to the most practical areas of translation in his or her language, namely, the areas in which translation work is mostly likely to be found, be it legal, medical and so on, and provides an opportunity to exercise in those areas, find out one's level of expertise, and gain an understanding of how to translate documents in those areas. Fifth, it explains the dynamics of the translation field, both career and freelancing, how one is to approach translation work sources, deal with translation agencies, and so on. Sixth, it discusses and lists hundreds of translation companies around the country, where most translation work is to be found. Finally, the newsletter provides a continuing, ongoing link between the participant and program, which continues to provide useful information and advice.

Another important feature of the program is that it provides a dual set of workbooks in each languages, namely, translation into English and from English. This is important to point out, because in the past two or three years there has been a major shift in the direction of translation here in the United States, from an emphasis on translation into English, to a new emphasis of translation from English into many other languages. This development, no doubt, has had an impact on all translators, positive for some and negative for others. But what has become abundantly clear is that translation in the United States today does not means, primarily, translating into English, but rather going either way.

The First Workbook: Spanish

The first workbook to be issued under this program was for Spanish translation. While Spanish is not necessarily the dominant language of international commerce, it is certainly growing fast in importance here in the United States, and it presents many problems which are yet to be resolved. During the past two years I've become aware of the fact that nearly every major company in the U.S. has, or is planning to have its business literature such as catalogs, brochures, manuals, promotion materials and so on translated into Spanish. Literally millions of people living in the United States at the present time are perfectly at home in Spanish, yet not so much at home in English. They represent a major segment of consumers of goods and services in this country, and have to be addressed in their own native language.

All of this may sound quite simple, but it really isn't. The fact remains that the Spanish spoken in Los Angeles, Miami, Washington or New York varies significantly, considering that the majority of Spanish-speaking people in those major urban centers hail from completely different parts of the Spanish-speaking world. In Los Angeles the great majority is of Mexican origin. In Miami, Cuban. In Washington, Salvadoran, and in New York, Puerto Rican. The problem is compounded by other factors. For one thing, the local Hispanic population in those places has begun to evolve local jargons, which are a mixture of Spanish and English. Names of everyday items, such as a can of spray, are given a new name in Spanish, namely, espray, rather than the original, more elegant Spanish name, rocío (incidentally, the new Oxford Spanish-English dictionary lists the word spray, as in "hair spray," as a Spanish word). Another problem is the fact that a great deal of commercial and technical literature uses highly technical terms, which are hard enough for most people to follow in their native American English, let alone in Spanish. Furthermore, we live in a time when new words are being coined almost every day, due to the fast rate of progress in nearly every field of human endeavor. We have to learn new words, terms and phrases in English, which is perhaps the leading language in the world today in originating neologisms. Then, in addition, we translators have to figure out how to translate those words into Spanish and other languages. Not to mention, of course, the vast differences between Iberian Spanish with all its varieties, and the various types of Spanish of the New World.

The problems I have pointed out are not limited to Spanish. They apply to other languages as well. When we translate into German, there is Swiss German as well as the official German of Germany. In French there is Haitian French, French used in parts of Africa, as well as the Metropolitan or Parisian French. The same holds true in regard to Arabic, Chinese, and many other languages.

Needless to say, the self-training program with its reference materials and exercises cannot solve all the problems of Spanish or any other translation. New words will continue to appear in the years ahead, and the conscientious translator will have to continue educating him or herself. But the advantage of the self-training program is that, hopefully, it points the translation student in the right direction, by explaining what the potential problems are, by offering some ideas on how to deal with them, and by providing information on where to find the necessary tools and assistance.

In my paper two years ago I discussed the problems of lexicography here in the U.S. and in general. Now, two years later, we still don't have a truly comprehensive dictionary of American English, similar to the great Oxford dictionary of British English. We lack good dictionaries in many languages and in many subjects. All of this means one simple thing: We cannot say to the beginning translator, "Look it up." Looking it up much of the time will not produce the necessary results. More direct guidance and assistance is needed.

In addition to Spanish, the other versions of the program now available are French, German, Russian and Japanese. Those languages were selected because the vast majority of technical documents in the U.S. are being translated from and into these languages. The five additional languages planned for next year are Chinese, Italian, Portuguese, Arabic and Hebrew.

I would like to reiterate that far too few colleges and universities in this country offer translation training. But academic studies can only go so far. Practical training is essential in this type of work. And this is precisely why this self-training program is an idea whose time has come. All of us in the translation business must be cognizant of and involved in the cause of translator training, which is designed to make our profession more professional, more respected, more systematic, and more effective.

Effective Freelancing

There is one more aspect to this self-training program. Most translators are freelancers, who work mainly directly with their clients or through a translation agency. A freelancer, by definition, is an independent operator, who is always on the lookout for new work. My own company provides work for over 500 freelancers on a fairly regular basis, ranging from 40 plus hours a week year-round, to a total of 200 hours a year. The amount of work we can provide each translator depends on the flow of material in that particular language, and on the translator's ability and availability. We can only keep a few translators on a full or near full-time basis. The rest have to maintain other active sources of work. This puts the freelancer in the peculiar position of having to market his or her services, in addition to performing them.

One important component of the self-training program is the discussion of how to market one's services, and where to find translation work. In recent issues of the ATA Chronicle there has been an interesting exchange of letters regarding the problematics of translator-agency-client relations. This is a complex area with which many freelancers have difficulties. The underlying problem, in my opinion, is the nature of translation itself. It is certainly not an exact science, and some of its built-in problems have been touched upon in this presentation. Most translations, to achieve the level of a high quality final product, must be reviewed by more than one knowledgeable pair of eyes. This is what translation companies most often do. They provide that additional pair of eyes, or even two or three pairs, as needed. The good ones have the ability to retain clients on a long-term basis, more so than an individual translator. They spend a great deal of time and resources on administrative problems, allowing the freelancer to dedicate most of his or her time to what they do best, namely, translate. And, most important, they provide a cushion between the translator and the client, often absorbing the criticism and the monetary losses, and dealing with costly legal issues. Some freelancers may not need translation agencies, and in effect, become a small agencies unto themselves. But the vast majority does, and therefore their relations with one or more agencies are critical. They have to know how to promote themselves to the agencies, how to price themselves, how to provide the languages and subjects most often needed, and how to continue to grow professionally. There are several hundred translation agencies or companies in the United States, who provide work to tens of thousands of translators. Over the last ten years they have grown in number and in size, and continue to grow. The cost and complexity of electronic translation tools will not diminish the number or size of those companies, but rather increase it. An organization such as the ATA, which has both individual and corporate members, is an excellent vehicle for cultivating freelancer-agency relations, and certainly any translator who takes him or herself seriously ought to belong to this organization. All of us together are engaged in a major ongoing campaign of educating the consumer as to the meaning, scope, limitations, problems and possibilities of translation, and all of us stand to lose or benefit from this campaign. This is why the self-training program puts a special emphasis on pursuing a better understanding of these relations, and on the marketing side of freelancing.

3. CONCLUSION

Having had several careers over the years, I can think of few that are as exciting and challenging as translation. There are no limits to the subjects a translator may cover. There is nothing boring about working with languages. The growing importance of translation in today's world is palpable, and the challenges facing translators are many.

At the same time, the training and reference literature available for translators have been extremely limited, and in some areas, particularly at the training end, practically not-existent. This is why the self-training program now being offered by Schreiber Publishing, a division of Schreiber Translations, Inc., is so vital for the growth and development of the translation profession. It may not have all the answers, but it certainly offers many, where none have been offered before. It is our hope at STI that this program will continue to grow and expand, addressing the individual needs of both new and experienced translators, not only in the area of self-training materials, but also in providing a variety of reference tools that reach beyond the traditional dictionaries into new concepts of flexible, expandable language and translation aids, most likely utilizing the new electronic technologies, so that translation may become not only more widely available, but also more accurate, more reliable, and better capable of transmitting information, ideas, and knowledge in every part of the globe where human languages are spoken.

Connections

Freelance Translating

THE TRANSLATORS AND INTERPRETERS GUILD: EVERYTHING YOU ALWAYS WANTED TO KNOW BUT WERE AFRAID TO ASK

Alan Gleason, Michael Goodmann, Elizabeth Harris and Sally Robertson
The Translators and Interpreters Guild

Keywords: The Translators and Interpreters Guild, Organizations, Referral Services, Mentoring, Independent Contractors, Health Insurance

Abstract: The Translators and Interpreters Guild, a labor union for freelance and in-house translators and interpreters, was chartered in 1991 as a nationwide Local of The Newspaper Guild (AFL-CIO, CLC). It operates a referral service to help members locate direct clients, as well as mentoring, dictionary sharing and collections support services. It offers a nationwide group health insurance plan, reduced-rate legal services and a variety of other member discounts. It lobbies on issues of relevance to its members and is working toward changes in labor law which would extend to independent contractors traditional union rights such as collective bargaining.

1. HISTORY AND PHILOSOPHY OF THE GUILD

As amply demonstrated at ATA conferences, the most important element of the translating and interpreting industry is the people involved in it. Even more exciting than the work itself is the way the people work together to improve their lot. The Translators and Interpreters Guild was created as yet another way for people to band together to help each other—and themselves. In some ways, the Guild could be viewed as one giant, nationwide cooperative.

However, rather than a cooperative, the organizational form chosen for the Guild was that of a labor union, a decision arising from a feeling among its freelance founders that freelancers had insufficient control over the conditions under which they worked. ATA adeptly fulfills its role as an umbrella organization uniting all sectors of the translating and interpreting industry, but we felt a need for an organization representing only "labor"—freelance and employee translators and interpreters.

The interests of translators and interpreters do not always coincide with those of the persons employing them. When the interests of workers and employers differ, unions have traditionally been the means through which workers band together to more effectively represent their interests and resolve conflicts with their employers. A union gives workers a voice, and we felt in need of a voice. Consequently, in April 1991, The Translators' Guild was formed as a nationwide Local of The Newspaper Guild (TNG), a union for newspaper employees founded in 1933. The Translators and Interpreters Guild (the name was changed in 1993 to more accurately reflect the membership) now has about 400 members nationwide (and several overseas), including 200 or so in the Washington, D.C. area and about 100 in California. It has regional groups (called "Units") in Washington, D.C., Northern California and Southern California. Its national headquarters are in Arlington, Virginia.

Our affiliation with The Newspaper Guild is serendipitous. When looking for a home within AFL-CIO, we were turned in their direction, and they agreed to take us in. They provide financial help (paying our phone bill, part-time office manager's salary and certain other expenses). We have access to their legal counsel to make sure we are operating within the confines of antitrust, labor and other laws. And they give us good advice and support on organizational issues such as building membership. At press-time, the Newspaper Guild membership was soon to vote on a merger with the Communications Workers of America which would make us part of one of the largest unions in the country. The merger would bring together, under one roof, workers in diverse areas of the increasingly interconnected information industry.

When we were first chartered as part of TNG, we dreamed of joining together to negotiate contracts with standard fees and working conditions, just like a normal union. We soon discovered that, since most of us are independent contractors, it is no more legal for us to set rates than for ATA to do so. We therefore decided on a two-pronged approach: establish programs by which we can do as much as possible within the current laws to improve our members' working conditions, and simultaneously work for labor law reform that would allow groups of independent contractors to enjoy some of the rights now reserved for employees, such as the right to bargain collectively with those who hire them. Both prongs are discussed below in more detail.

Many people have asked what makes the Guild different from ATA. The main difference in practical terms is our eligibility policy. Whereas ATA membership is open to anyone involved in the translating and interpreting industry, Guild membership is limited to labor, i.e., freelance and in-house translators and interpreters, or those aspiring to become such, and excludes management. Although definitions are very difficult in this complex industry, we define management as the owners and operators of translating and interpreting bureaus, or persons with in-house management positions in which they supervise translators or interpreters.

One point should be emphasized in order to clear up a misconception that has dogged the Guild since its inception. It is definitely not the Guild's position that running a bureau and subcontracting to freelancers is a bad way to do business, only that it puts the owner or subcontractor in a position in which his or her interests are no longer identical to those of the freelancers doing the work. We exclude bureau owners not because we don't like them, but because it is the only way we can unequivocally promote the interests of the translators and interpreters who work for them. A translation bureau may treat its contractors and employees impeccably, but it is unlikely to put their interests above its own. The Guild's mission is to put them first.

One of the strongest pillars on which the Guild is built is its dedication to the principle of democratic operations. Members are encouraged to make their views heard and to propose changes when they are dissatisfied with the way a program works. We hold referenda on important issues when it is perceived that there is significant discontent with some aspect of our operations. The most notable example was the reversal of our original policy of allowing only one target language for Referral Service work. Although the founding parents of the Guild, and the leadership at the time of the referendum, still felt strongly that work obtained through the Referral Service should be performed only into one's native language, the vote came out the other way and the policy was changed. We hope that the ideal of democracy is not just a luxury affordable only by small, young organizations; we intend to ensure that it persists. We have been encouraged to observe that The Newspaper Guild shares our dedication to this principle. If we truly believe our members are worth protecting, we must also believe that their views are worth respecting.

2. LOBBYING EFFORTS AND THE LONG VIEW

Like any good labor union, the Guild lobbies legislators on issues that affect our members. Our Policy Issues Committee follows federal and state legislative developments and informs our membership through the Guild newsletter about issues that may concern them. In 1993, Guild representatives knocked on doors in Washington urging members of Congress to reinstate the tax deduction for health insurance for the self-employed. We have written letters to Congressional committees about workplace issues that affect in-house translators, for example protesting withdrawal of regulations on repetitive stress injury by the Occupational Safety and Health Administration (OSHA). We use the resources of The Newspaper Guild and the AFL-CIO to stay informed and to make our voices heard.

Like ATA's PALRAC, we have closely followed the recent efforts by many state labor departments to regulate the employment status of translators and interpreters. Unlike many industry representatives, however, the Guild does not view this as a simple black-and-white, us-against-them issue. Rather, we recognize that some members of our profession are entitled to, and would actually benefit from, employee status under labor law. Prime examples of this are the groups of court interpreters in many parts of the country who are demanding the right to negotiate their contracts as employees of the courts. Many interpreters have reported to the same courthouse every day for years, yet are denied the benefits and bargaining rights granted to other courthouse workers who happen to be classified as employees. Many are clearly underpaid, such as the Milwaukee County court interpreters, who receive the same pay as bilingual clerical staff. Nor are the consequences of employee classification necessarily as dire as some in the industry would have us believe. In the July/August 1995 issue of *Translation News*, ATA and Guild member Stewart Colten made the point that translators and interpreters classified as employees could still get deductions for business use of the home on a par with those granted to independent contractors.

Unfortunately, the Guild's attempt to present this point of view has prompted some to label us as "pro-employee-status." This is not the case. The majority of our members are freelancers who do not wish to be arbitrarily classified as employees. We do believe, however, that freelancers should be extended rights now enjoyed only by employees, such as the right to bargain collectively with users of their services for reasonable rates and working conditions. There are precedents for this application of labor law in the garment, construction, and entertainment industries. The American Federation of Musicians, for example, has traditionally been able to negotiate contracts on behalf of members who work at different venues from night to night. The Guild supports legislation that would encode this right in labor law.

Ironically, we face an uphill battle in convincing some in the labor movement that they must sign up freelancers like ourselves and find a way to protect our interests if they are to survive in the U.S. workplace of the nineties. Unions are slow to acknowledge that the widespread corporate tactic of firing employees and "outsourcing" their work to independent contractors has accomplished its goal—to save the company money by avoiding payment of benefits—with the corollary effect of weakening the union. For this reason, we have initiated contacts through The Newspaper Guild with other unions who represent freelancers, in the hope of working together on independent contractor issues. We have joined a new group called the Creators Alliance that unites us with other organizations of professionals who deal in words and images—the Screenwriters Guild, the National Writers Union, the American Society of Media Photographers and others.

Meanwhile, what can the Guild do in the workplace for its freelance members? Well, plenty. We can lobby for improved working conditions and fair compensation, as we did this year on behalf of the court interpreters in San Francisco and San Jose. Through our Member Advocacy program, we can write letters in support of members who are trying to get their clients to pay or adhere to other contract terms. We can teach freelancers skills to help them help themselves, such as negotiating effective contracts with clients. We can take advantage of The Newspaper Guild's resources and expertise to teach potential employees how to organize and collectively bargain, as we did at a forum in the San Francisco Bay Area this fall. And we can join forces with freelancers in other professions as we are doing through The Newspaper Guild and the Creators Alliance.

The one thing we can't do for freelancers is establish standards of compensation. But we can support the efforts of independent contractors to improve their livelihood. And we can continue to lobby at the federal, state and local levels for legislation that gives freelancers the right to organize and demand better treatment from those who hire them.

3. REFERRAL SERVICE

A distinctive component of the Guild (with respect to most other associations of translators and interpreters in North America), and a factor which has doubtlessly stimulated its growth, is our Referral Service for translators (in operation since 1992) and for interpreters (new this year). The goals of this service are comprehensive and lofty and not inconsistent with our mission of enhancing the economic and professional profile of those who labor in our chosen occupation. These goals include:

- encouraging direct transactions between our members and their clients
- providing to both freelance and in-house translators and interpreters a practical alternative to exploitative or otherwise unfavorable working arrangements
- promoting free enterprise by enabling translators and interpreters to compete on a more equal footing in a truly free market, charging what that market will bear
- sharing the cost and effort of marketing
- building a reputation for high quality and spurring translator productivity
- encouraging clients to expect and to demand high quality
- highlighting the value of translation review as an effective vehicle for quality control, and making responsible quality control practices affordable and worthwhile for translators
- eliminating contractual restraints on competitiveness
- improving the translator's or interpreter's return on investment and work by distributing—and thus reducing—overhead costs and administrative time
- offering clients the quickest and most efficient route to the linguistically and topically best-qualified translator or interpreter for a given assignment
- enabling translators and interpreters to remain in the field long enough to acquire mastery and to build a reputation without undue hardship.

Implicit in a number of these principles is the overall desire to increase the opportunity for the translator or interpreter to work directly for the end-user client and, accordingly, to decrease drastically our dependency upon agencies and other profit-making intermediaries. Unfortunately, the latter relationship is one in which, not infrequently, the interests of the translator are a matter of minimal concern.

Although Guild membership is an absolute prerequisite for participation in the Referral Service, the Referral Service is by no means an automatic benefit of Guild membership. Rigid standards of eligibility exist for inclusion in the Referral Service, whether as a translator or an interpreter. After having joined the Guild, applicants interested in the Referral Service for translators must complete a comprehensive application which requests data on the applicant's source and target languages, a detailed indication of the topics and sub-topics of specialization which the applicant is competent to translate, and the specific types of translation and related services which the applicant is prepared to offer. Comprehensive data is also sought regarding the applicant's equipment and technological resources. Most importantly, references must be provided for each of the target languages for which the applicant seeks to be listed. The applicant must provide both peer references (translators who have reviewed or supervised the applicant's work, are native speakers of the target language and can attest to the applicant's fitness as a translator) and client references. For purposes of assuring a greater degree of consistency and fairness in the review process, a recently introduced requirement calls for every applicant to complete a sample standardized translation, provided by the Referral Service. Those applicants claiming a second target language also submit a writing sample of roughly 500 words in length on the subject, "How I Acquired my Second Target Language Fluency." In instances in which an applicant is unable to provide the required number and type of references, translation samples prepared by the applicant may be submitted, in lieu of the missing references, for review by the Referral Service Committee.

Applications undergo a process of review by the Referral Service Committee, which is made up of several Guild members already approved for the Referral Service. This process includes contacting the references provided and asking each a series of standard questions pertaining to the applicant's translation abilities. If the application is approved by the Committee, the applicant is listed with the Referral Service for the language pairs given on the application.

The process is by no means an easy one, and for some 12% of the applications reviewed to date, the Committee has found itself unable to approve the application. In such cases, the Committee will, upon request, review additional information and reconsider an application one further time. The second decision is final. After 6 months, a totally new application may be submitted. Unsuccessful applicants are advised to use the 6-month period to further polish their skills, possibly by obtaining assistance from another Guild member through the Guild's Mentoring Program, which will be discussed below. To further illustrate the exacting nature of Referral Service standards, only *three* applicants have been approved to date for translation into more than one target language.

In the case of interpreters, a similarly rigid set of Referral Service guidelines has been established. Applicants are asked to list their A, B and C languages and the interpretation modes (simultaneous, consecutive, court/legal) for which they seek consideration. Two peer references must be provided per language direction and mode, as well as at least one client reference per mode. Applicants must be able to demonstrate at least 50 days of experience for each mode. However, for simultaneous mode only, 50 days' experience into A, 25 into B and 25 from C are required in order to be considered for those languages.

Job requests received at the Guild office from potential clients are referred on a rotational basis to Referral Service members with the desired language pair and subject specialization. Priority of assignment rotation is accorded to the Referral Service member having the earliest date of application. Members revert to the bottom of the list once they have accepted an assignment or received two referrals which did not result in work.

When a potential client contacts the Referral Service, the pertinent information regarding the assignment is recorded and members with the desired language pair and subject specialty are contacted by the Guild office (in the order of the rotation), until one can be reached. The member then makes contact directly with the client and discusses the assignment. All negotiations regarding the job are thus conducted directly between the member and the client, and the client makes payment directly to the member. The Referral Service merely plays the role of bringing the parties into contact with one another. After the initial contact is made, the client may contact the member directly for further translation assignments.

An important aspect of the Guild Referral Service is its emphasis on quality control. This is achieved through a probationary period, which continues until, in the case of translators, the "rule of 5 & 10" has been satisfied. This means that, for each language pair for which a translator has been approved, the first five jobs *and* 10,000 words referred by the Referral Service (whichever comes *last*) *must* be reviewed by a reviewer. Wherever possible, this must be done *before* the job is sent to the client. Following the initial probationary period, each translator remains subject to periodic spot checks for each of the language pairs in which he or she practices. Interpreters must provide one new client and one new peer reference for each language direction and mode, from a period no earlier than 18 months prior to and no later than 24 months following the date of acceptance into the Referral Service, and this information must be updated once every five years thereafter.

The not inconsiderable administrative expense incurred in running the Referral Service (budgeted at some $30,000 for 1995) is covered through an annual maintenance fee of $48 for translators and $60 for interpreters, required to maintain one's listing. This amount is payable in addition to members' Guild dues of $72 per year. Following an initial period not to exceed two years, this maintenance fee may be waived if the member has received no work at all through the Referral Service. In addition, members pay a 10% commission on all work received through the Referral Service. (For translators, this rate of 10% is payable only on the first 50,000 words of translation from each client.)

Additionally, the Referral Service enjoys an invaluable measure of support from The Newspaper Guild, in that TNG subsidizes the salary of the office manager at Guild headquarters. An appreciable portion of her time is dedicated to providing the administrative support necessitated by the Referral Service.

A program of marketing activities for the Referral Service is carried on through a separate Guild committee which spreads the word through such activities as advertising, mailings, attendance at networking events, phone calling campaigns and presentations to groups of translation users.

So far, according to all available indications, the Referral Service—despite its relative youth—has been a remarkable success. Over 600 job requests from clients have been received since official operations began in November 1992, of which 42% resulted in jobs for Referral Service members. In the first eight months of 1995, we received about 125 such requests, of which 54% yielded remunerative assignments. Some of these projects have been quite extensive, with 18 representing jobs of 10,000 words or more and even 2 of over 50,000 words. The amount of work a member may receive will, not surprisingly, vary considerably according to language pair. Referral Service membership rolls now stand at around 115 translators and 3 interpreters (the interpreter program being still in its formative stages), with a wide assortment of language pairs (48 as of earlier this year) and fields of specialization represented. Client feedback, solicited on all jobs referred, has been consistently positive. Peer review of translations has afforded a unique opportunity for direct interaction between colleagues, a valuable tool seldom available under other circumstances.

Consistent with our mission as a labor union, the Referral Service represents a true banding together of translators and interpreters to improve collective working conditions. In so doing, our profession may achieve progressively higher levels of recognition and compensation, under ever-improving working conditions.

4. OTHER GUILD PROGRAMS

The Referral Service is thought by some to be the biggest drawing card of the Guild, but it is by no means the only benefit that comes with membership. The beauty of Guild programs is that they, like the Guild, are centered around one underlying principle: that we, as members, are all trying to make a living in the language profession we have chosen, be it out there on our own or working under someone else, and we need each other. The prime example of this is the Mentoring Program, but the other programs, such as the Dictionary Sharing Program, the Member Advocacy program, the Referral Service, our newsletter *The Voice*, local Unit activities, the Union Member Discount Program and other AFL-CIO Union Privilege benefits, all show how solidarity can help improve the lives of independent translators and interpreters.

Mentoring, as defined by the Guild, is a translator or interpreter helping a colleague by sharing information. We recognize that while a translator or interpreter may be very good at translating or interpreting, he or she may have trouble operating a modem. Or preparing taxes. Or finding clients. The Guild keeps a listing of all members who have volunteered to give advice on these and other subjects. Members who request help in a certain area have access to the information they need to contact any of these individuals. The program is loosely structured to allow people to establish mentor-protégé relations at the level they need. They can call a mentor and ask one question or a thousand. We hope to expand the Mentoring Program by identifying "Macro Mentors" who would work with beginners on every aspect of running a translation or interpretation business, teaching the kind of quality practices and professionalism the Guild espouses.

The Dictionary Sharing Program was created in response to members' suggestions and comments about how expensive reference materials are and how much can be gained from pooling our resources. We keep a listing of all the dictionary and reference works owned by the members who have agreed to participate in the program. Other members may then call and request the list for their language pairs. The program has been a lifesaver for members who have called the office to find a certain term or group of terms they needed for a job. In all cases, the caller was given the name of someone to call who would look up terms or provide specific subject information. The program also has listings of glossaries or dictionaries that members have volunteered to loan or sell to interested parties. The next step for the Dictionary Sharing Program will be to enter all the information in a database and put it on the Guild's Bulletin Board Service (BBS) for easy, self-service access at all hours.

The BBS was recently established to allow for such easy access to all Guild programs. Members can send each other messages and receive information on Guild activities. Plans for the BBS are numerous. In the future, Guild members will be able to log on and find mentors or protégés, find or sell dictionaries or equipment, read and contribute to *The Voice* and find contact information on other Guild members. Prospective members can now sign on and read about the Guild and its ideology, programs and services. And, of course, those who don't know how to use a modem can still call the office and find a mentor who can teach them.

For those hesitant to embrace new technology, we will always have a printed source of helpful information. Our newsletter, *The Voice*, has won awards for excellence two years in a row in the Journalism Awards Contest of the International Labor Communications Association. ILCA judges said *The Voice* was "full of good reading that obviously must be pleasing to the members of this particular union of independent translators." It is our forum for discussion of issues pertaining to independent contractors, ethics and standards in the language professions, technological advances, political events and legislation, and it is the distribution center for our own home-made news, such as workshop reviews and members' activities. It is distributed to Guild members, prospective members, and, through newsletter exchange agreements, other organizations of translators, interpreters and linguists. *The Voice* is sent to certain members of our parent organization, The Newspaper Guild, and every member of The Translators and Interpreters Guild receives TNG's newsletter, *The Guild Reporter*.

TNG not only gives us their newsletter, but allows us to share in the Union Privilege programs they enjoy as part of the AFL-CIO. The programs vary in complexity, from simple discounts on rental cars, theme park admissions and prescription drugs to the low-interest Union Master Card. Other popular programs include a travel club, mortgage and loan programs and life insurance. Perhaps the most popular with our membership is the Legal Services Program, which allows for a free half-hour consultation with a Union Privilege lawyer, a free follow-up letter and a 30% discount on additional services. Members have used this service, for example, to deal with clients who refuse to pay.

The Translators and Interpreters Guild also does its part to help members get paid what they are due. For documented cases, the Guild will write a dunning letter to a client stating the Guild's support for the unpaid member. Members have helped each other deal with mutual problematic clients, either informally, by comparing notes about what works and what doesn't, or in a formal setting, such as a workshop.

In addition to the three existing Units in Washington, D.C., Northern California and Southern California, plans are underway for the formation of other units in areas where the Guild has a high enough concentration of members, such as Chicago, New York and Miami. The D.C. Unit has held several workshops on various aspects of running a freelance translation or interpretation business. Participants have learned good accounting practices, how to find clients and other nuts-and-bolts techniques. The D.C. and Northern California units meet regularly, both for business meetings and for social events. These meetings, like the workshops, are good for seasoned professionals and newcomers alike, as they provide excellent networking opportunities and allow for the sharing and comparing of experiences.

Our unique position as a "nationwide Local" has its geographic difficulties, but we nevertheless look after the needs of those in areas not previously mentioned. The Newspaper Guild helps us subsidize travel expenses for those outside the Washington area who wish to attend our Annual Planning Retreat. Members also see each other at other Newspaper Guild functions, such as the annual New Officers Seminar. Committee meetings use conference calls to allow "out-of-town" members to participate.

One prime example of our commitment to geographic diversity is our new health insurance plan, available nationwide through the Communications Workers of America. Members, regardless of location, may choose from four plans with group rates and a network of physicians across the country. The four plans offered are: a Health Maintenance Organization, Preferred Provider Organization, regular indemnity insurance and a low-cost option with high deductibles. Members in Northern California are also offered a plan by special group arrangement with Kaiser Permanente.

5. CONCLUSION

The day-to-day reality of the Guild is like that of many young organizations: lots of good ideas and too little money and/or volunteer effort to implement them. Some members are true believers and put in a lot of time to make the dream work. Others are disappointed when they don't get the immediate volume of work they were expecting from the Referral Service. At this point in the Guild's development, we still have far to go before reality catches up with our vision. For that reason, the Guild seems to works best at this point for people who are willing to work together (or at least hang in there) until we reach our goals. But in moments when the ideal of working together for our common good coincides with reality—as when members leave a workshop raving about all the information they gathered, or when Referral Service marketing efforts hook a particularly juicy client, then we know we're on the right track.

We are making slow but steady progress toward the goal of bettering the working and living conditions of translators and interpreters. We have broken through some barriers already—by providing nationwide health insurance for our members, and by helping them connect with new clients. We intend to continue breaking through barriers until we establish the clout and reputation necessary to create a professional environment for our members commensurate with the highly skilled work they perform.

For more information on The Translators and Interpreters Guild, call 1-800-992-0367. Or try out the Guild's bulletin board at 703/524-3157.

SETTING YOUR PRICE: COSTING PRINCIPLES FOR FREELANCE LANGUAGE SPECIALISTS

Jonathan T. Hine Jr., CRA
American Translators Association

Keywords: Cost Analysis, Freelance, Pricing, Translation Business, American Translators Association, Home Office.

Abstract: Freelance translators and interpreters are in business. Pricing one's product or service correctly is crucial to business success. The subject of prices has stirred much controversy in the ATA in recent years, but the principles to follow in deciding what to charge are basic and need not cause problems. A good business person examines needs and costs and sets the price required to obtain the needed return.
Service-providers charge for their expert time, not a physical product. Unit pricing based on word or page count has confused beginning translators and even experienced ones. It is not necessary for a translator to confer with colleagues to convert a required labor rate into a word count. Freelance translators share the working environment of freelance writers and can learn from the large body of experience of that community.
Using basic costing principles, the author will step through a non-technical procedure for choosing a minimum price. Two sample fill-in forms will be provided. One may be used for figuring what to charge. The other can help the lone translator or interpreter track translation volume and revenue. The latter form is important for staying in business and for tax reporting.

1. INTRODUCTION

This paper is not about language. It is not about translating or interpreting. It is about business -- any business. As translators and interpreters, we can learn a lot from other professionals and self-employed entrepreneurs about the subject. Freelance language specialists, in particular, need to apply basic business principles to their work to avoid costly mistakes. Fortunately, the principles are few and relatively straightforward.

1.1 Principle: We need the money.

There is a framed sign in the office of the Director of the Small Business Development Center in Charlottesville, Virginia. Charles Kulp, a slender, pleasant man sporting thin-rimmed glasses and the ability to listen intently, keeps it on the wall behind him, in plain view of his clients. Most of them are there for advice, and the sign keeps the counseling session focused and pointed:

If you are not in business for the money ...
You are not in business.

The point of this paper is that freelance language specialists, no less than their customers, are in business. They may be working part-time to supplement their income or trying to support their life's passion. They may be trying to make a living or better, working forty hours or more each week. In any event, if they hope to break even or make a profit, they are in business.

There is a labyrinthine corpus of laws and regulations affecting the definition and conduct of business, and a Byzantine plethora of government entities often stumbling over themselves to enforce it. As Jan Maier so aptly illustrated when she took on a two-year battle with the Colorado Department of Labor, one can run afoul of these entities even when doing everything correctly.[1] Woe, then, to the dilettante who naively takes money for language work without a business-like respect for the consequences.

One consequence stands out more starkly than the others: going broke. It is easy to avoid, because it involves the application of some business sense to one's situation and making sure that the money coming in equals or exceeds the money going out. It involves no statute, violates no Federal or state regulation and requires no fine-tuned analysis of contractor-employee status.

This article will not make anyone rich. It will not define "the going rate." It will provide one way of looking at our personal needs and the needs of the business and determining how much to charge for language services. In other words, it finds the minimum rate. It is my premise that if we total our needs accurately and completely, the rate obtained by this method will tell us how much revenue is enough. That will drive other business decisions, such as whether to get a better-paying day-job, increase our marketing efforts, redirect our efforts to profitable areas and away from losers.

One of the most important business decisions we can make is to get out of the business. This is not meant harshly. A proper analysis of our needs may reveal that we will never break even. If we truly love our language work as passionately as we think, we may want to devote ourselves to it as a serious avocation. We could take on the translation of the family records or that fascinating book discovered on vacation or re-translate a beloved author for our own satisfaction. If we are interpreters, we could do volunteer work with social agencies or non-profits. Translation is a noble and scholarly profession and there is much satisfaction to be had in pursuing it for its own sake.

Yet even the literary translator negotiating a book contract would do well to know just what this labor of love is worth and negotiate from a position of strength. It would be no fun to commit oneself to a project and a deadline that would sap the family's resources beyond their limits.

1.2 Principle: We charge for expertise, not product

The difference between a professional and a semi-skilled wage laborer is expertise. Not everyone can translate, and customers pay to make that expertise available to their document -- both in language pairs and subject matter.

1.3 Principle: We are really selling time-based units of our expertise.

Since we are selling a skilled service rather than a product, we must measure the service appropriately. Professionals usually charge for their expertise in units of *time*. For that matter, any form of labor only makes sense economically if measured in time units. That is why salaries and wages in industry are usually calculated in monthly, weekly or hourly units. Even when the worker is paid by the piece -- for example in the sweat shops of New York's garment district not so long ago -- the amount paid depends on the expected speed of the worker's output. Speed is a time-based unit.

1.4 Principle: We need to separate the business from other sources of income and expenditure.

This is just good advice, but it may trip up more would-be business people, especially home-based freelancers, than almost anything else. The records, the money and the identity of the business in the owner's mind must be distinct. A business by definition must be a "going concern," which implies a certain personality of its own. Incorporation is not necessary, but there are several characteristics of a business that identify it as such.

One of the most important actions the free-lancer can take to give the business an identity is to set up a separate account for the business at the bank. Money from non-business sources (like the other job) goes into this account if the business needs it. The owner takes money from the business account when necessary. These must be clear, traceable transactions in and out of the business.

There are other ways to show that we are seriously in business. We should register a fictitious name at the local courthouse, even if it is simple as "Trading As [My Name], Translator." If our town requires it, we should apply for a business license and pay taxes. A separate credit card for business charges is a good idea. It can be a personal card we keep for the purpose.
A separate business checking account and credit card can make for less bookkeeping. When we write a check, we have to record the transaction in the register. The combination of a well-kept checkbook register, credit card statements and a small notebook may be all a self-employed professional needs for a long time.

2. CALCULATING OUR NEEDS

We cannot calculate how much to charge a customer without first knowing how much money we need. This means we must build a budget for the business. Tax time presents a good opportunity to do this budgeting. We have to assemble so many records for our taxes, that it is cost-effective to extend the process to all our expenses and see what life costs.

Preparing a business budget highlights the importance of separating personal and business finances. Those doing it for the first time are often surprised to learn how much they have been subsidizing the business from other income.

A business budget has three major components.

1. Personal Services Budget (Direct Labor). What the owner needs to support self and family. The personal budget becomes the labor cost of the business.
2. Other Than Personal Services Budget (OTPS). A business needs money of its own to survive.
3. Growth and Capitalization. Any business must earn enough money to buy the equipment and materials it needs to expand.

2.1 The Personal Budget

The personal budget includes all the expenses the translator or interpreter needs to live comfortably for a year. Whether full-time or part-time, the business professional needs to calculate a complete budget for self and family. "Need" in this context is highly individual and has nothing to do with luxuries versus necessities. If making enough money to invest in the stock market would be nice, it is a need as far as this budget is concerned. So is buying a new car or accumulating the down payment for a home.

2.2 The OTPS Budget

The OTPS budget includes the business phone bill, paper and other office supplies, vehicle and travel expenses, advertising, postage stamps, the cost of periodic mailings, etc. Schedule C from the 1040 package for Federal Income taxes can serve as a convenient starting list.

There are some areas that create pitfalls for the unwary. One of them is the tendency to shortchange the business in building the budget. When first starting out, the amounts of money may seem so high that they tempt one to economize during the budgeting process. One should save the corner-cutting until after the needs are clearly calculated and compared to potential income.

Another mistake is to overlook expenses. For example, business cards are not cute: they are an advertising expense. Postage should be a noticeable number. If not, marketing may be inadequate or family stamps may be mixed in with the business stamps.

Decide how to report car & truck expenses on your taxes and use the same system for the budget. The IRS allows a deduction for mileage or actual expenses. Watch out for the difference between business travel and commuting. It is business travel if the cost is included in the amount we bill the client, especially if it shows on the invoice. It is business travel if we do not translate or interpret on site, but simply pick up assignments or meet people and bring the work back to the home office. But it may be commuting if we sign a contract to work on-site for an extended period of time, the client does not pay for the daily trip to the site and we are not making rounds elsewhere during the day. Commuting comes under the personal budget. Get professional tax advice if you find yourself in this situation.

2.3 Growth and Capitalization

Think of the business as an organism. An organism that does not grow dies. Even maintaining a steady personal income requires growth planning because customers do go away and have to be replaced. The cost of things usually rises, and business growth must keep up with inflation.

The growth budget contains the bigger computer, the communications hardware and the new software. The cost of moving to a larger office goes here. So does the fax machine, the scanner and the dictionaries we borrow from the library but hope to buy some day, an extra telephone line to the office, an Internet service or additional on-line services -- anything we need or want for the business but we do not have now. Once we have the list of what we need to grow, we price it out and figure out when we need to buy each item. Then it becomes clear how much money we need to amass each year towards the purchases.

There are some things a modern translator needs to perform competently, like a computer, fax, modem, dictionaries and software. To avoid letting the growth plan become too complicated, budget for those and include a growth factor for the rest. After setting aside money for taxes at the end of the year, buy growth items as the cash reserve allows.

The business plan depends on the data in this part of the budget. Every business should have a plan. Freelance translators have much in common with freelance writers when it comes to business, and there is much help available for the home-office operator. The local library should have books about running your own business. The Small Business Development Center or the local community college may offer courses or consultation for starting a business. They may make the business plan sound very complex, but it is simple for the free-lancer. We figure out where we want to be in, say, five years. Then we write down what we have to do and buy to get there. And that tells us how much we have to earn.

2.4. Pulling It Together: The Business Budget

Table 1 combines the three considerations discussed above.

Business Budget Worksheet		
ITEM	**MONTH**	**YEAR**
Personal Budget:		
Rent/house payment	525	$6,300
Groceries	400	$4,800
Clothing	125	$6,300
Vehicle fuel	60	$4,800
Charitable pledges	90	$1,500
Eating out	120	$ 720
Vacation		$2,000
Utilities	250	$1,440
IRA	188	$2,250
Other (hobbies, school, etc.)	220	$3,000
SUB-TOTAL: PERSONAL BUDGET		**$33,110**
Other Than Personal Services		
Advertising	10	$ 120
Vehicle (mileage)	120	$1,440
Fees	15	$ 180
Depreciation (179 expense)	350	$4,200
Office expense	10	$ 120
Rent	150	$1,800
Supplies	100	$1,200
Utilities	110	$1,320
Dues		$500
Other (Postage, books, training, etc.)	75	$ 900
SUB-TOTAL: OTPS		**$11,780**
LABOR BUDGET		**$33,110**
GROWTH (3%)		**$1,347**
TOTAL REQUIREMENT		**$45,790**

Table 1: Business Budget Worksheet.

The figures are completely hypothetical. This is only one way to assemble a business budget. One could also replace the Personal Budget portion of the worksheet with a "desired salary," which could be either the full personal income needed for the family or only the portion that the business must supply.[2]

While the Schedule C of the Form 1040 of the Federal Income Tax return is a helpful starting point, it cannot adequately capture all the categories of expenses in a business. The business person should add whatever accounts are needed to track the expenses of the firm.

3. DETERMINING THE HOURLY RATE

The next step is to put a value on the productive hours that can be billed. Whether expressed in an hourly rate or not, the amount on the language specialist's invoice should be a function of the expert's time.

One can only bill the hours actually worked on a client's project to that client. The label "productive hours" is misleading, because the hours spent on the business for the benefit of all clients (research, travel, marketing, training, etc.) are just as important to the business and each client as the time in front of the document or in the interpreters' booth. For this reason, I prefer to refer to billable and non-billable hours.

The service-provider bills the customer for all billable hours and a fair share of the non-billable hours. The latter are often called "indirect costs" or "overhead." To set the billable rate correctly, divide our total income requirement by the number of billable hours in a year. Table 2 illustrates the process.

Conditions	Hours/ year
52 weeks @ 40 hours/week (full-time)	2080
Less a two-week vacation	2000
Less Eleven State and Federal Holidays (8 hours/day)	1912
Less Allowance for sick time (10 hours/month)	1792
Less Overhead (e.g., 30%)	1255

Table 2 -- Finding the number of billable hours in a work-year.

Part-timers enter Table 2 with a lower number: whatever the total hours they can devote to the business in a year. They calculate what percentage of full-time the total is, then they apply that percentage to the holiday and sick time they budget. For convenience, they could choose the allocations for holidays and sick time used in their other job(s).

Those who have had to employ others will recognize that the considerations on this chart apply to the costing of labor in any setting. Self-employed business people have control over how many holidays they take and how many hours of sick leave they need to budget. They can fine-tune these lines better than an employer or personnel expert can.

One of the common mistakes of the self-employed is to work around the clock. Those whose businesses survive learn to take their holidays and vacations. Working holidays and quaffing cold medicine to a deadline are part of life. But the language specialist who does not budget holidays, time off to be sick and non-billable overhead in setting rates is charging less than the time is worth.

The arbitrary thirty percent used in Table 2 would equate to working one hour on non-billable activities for every two hours spent actually translating or interpreting. Unreimbursed travel time is a typical indirect cost. So are errands for office supplies, the bank, paying bills and other activities that apply to the whole business, not just a particular job.

The words "overhead" or "indirect costs" have a dirty name in some circles, but it is a fact that the non-translating part of the business is as much a part of it as the translating. The trick is to track the non-billeable time carefully and make sure that it stays reasonable and on task. Anyone running a business, including customers and the agencies, should understand indirect costs and expect them to be in the rates. Nevertheless the subject is almost never mentioned in rate negotiations.

Using the formula

Labor Rate = (Income Requirement) / (Billable Hours)

we find that the mythical translator in Tables 1 and 2 must bill $36.49/hour. It is a common practice to round such numbers up to the next higher dollar.

Clearly this is a very individual exercise, so there will always be a wide range in rates. A translator who just finished school may not need to charge as much as someone with thirty years of experience, two college degrees, a mortgage and teenagers. The client reading a résumé should be able to see the difference and should remember the business motto, "You get what you pay for."

Needs-based analysis of this sort can help the free-lance contractor with the plans and the decision-points in life. For example, one can divide the money needed between the free-lance revenue and other work (the "day-job"). Part-timers will typically start with something less than forty hours per week in computing the hourly rate. One can set milestones in the business plan to have the translating pay for itself over a given period of time and grow X% each year after that. Customers should not see a rate change as a language specialist moves from part-time to full-time or back.

Do not make the mistake of having another source of income pay for business operating expenses like postage or capital equipment like the computer. In terms of cash, that may be exactly what happens of course. But on the business books, there should be equal and opposite entries: one for "owner's contribution" and another for the business expense.

The money the owner spends for things the business cannot afford yet needs to be recovered in the rate. Failure to recover that money amounts to giving away free services. Clients may love an artificially low rate and the illusion of getting lots of business may be quite infatuating. But sooner or later the business will go broke, or the family will wonder why there is never any money left over from all the translation work. These conditions may also invite an IRS audit which could determine that the language specialist is engaged in a hobby rather than a business for profit and disallow several years of deductions.

4. THE PIECE RATE

Translators in the United States are usually asked to quote their price and bill their services in dollars or cents per word. In other countries, quotes by line or page are also common, but it all involves the same exercise. In labor management, we call this kind of pricing a piece rate. Most interpreters, editors, FBI contractors and such are paid by the hour, so they do not need to convert their hourly rate.

Conversion from hourly to piece rate involves estimating how many hours the job will take, multiplying by the hourly rate to estimate the total quote for the job. Then divide by the estimated number of words, lines, pages or whatever to arrive at a piece rate to quote to the customer.

The speed of a first draft should be higher than the actual word-per-hour rate of the finished product. One can easily spend twice as long in proofreading and terminology checking. Each additional pass through the text will lower the final translation speed. Failure to take that extra time with the work will allow mistake-ridden translations to go to the customer. That does not encourage repeat business.

The key to making a good estimate is knowing how long the job will take. This is why translators must insist on seeing a sample before quoting a job. Each translator's work habits and subject familiarity are unique, so each has a different output in words per hour.

Not only is the basic production rate highly variable, but many other factors affect the time required for a job. Here are some of reasons why the piece rate may go up or down:

- Research
- Finding special references or materials
- Typing
- Language expansion factor
- Graphics or formatting
- Delivery
- Travel requirements

The material could be extremely technical or the fax could be nearly illegible. The translation project manager on the telephone may not be able to read the original and may be guessing at the subject. Sometimes a "simple letter" turns out to be a detailed and technical document between two specialists. The material may concern cutting-edge technology or novel research by a scholar. Printed dictionaries cannot handle that material. It takes extra time to go through library collections, on-line databases and periodicals to verify the terminology.

The translator should not expect one customer to pay for all the research that may apply to several jobs. If one can reasonably foresee some repeat business that will use the research, one should spread out the charge for the research hours. This may lower the word rate considerably on the first job or so, but never to the point that the research is free.

Formatting and graphics can add massively to the cost of a job. Free-lancers should be frank with customers and warn them about this time-consuming exercise. Often the translation project manager will accept text translations with call-outs and have less-expensive clerical employees of the agency do the paste-up. Sometimes, however, the setup to match the original is easier on the PC than the extra complication of explaining where the translated text goes.

The subject of whether to charge for target language or source language word count is a popular topic of conversation among translators, but the subject may have been overtaken by technology. Agencies and translators alike seem to prefer word counts generated by computer software, and that is often available only in the target language, because the translator generated that document on a PC. In those cases where the source text is in an electronic format, there is some room to discuss which word count to use.

There is an expansion factor in language pairs and it affects the price quote. The job will often be estimated on physical source-language word count, but be billed on a computer-generated target-language word count. This is because that is what is available at the time of each count. Some languages use many more or many fewer words than English to say the same thing. Translators may need to apply appropriate percentage factors for their language pairs to the estimated number of words in the target document before converting their hourly rate to a word rate.

Beware of over-generalizing the expansion of word counts. Translators have individual styles and word usage that may be more heavily dependent on the subject matter and the culture of the community of readers and writers than is apparent from the literature. Tracking individual word counts may reveal that a particular subject or type of document reverses the usual expansion factor for that language pair.

5. RECORD-KEEPING

Clearly, estimating a job correctly requires knowing how fast one translates. A beginner can only guess, but by keeping consistent records, the translator can immediately begin refining estimates to match personal realities.

Tracking word counts, revenues, accounts receivable and overdue payments can be done on one form. It can be a record book divided in columns, an unused checkbook register or a spreadsheet kept on the PC. It is important to keep the design simple and easy to use.

Job #	Date Sent	Rate	Count	Revenue	Hours	$$/hr	wph	Date Paid	Remarks
URI-01-95	15/1/95	$0.13	15,568	$200.00	26.30	$76.95	592	22/3/95	Marfiak
DDB-09-94	31/3/95	$0.08	135,344	$10,827.52	235.20	$46.04	575	15/4/95	CAT-01-84
SPR-07-93	12/4/95	$0.08	253,889	$20,311.12	789.00	$25.74	322	28/7/95	Barbui
XCT-01-95	02/2/95	$0.13	7,895	$1,026.35	14.50	$70.78	544	15/3/95	Tradex
URI-02-95	15/2/95	$0.13	11,250	$1,1462.50	22.30	$65.58	504	21/3/95	Marfiak
...	...	...	...	...	...	...	...	...	...
SUMMARY			**718,939**	**$61,463.33**	**1725.50**	**$35.62**	**417**		

Table 3 -- Section of a Translators' Sales Record.

Table 3 is one suggestion using a computer spreadsheet. Its column headings could be also used in a manually-maintained record book. Interpreters and editors could use a similar record sheet.

The summary line tracks the performance to date. It contains sums of the "Count," "Revenue" and "Hours" columns. The "$$/hr" and "wph" cells on the summary line are generated from the appropriate sums on the summary line, not the data in the columns above them.

Note that for each job, there is an entry for words per hour and dollars per hour. These numbers allow the translator to monitor sales and refine the word-rate for different types of material. The summary line answers questions like "How many words do I translate per year?" or "How fast do I translate medical journals or lease contracts?" It contains the gross sales revenue numbers needed for the Schedule C of the Federal income tax return or local taxes.

6. CONCLUSION

The costing method outlined above should help language specialists determine what they need to charge for their work to break even. Many variations are possible and individual translators or interpreters should adjust the forms to suit their work style and personal requirements.

The sales record allows the language specialist to track sales volume and revenue, as well as hourly rates and piece rates coming in from the work. This will support tax reporting and business decision-making.

Most professional translators in the United States also work at something else. They enrich their lives by pursuing multiple activities in parallel. The technical specialization of their other work may be the source of their expertise as translators. Regardless of other sources of income, freelance language specialists are in business. To succeed, entrepreneurs must have a passion for their work. Like entrepreneurs, freelance language specialists must pay attention to the bottom line to ensure that they can continue to do the work they love.

NOTES

---

[1] Jan E. Maier, "Translators Ruled Independent Contractors in Maier v. Colorado Department of Labor Case," *ATA Chronicle*, Vol. XXIV, **6**: 9 -10 (June, 1995).

[2] Claude Whitmyer and Salli Rasberry, *Running A One-Person Business*, Berkeley: Ten-Speed Press, 1994.

Connections

French

FRANGLAIS: THE INFILTRATION OF ENGLISH WORDS AND PHRASES INTO THE MODERN FRENCH LANGUAGE

Jeanine A. Normand
ATA and AWIL

Keywords: Franglais, *Académie française*, Anglicism, French Purity, Language Committees, Infiltration, language decrees.

Abstract: English words and phrases have infiltrated the modern French language for centuries. But some borrowings and neologisms, forming a jargon called Franglais, are accused of being illegitimate by defenders of French language purity. The infiltration is on two major levels: the lexicon and the grammar. These two processes are examined on a general level with illustrative examples. The *Académie française* (the French Academy), the official institution that regulates correct usage and produces a dictionary, is supportive of recent legislation to prohibit Franglais in public in France. While the decree covers all "foreign" influences, the main target is English. This is the second time during the last half of this century in France that legislation has been levied against Franglais under the guise of being against any foreign influence in the French language. The banished words, phrases, and grammatical patterns include even some previously long-accepted examples that are now under fire. There is a new Dictionary of Official Terms of the French language. The law has provoked serious debate on the role of linguistic infiltration, and if there is a connection between it and the development or decline of a culture.

1. PROLOGUE

The French have throughout their modern history been very nationalistic in spirit. This includes chauvinism (indeed, the very concept is French) about the country, the culture, the compatriots, and the language. Ever since the Seventeenth Century, when Boileau and Malherbe set the standards of French Grammar and usage, adhering to such standards, and the ability to use the French language cleverly have become admirable attributes highly sought after in France.

The French believe, in spite of what they may know about other languages, that their *own* language is still the clearest, the most concise, the most logical, the most mellifluous, and even the most poetic.

English, however, is the most widely used language in the world today, and it is taught as a foreign language at some level in the educational systems of almost every country, according to Jane M. Alden of the U.S. Department of State, in a 1973 address to the Georgetown University Roundtable on Language and Linguistics. And now, twenty-two years later, when U.S. culture, business, politics, and technology have become evermore prevalent forces, this is certainly a increasingly intelligent discussion. In today's world of the information superhighway, and of instant, world-wide satellite transmissions, it is an economic, educational, and professional advantage to individuals and nations to be able to use English.

It is also thus inevitable that English words and phrases would make their way into other languages, especially in fields where the USA has so many fast-breaking new developments.

This study looks at the progression of a language believed to be "pure" and worth defending (an ambiguous notion at best) from the assault of "foreign languages." The French language is most vigorously defended against *English,* the paramount assailant.

2. THE RAVISHMENT OF THE FRENCH LANGUAGE

The recent modern French language is changing very swiftly. Many French linguists blame the English influence as the most dangerous agent threatening the breakdown of the traditional, immaculate French language. Many French people are concerned about the state of their language, and some have expressed their feelings of chauvinism in dramatic ways: "National degradation is always announced by the degradation of the language."[1]

The infiltration of English into French is reflected widely in the media, and it affects the vocabulary in a comprehensive cultural spectrum. More and more, English words and phrases are becoming popular in place of their own perfectly adequate French equivalents.

Such a complex infiltration can be traced on two major levels: the lexicon and the grammar. The infiltration is the most noticeable in the lexicon: neologisms and borrowings; and the non-standard spellings and pronunciations thereof. There is also much evidence through linguistic analyses that in specific cases, the grammar is changing pattern to imitate some English usage. Such important grammatical adaptations reinforce the significance of the lexical enrichment.

The process of English neologisms and borrowings infiltrating the modern French language is essentially an enrichment of the vocabulary. This is especially true as this lexical process involves curious pronunciations and spellings somewhat gallicized. And on a more complex level, the handful of translated expressions that seem to influence certain aspects of the grammar, such as adjective placement and verb creation, undeniably indicate a deeper penetration. Some French linguists are convinced that a "new look" French is in progress.

Many French defenders and institutions treat this development as a substandard dialect widely recognized unofficially as *franglais*, coined from combining the French word for English (*anglais*) and the French word for French (*français*). On occasion, this development has also been termed *le gallo-ricain*, with *ricain* as French slang for American (*américain*). *Le sabir atlantique*, another term, was coined by René Etiemble, one of France's leading experts on *franglais*, where *sabir* means disparate foreign elements in the language, and *atlantique* is an understated reference to American and British origin of these elements. A few linguists have even labeled this jargon a dialectical decoration. For the purposes of this discussion, the popular term *Franglais* is chosen.

The process of borrowing anglicisms into French has a long and distinguished history although the official and the public reactions to it have varied along the way, and there is still a wide range of opinion on the subject. Today, educational and economic groups of all ages in France come in frequent contact with briticisms and americanisms, otherwise known as anglicisms.[2]

It could be said that for as many French people who speak with aplomb a near dialect that is chock full of anglicisms, there is an equally large group who are appalled at a so-called debasement of the language.

3. LEGISLATING AWAY THE ENGLISH INFILTRATION "PROBLEM": LINGUISTIC LIBERTY LOST?

Indeed, English infiltration as a language "problem" is considered so grave that the French government finally had to intervene as a last resort to save the language. The first time this "last resort" was used in recent times was with the Law Number 751349, passed in January 1973. It was structured to cast out from France the foreign, and especially English, words and phrases from all advertising, business contracts, TV, radio, and all other government and public functions (parts of the media in France are government-owned). This is not only the first time in recent times, but it's the first time that the French government made a language decree since it passed the infamous 1537 edict that convincingly outlawed the use of Latin in the courts and in official documents.[3]

And again since the late 1980's, much scholarly debate has focused on whether the level of performance of French high school graduates had deteriorated; and if so, is it the influence of Franglais. Comparisons were made with the levels in 1926, and even more detailed comparisons with those in 1975, against 1994. The verdict is unanimously this: expertise in [standard French] spelling, grammar, and language has plummeted.[4] The French Ministry of Culture and Francophony (where francophony implies all matters concerning the French language) had no trouble at all coming up with a remedy to recommend to the government.

Thus, the French government was forced a second time to make a language decree, passed by the National Assembly on 30 June 1994, and made effective on 4 August 1994. This law banishes of all foreign words and phrases in public and in private French usage, with fines levied on the transgressors. (Does this sound familiar?) And a dictionary of approved words, the Dictionary of Official Terms of the French Language, which lists approved French equivalents for the exiled words - mostly anglicisms, was widely circulated.[5]

The outcry against this decree, even among government functionaries with a reputation for being stuffy and pompously chauvinistic, caused the Constitutional Court to straightaway mitigate the effects of the decree. The decree was watered down to have jurisdiction only over government bodies and some state-owned companies.[6]

Everywhere in the French press since mid-1994, even in French publications known for being dry, there are serious discussions and humorous articles about this situation. It seems that everyone wants to defend freedom of speech along with the purity of the French language and culture. While at the same time, they all keep speaking Franglais, probably because it's chic and voguish; what the French call *cool*.[7]

And anglicisms have been considered "cool" and fashionable among French society for quite some time. In the seventeenth and eighteenth centuries, when English and the English way of life had a great influence on the French, much military and political terminology was borrowed from English just before and during that period. English manners were socially desirable in much the same way that American fads and customs are today.

English words and phrases, borrowings, neologisms, and anglicisms, which sometimes replace French expressions, are flagrant in subject areas concerning music, dance, fashion, sports, politics, economy, society, science, and technology. They are widespread in the mass media. For instance, more than five hundred regularly-occurring anglicisms were counted in the popular comics alone in the 1960's.[8]

4. PURITY AND CLARITY OF THE FRENCH LANGUAGE

The French have always been concerned with the purity and clarity of their language. DuBellay, writing at a time when the majority of the French population didn't even speak the national language, and when formal education was in Latin and Greek, sought to minimize the influence of Latin by publishing the Defense and Illustration of the French Language in 1539.[9] French was spoken and written, in fact, by "elites abroad before the French nation learned it. In 1296, the Italian adventurer, Marco Polo, wrote the entire account of his Chinese travels in French."[10]

With the dawn of the seventeenth century, the French language became stabilized. The grammar was written and engraved in the hearts of French grammarians forever by Malherbe and Boileau who set the standard never to be tampered with again.

Thanks for this standard also go to the ambitious Richelieu. A club of nonwriting writers had begun meeting informally to discuss philosophy as a reaction against female domination of the literary salons frequented by the educated nobility and hosted by trendy lady aristocrats. They caught the watchful and suspicious eye of Richelieu, who pressed them into government service by constituting them into the now-famous *Académie française* (French Academy) in 1635. The *Académie française*, this official agency of linguistic formalism, was forged out of "the absolute will to control and discipline language." Remember that the *Académie française* also came to government duty in part specifically to protect the french language from feminine influence. In any case, it was forevermore deemed responsible for the eternal tasks of cleaning up the language, and of writing a dictionary authorizing correct grammar and pronunciation. Since that time, the *Académie française* has taken itself very seriously as the defender of the purity of the French language. Some, however, consider it an outdated and uselessly slow organization that serves little purpose to the problem of the rapidly changing language in France today.[11]

5. A CRUSH OF COMMITTEES

The 1973 language decree may be the first official one since 1537, but the French government has always considered itself responsible for the maintenance of good usage, and for control of the development or changes in the language. In the times of the French Revolution, they invented new titles that purportedly equalized all citizens, a new calendar, and much new vocabulary required by law to replace aristocratically-biased expressions. This repressive language control did not last, and the French quickly returned to the regular calendar during that age of political turmoil in Europe. Nonetheless, the far-reaching effects of this great Revolution in the linguistic context was that of enriching the general popular vocabulary with gallicized words from essentially Greek and Latin roots, French grammar remaining intact.[12]

The French government has formed committees to protect the language in a way similar to that in which the Americans form them to save wildlife threatened with extinction. There are the committees of French as the Language of Europe, the French Language in the Beaux-Arts, and the International Association for French-Speaking Parliamentarians, and countless more, with new ones popping up continuously. By 1972, Georges Pompidou, the president, had set up fifteen additional official committees to find French alternatives to English words.[13]

In 1958, the Academy of Sciences pledged to see that French kept its place in the scientific world by calling on its government to require French at the international scientific and cultural conferences. DeGaulle had announced, "it is of national interest that our scientists and technicians, and all of our scholars, use our language with all the respect due to it."[14] The linguistic law of 1994 requires that conferences, meetings, presentations, and exhibitions that take place in France to be in French.

In 1972, as a preparation for the linguistic law of 1973, President Pompidou spoke for the French nation. They were "seeking 'a way for us [French] to distinguish ourselves from the U.S.' [and he] authorized a slew of committees to draw up a master list of government-approved anglicisms. Unapproved words are henceforth forbidden to any decree, circular, instruction, letter, or other document, including a lunch order for *un sandwich*. The [mandate] contained not one word of Franglais, which is what we Americans call a tour de force."[15]

In effect, scholars who were assigned to the task of making this list "worked long and diligently on more than 15,000 potentially 'barbarous' words." This long list of the most offensive anglicisms proscribed in government agencies, which includes radio and TV, made employing their French equivalents - some of them newly coined - ***mandatory***. The early list of anathematized English terms covered only those theatrical and technical. Those scholars were to next "turn to the fields of finance, computers, and tourism."[16] And now the scholars are put to the task again in a new committee to find acceptable French equivalent to the anglicisms with the language law of 1994.

6. ANGLICISMS BOTH POPULAR AND INFAMOUS

The Office of French Vocabulary conducted an inquiry to find out some of the most unpopular anglicisms. This vocabulary may seem perhaps superficial or specialized at first glance, yet the extent of the population exposure through these channels has contributed in fixing anglicisms into the language, especially among the youth. And many of the anglicisms from the following examples are in the French dictionaries!

The following terms, all taken from that inquiry, have adequate equivalents in French: *le business man* (*l'homme d'affaires*), *le leader* (*le chef*), *le living room* (*la salle de séjour*), *le rush* (*la ruée*), *up-to-date* (*dernier cri* or *à la page*), le fair-play (*le franc jeu*), *le star* (*l'étoile* or *la vedette*), and *le speaker/ la speakerine* (*l'annonceur, l'annonceuse*).[17]

A certain amount of English expressions have become popular through commercial channels: *Balenciaga for men* (an aftershave brand), *la salade mixed*, *le* or *la t-shirt*, *le Scotch* (adhesive tape), *le scotch* (the drink), *le kleenex* (paper tissue), *le reporter*, *le one-man-show*, *le lipstick*, *time is money*, *les toys*, *la cover-girl,le play-back*, *le blue-jean*.[18]

The generation gap among Franglais speakers is spanned when grandchild and grandparent enter a toy store in France. Since the last 30 years or more, probably two-thirds of the toys the child finds most attractive are made in the USA, or are promoted by US companies. The brand names, instructions, parts, labels, and promotional brochures are usually in English. Note the following examples, which represent but a few: *Le teddy-bear, le bunny, les dinkey-toys, le scooter-ball, le Kentucky-rifle with operating hammer and trigger assemble a full size rifle from this kit, Human-Skeleton assembly kit,* and *le baby-doll.*[19]

Other significant English influences on French grammar developments include an overall leveling of gender in anglicisms; and verb, adverb, adjective, and noun neologisms. There are even some "rules" for the formation of Franglais. "In order to create gallicisms, it only takes sticking a hyphen between some English words." This is certainly the most frequent practice, as seen in previous examples, and more: *le dead-heat, up-to-date, le chewing-gum.*[20]

In general, foreign words and phrases in French usage are invariable in gender (masculine) and number (singular). With anglicisms, there are some exceptions: *les toys*, *la speakerine*, *le tennis-mens*, *la nurse, la camera*, *la pin-up, la jam* (jam session), *la surprise-party* (any kind of party), and *le* or *la t-shirt*. Notwithstanding these exceptions, gender is almost always masculine, and that's always the case of those anglicisms ending in *-ing*: *le yachting, le camping, le parking, le kidnapping, le high-jacking, le marketing, le shopping, le looping.*

Verbs are formed by adding the first conjugation group infinitive ending *-er* to the anglicism, and conjugation is thus "regular" and simplified: *shooter* - to shoot the ball; *smasher* - to smash the ball; *window-crasher* - to window shop; *interviewer* - to give or submit to an interview. A few adverbs are corrupted by taking the English adverb patterns, and adding the common adverb ending from French (*-ment*) corresponding to the English "-ly": *pratiquement*, on the pattern of "practically," with an added new meaning of *presque* or *quasiment*, when the standard meaning is *en practique* or *en réalité*.[21]

Adjective placement, by analogy with English, is reversed in many new expressions: *une urgente mission, le très moderne style, l'entière population française*.

With all the brouhaha about anglicisms corrupting the enforced authenticity of the French language, it is noteworthy that ***some anglicisms originally came into English by way of Old French***, and they are now re-entering the French language. Here are some noteworthy examples: *tennis* - from *tenez, tenetz* of OF; *warrant* - from *warrant* OF, variation of *garantie* OF; *la relaxation, se relaxer* - originally from a religious, then juridical OF term; *gentilman* - from *gentil* OF; *le baby-beef* - from *boef* OF.

And let's not forget that among the neologisms are the anglicisms that are being invented that don't exist in English: *le turf, le turfist* - for horse racing, and horse race fan; *le tennis-man, le tennis-mens* - for male tennis player(s); *le pressing* - for the cleaners; and *le looping* (airplane acrobatics). Semantic differences occur when some English words are adapted with a different meaning specific to French culture, as in the case with the following expressions: *le drugstore* - for a variety store that often includes a bar or restaurant; *la nurse* - for nanny; *le flirt* - for a brief romance; *le playback* - miming or lip sync; *le parking* - for parking lot; *la surprise-party* - for any kind of party; *le window-crash* - for window shopping; *lifté/ liftée* - for someone who's had a facelift; *le pull* - for sweater; and *le smoking* - for dinner jacket.

There is also a tendency to translate popular English expressions and substitute a word or verb in pattern. An example of this would be the verb *admettre*. Standard usage of the verb is in the same sense as *avouer* or *reconnaître* (to avow, to acknowledge, to be grateful for), as in *J'admets qu'il a raison; admettez que c'est bon*. Following the English pattern, *admettre* can also mean to admit in the sense of to confess.

7. ANGLICIZED DIALECT OR THE NEW MODERN FRENCH?

The fact that languages are constantly changing, and that words are taken "from many sources is indicative of a cosmopolitan attitude that is the very opposite of, say, Icelandic, and to a lesser extent, German ... Esthetic preferences are [unequivocal] matters of style, and they have nothing to do with the subtle patternings that make one language different from another."[22] And in the history of all languages and language development, there has always been a discrepancy between the written and the spoken, between established style and popular style, in spite of the *Académie française* or the law.

And like the equally ambiguous notion of keeping a race of people "pure" to prevent it from mixing with other "races," this notion that languages can be contained by government rule becomes a moot argument as well. Languages are living, growing communications tools that do not exist to obey the laws of State, but rather to serve humankind and to adapt with us.

The dialectical differences, the slang phenomena, the borrowing and dropping-out of words and phrases, and many other cultural and political influences have always contributed to what makes living languages evolve. Yet does it have to go all the way to *l'anglofolie* (a craze for that which is English - language, people, and culture) or *l'americanolâtrie* (worshipping that which is American - language, people, and culture)? Are the processes that have placed these anglicisms so firmly in the French language inherent or environmental? Is it deterioration or enrichment? Is it Franglais or the New Modern French? And what does it all mean for what we cherish as French culture? *Mes amis*, this discussion brings up more questions than it does conclusions, so I guess we'll just have to wait and see.

8. End Notes

NOTE: Many sources were in French, and where they have been quoted and paraphrased, the translations are all those of the author of this study.

1. Sanche de Gramont, *The French: Portrait of a People*, (New York: G.P. Putnam's Sons, 1969), p. 280.

2. Some are distinctly americanisms, and some are distinctly briticisms, while many anglicisms are both. "Anglicism" is chosen in this discussion to refer to the English words and

phrases of either or both cultures found in nonstandard and standard French.

3. Pierre Honneger, "Non-Bon Mots in French: Franglais," (*Time*, 19 January 1976), pp. 62-63.

4. *Le Nouvel observateur*, 1546, 23-29 June 1994, pp. 4-13.

5. *France-Amérique*, 1182, 4-10 February 1995, pp. F-H.

6. Ibid.

7. *France Magazine*, 32, Fall 1994, p. 8.

8. René Etiemble, *Parlez-vous franglais?*, (Paris: Gallimard, 1964), p. 89.

9. Sanche de Gramont, p. 249.

10. Ibid., p. 245.

11. Ibid., p. 251. About thirty years ago, when the *Académie française* reached the letter "d," they decided to include "drugstore," but with its adapted French pronunciation and definition.

12. "Tongue-Twisting," *Newsweek*, 79, (April 1972), p. 73.

13. "Tongue Twisting," p. 73.

14. René Etiemble, *Le Jargon des sciences*, (Paris: Hermann, 1966), pp. 172-178.

15. M. C. Wrenn, "Le Big Flap Over Franglais," *Life*, 73, (April 1972), p. 20.

16. "Le Come-Back de Savoir-Faire." *Newsweek*, 81, (January 1973), p. 35.

17. Ibid.

18. René Etiemble, *Parlez-vous franglais?*, pp. 230-247.

19. Ibid., pp. 230-247.

20. Ibid., p. 62.

21. Ibid., pp. 78-80.

22. Thomas Pyles, *The Origins and Development of the English Language*, 2nd ed., (New York: Harcourt, Brace, Jovanovich, Inc., 1964), pp. 340-341.

Connections

General Interest

THE GENESIS AND METHODOLOGY OF *A TRANSLATION TRIPTYCH*

Alex Gross
Cross-Cultural Research Projects, ATA

Keywords: Famous Quotations About Translation, Translation History, Translation Errors, Language As Evolved Animal Spray, Computer vs. Print Presentations

Abstract: This paper discusses *A Translation Triptych*, an electronic presentation about translating and interpreting. As its name implies, the presentation consists of three principal parts:

1. *Translation and Translators--Quotations Through the Ages*, comprising approximately 150 observations on both the spoken and written forms of our profession.
2. *Famous Translation Bloopers, Real or Imagined*. This is a collection of some remarkable translation errors, most of them first noted in the pages of the *ATA Chronicle*.
3. *An Unofficial History of Language and Translation*.

The presentation takes the form of an electronic slide-show lasting approximately eighty minutes. It will be continually recycling throughout the Nashville Conference, so that sooner or later most people in attendance will have a chance to see most of its parts. In the paper the respective advantages and drawbacks of the print and electronic media are discussed, the contents of the presentation are analysed in some detail, and the strategy for creating a scholarly apparatus for electronic works is examined. *A Translation Triptych* will be located in a central area of the Conference site.

In their preface to *"The Craft and Context of Translation,"* William Arrowsmith and Roger Shattuck observed in 1964:

> "...intelligent comments on translation...tend to be unavailable or scattered, tucked away in odd corners, and their arguments diffused."

Although many excellent general treatments and specialist studies on translation have been published during the thirty years since they wrote, on the whole their observation rings as true today as it did then. Only a few *"traductologues"* or "translatologists" (the French term being somewhat more common than the English one) are likely to have access to such information, and even they might be hard pressed to separate "intelligent comments" from the far greater body of less striking ones. How much truth their observation contains may perhaps be gauged by the fact that it was, for all its centrality, encountered relatively late in the research for the current project.

A Translation Triptych represents a first attempt to break through this specific language barrier and present such comments, along with related material, to a wider audience. Its three main sections comprise a compendium of about 150 translation-related quotations through the ages, a selection of outrageous translation errors, and a semi-humorous sketch on language and translation, with a brief "quiz" on translating and interpreting as an addendum.

This work differs from previous studies of the subject in two important ways. First, it is directed mainly towards two audiences: a somewhat new genre of specialist audience in the first instance but in the long run towards the general public as well. So far neither potential audience has had any extended opportunity to examine this material. Even the specialist audience in question, though an entirely logical one for these texts, has for some reason rarely been addressed by the scholarly community in the past. It is none other than the community of working translators, including not only scholars, literary specialists, and teachers of translation but also conference and courtroom interpreters as well as business and technical translators. The presentation has in fact been designed to be premièred at our very own 1995 National Conference of the American Translators Association in Nashville.

Aside from this somewhat atypical venue, *A Translation Triptych* differs in one other important respect from previous studies on translation. Its chosen medium is not a book, a monograph, or a lecture--it has in fact been organized on a disk as a computerized slide show, though it will also be available in printed form. The very nature of the electronic medium has of course influenced the manner of presentation, as have engraved stone or clay, papyri, incunabular and printed books, films, and audiotapes in their respective eras. Each of these media has its own advantages and drawbacks, and the computer is no different.

Since the presentation will be viewed during a busy conference, certain liberties have been taken with the manner of displaying some information on the screen, though fail-safe mechanisms have been set in place to ensure the maintenance of scholarly rigor. It is perhaps useful to remember that those viewing this work will be balancing food and refreshments, consulting programs, and dashing off to meet friends or attend sessions, all of these activities necessarily encouraging a short attention span even among the scholarly. On the other hand, unlike most conventional presentations, *A Translation Triptych* can be continually recycled for twelve hours every day, perhaps on more than one monitor, so that all who wish can return and review its contents more reflectively.

One further factor surrounding the development of this project actually had the effect of encouraging more rigorous treatment rather than the reverse. Although *A Translation Triptych* was intended primarily as a Conference event, from the outset it was also conceived as part of a more ambitious proposal for a museum exhibit devoted to translation and interpreting, which might hope to reach and educate a more extended audience about this field. Within the fifty separate displays of this larger exhibit, *A Translation Triptych* would comprise no more than one. Such an exhibit would echo to some extent the excitement and "information overload" of the conference setting, making it feasible to convert the one into the other with at least some degree of ease. But since both projects went through their initial research phase within a similar time frame, more extensive work on the larger presentation could fertilize the smaller one to some extent.

Before reviewing the actual content of *A Translation Triptych*, a discussion comparing and contrasting print and electronic media as vehicles for presenting ideas may be in order. It should not be supposed that any simplistic dichotomy between these two media can exist. There are many different styles and approaches in using computers to display information, just as there are many different ways of employing print. And in each case both creativity and native presentation skills are certain to play a role. Moreover, the available range of authoring software--even at this early stage of computer history--is more than broad enough to allow a veritable gamut of creative choices.

Only after carefully examining some twenty different shareware and commercial products of this genre did the author finally settle on one program called *Automessage*. That so thorough a search was needed reveals much about the many faces of this medium. Some programs permit instantaneous flashing of a message on the screen, but this can induce boredom or even hypnosis over the long term. Others enforce strict message boundaries and/or uninspired typefaces. Still others boast an infinity of typefaces and colors but lack crucial editing tools. *Automessage* was chosen because it offered the best of many possibilities for this particular purpose and also because of its engaging manner of painting messages on the screen as one watches, a sort of latter-day "handwriting on the wall." The options of using a hypertext or expert system environment were also considered but rejected because they limit the number of people who can use the computer at once.

The print medium uses the "page" as its model, while computers are ineluctably bound to the "screen," assuming the message is meant to be viewed on a monitor. This limitation is actually less restrictive than may at first appear. In order to grasp more clearly how "screen" and "page" can differ from one another, let us look at an excerpt from this presentation and compare the two versions. One of the quotations from *A Translation Triptych* looks something like this on the screen:

> Now it is words and
> their associations which
> are untranslatable, not
> ideas...there is no
> idea...which cannot be
> adequately produced as
> idea in English words.
>
> --SIDNEY LANIER, 1897

The quotation is placed within a box to simulate screen borders, with the bottom line in capital letters to represent the different colors on the screen. This is an ideal passage in terms of this particular program because it is exactly seven lines long or, rather, has been "segmented" to fit into seven lines (no actual editing has occurred *per se*). A blank line is inserted above and below the quotation simply to make it more readable, and the type has been centered for the same reason.

Thus, it is not merely by "screens-full" that messages on monitors are to be measured, but by "readable screens-full." This important criterion also has its print counterpart, as we shall see. The reason the number of lines is important is that the *Automessage* program permits exactly ten lines on each screen (or only five if one were to choose the "double-height" option for the message). But this is not a general computer standard or even a DOS one--it applies only to *Automessage*, and other programs will have their own internal limits and/or drawbacks. It is necessary to work within the limits of each program to achieve appropriate results.

Now here is the very same passage as it might appear in an academically "correct" printed format:

> Now, it is words and their associations which are untranslatable, not ideas...there is no idea...which cannot be adequately produced as idea in English words.
> **--Sidney Lanier, 1897**
> *The English Novel.* New York: Scribner's & Sons, 1897.
> pp. 190-91. *(cited by Morgan, 1959. Original text examined and one missing word restored.)*

It should be clear from the beginning that the printed form of this citation is intrinsically less easy to read than the screen form, even if we ignore the several lines needed for a correct scholarly reference. Although more information has ostensibly been transmitted, the overall effect could actually become one of less information. The other lines add what might be termed a "noise factor" by the standards of Shannon's Information Theory. If these last three lines were included in the screen version, where six lines might be required----and if this model were followed throughout the presentation--overall readability would be greatly reduced and the willingness of conference or museum goers to keep watching would be sorely taxed.

This extended form is indeed closely related to the one which appears in the printed List of Sources, with the exception that the book reference is shown simply as "Lanier, pp. 190-91," with the full details present in the Bibliography. Here we also see the typographical counterpart of contrasting screen colors: in this case bold, italic, and slightly reduced fonts are used for the attribution and notes.

But what if the message is more than seven lines long--does this mean it must be "procrusteanized" into a shorter form? This is by no means the case, as the program also permits scrolling the screen down for longer text, or splitting a message into two segments, or setting up top and/or bottom lines that can announce a continuity of theme, as in the "Tale of Moses Ben Ezra" or the examples of Machine Translation. It is theoretically also possible to employ 256 different color combinations on each line of each screen, though only a few have been chosen here: the ATA colors of light blue and white for the "body font," black on white for attributions and comments, and yellow on red for occasional emphasis.

Actual editing of these texts has been avoided in almost all cases, and only two are marked as having been "adapted." It nonetheless remains true that in a conference or museum setting, shorter messages are more likely to be read than longer ones. For this reason the author has sought out citations between three and thirty words long, though a number of exceptions have been made.

There is probably a lesson here for intellectuals and scholars in general--the assumption that longer treatments of a subject are necessarily more accurate or convey more actual information than shorter ones is possibly in error. A popularized magazine article can indeed be more informative than a scholarly book, and even a bumper sticker can on at least some occasions say more than the thickest Germanic "*Prolegomena zum...*" From Ancient Greek epigrams down to *Quotations from Chairman Mao*, people have sought out simplified summations of knowledge in the hope of understanding more of the world around them.

As for the quotations themselves, what is perhaps most remarkable is that only a few recurrent themes seem to emerge over the span of more than two millennia. Aside from those passages offering specific technical advice or singling out specific works for criticism, these themes are really only three or four in number. By far the most frequently mentioned issue is the ongoing debate between "literalists" and "liberalists," between translating "words" vs. translating "sense." And by far the majority--though not all--of the most famous authors and commentators come down decisively in behalf of the latter. Granted, these are for the most part translators of literature rather than technical or scientific specialists, but a general feeling of consensus is nonetheless present. (It should perhaps be added that the author did not expressly seek out passages favoring this opinion and would have gladly included those expressing the opposite view, but they were distinctly hard to uncover).

There is also a clearly recognizable counter-current of those who suppose translation is nearly impossible, is practiced mainly by self-deceiving incompetents, or even constitutes a breach of human or divine law. Typifying this school of thought through the ages, one may encounter the added suppositions that one language is "inferior" to another or that the crux of the matter lies in the great number of "bad translations." When the problem is thus defined, the solution often emerges that all translators must be trained in a single translation method with clearly set rules, an argument heard more than once over the centuries. These rules frequently belong to the person advocating such a system, even though it remains far from clear that the problem has been correctly defined to begin with.

How are we to explain the continuous reemergence of these same disagreements through so many eras and cultures? In the sketch comprising the third part of the *Triptych*, it has been suggested that language might be an actual biological process analogous to--and evolved from--primitive spraying by animals. Because language is such an instinctive process, most people are quite unable to view it as a complete system or make meaningful generalizations about it beyond a certain limited point. This was indeed the argument advanced by Leonard Bloomfield's *"Secondary and Tertiary Responses to Language,"* which the author has discussed elsewhere[2]. Bloomfield concluded that many of our attitudes to language might turn out to be "a matter of psychology and sociology." Although much of his work is now forgotten, this particular observation perhaps deserves further attention.

The notions of "culture shock" or "future shock" are both familiar to us. A comparable emotional confusion may affect many people when they first encounter a foreign language or even a translation from a foreign language. Perhaps it should be called ***"language shock"*** or even ***"translation shock."*** Depending on the circumstances, such a state can prove more or less serious for those suffering from it. Suddenly the familiar furniture and surroundings of one's own language, which one had supposed were synonymous with reality itself, are ripped away. One is invited--in some cases even forced--to assume that a totally new way of categorizing the surrounding world is possible. It is a break in the fabric of life that most people--and even quite a few translators--are unable to fully account for. Add to this the virtual certainty that no two translators will ever translate the same text in exactly the same way and that inexperienced translators can argue endlessly over which is the so-called "correct" translation of almost any passage. Considerable grounds for confusion are quite clearly present.

Such an experience can cause a considerable jolt, especially if one had never really devoted any thought to language before. The immediate reaction of those so afflicted is familiar to all of us and may provide a concise symptomology of "language shock." Those suffering from this condition profess to disdain the other language--and perhaps the other culture as well. They may reject any influence springing from that language or culture--including the work of a translator or interpreter. Alternately, they may choose to blame their own language or culture. What we may be seeing in many of these quotations may be nothing less than a repeated reinfection by "translation shock" or "language shock" through the centuries.

But can there ever be any reliable cure for this syndrome? If so, it can only come about when human beings become more sophisticated about the true nature of language. To what extent are we all like moles, happily burrowing through our own networks of tunnels and patiently laying down our own scents upon our domain? And how do we differ from such moles as they suddenly come upon a neighboring set of burrows with totally different scent marks they can neither recognize nor understand? Let us now suppose that a "translator mole" came along and tried to explain these foreign aromas to the others. How would moles react in such a case, and how do we?

We know the answer only too well. The mole's reaction will be to defend, to repulse, to kill. Humans are not usually allowed this option, and so we settle for criticizing the intruders' spray marks--their spelling, punctuation, accent, grammar, or choice of words, not to mention their appallingly "bad translation." And yet, at the very moment of our utmost arrogance about language--just as we assert our superiority over animals for having invented it and boast of our scientific methods for studying it--at that very instant perhaps we should recall that we too are at work as animals, spraying everything around us in the hope of protecting our environment or effecting a magical change in its nature.

Another major theme present in these quotations is of, course the practical value of translation, with such major voices as Bruno, Goethe, Roget, and Kelly summoning us to recognize the crucial connections between translation, science, and the entire world economy. One further frequently voiced attitude--here exemplified by Fray Ponce de León, Cowley, Ladmiral, Meschonnic, Levy, and Fauchereau--is that the academic establishment--and perhaps even those who purport to study translation--penetrate only marginally into its real nature and produce little beyond "chatty essays."

This position is taken even further by the striking observations of Steiner and MacFarlane:

> However, despite the rich history, and despite the calibre of those who have written about the art and theory of translation, the number of original, significant ideas in the subject remains very meagre.[3]

> [My intent is] to underline the need for some new provisional theory of translation--new in the sense that it should be diagnostic rather than hortatory...concerned... with actualities...It is not the principles of translation that need re-adjusting...but rather our ideas about them.[4]

There is also a fair consensus through the ages, shared by Cicero, Quintilian, Iamblichus, Roger Bacon, and Tytler, that no so-called "universal grammar" can account for the many differences between languages and that only concerted creativity and inventiveness can bridge the scarcely trivial gaps between them. One author goes even further and insists that there may be circumstances where no such bridge can be built at all:

> True, translation may use the value terms of its own
> tongue in its own time; but it cannot force these
> on a truly alien text.[5]

This observation necessarily brings us back to Sidney Lanier's assertion in our first example, that any idea, as idea, can be expressed in English. Such a claim remains essentially unprovable--if an idea did exist that could not be expressed in English, precisely how would we learn about it? We are confronted here by a genuine paradox, one which poses for language the same order of difficulty that Gödel's *Entscheidungsproblem* posed for mathematics.

As noted, research procedures for this study have been rigorous, though this has been less necessary for the two specially composed sections and the collection of translation errors, most of which were printed in the *ATA Chronicle*. But the compendium of quotations has required and received special attention. Although its actual text is a mere 3,000 words, about the same length as many magazine articles, this section comprises a veritable mine field of attribution problems, covering many eras and cultures. Unlike typically journalistic "Thoughts on Spring Through the Ages" (or "Thoughts on Love, Friendship, Wine, etc."), its citations could by no means be found simply by picking up one's desktop *Bartlett's*. Almost all these passages have indeed proved Arrowsmith and Shattuck's words: they are truly "scattered," "tucked away in odd corners," or sometimes close to "unavailable."

From the outset it also became clear that any usable citations must meet three highly selective criteria:

1. They should be relatively brief, ideally no longer than 30 words (in fact, the average length is 21 words).

2. They should be intrinsically interesting and contain some useful insight for today's translation professionals.

3. They should, wherever possible, come from the pen of well-known authors or authorities in the field.

Thus, the aim of the project was from the outset an extremely specialized and selective one. This goal made a careful search strategy essential from the very beginning. The work that ought to have provided many solutions, Lefevere's *Translation--History, Culture: A Sourcebook* is entirely composed of quotations about translation through the ages and is doubtless extremely useful for some scholarly or classroom purposes. But almost all its selections were too long or diffuse for the purpose of this presentation, and none in fact has been used (though one or two may overlap from other references). The most valuable resource encountered was not a book at all but Bayard Quincy Morgan's twenty-page bibliographical section from the 1959 collection of essays *On Translation*, edited by Reuben Brower.

This bibliography provides some 280 listings related to translation, about half of them containing quotations, partial quotations, seeming quotations, or summaries. Of these 140 listings about 32 proved of immediate interest and were in almost all cases traced to their source or sources. The well-known books by Steiner and Ballard have been two other useful resources, and to a lesser extent the volumes by Rener and Kelly have also been consulted. Given the scope and complexity of this subject and the relative scarcity of useful leads, some 40 existing dictionaries of quotations, together with a few electronic resources, have also been consulted and compared. A final source has been the author's own general reading over several decades, though this became more directed in nature as the project matured. As of this writing, a more diffuse "fishing expedition" has also been launched into the pages of the profession's many journals in the hope of discovering yet other quotations.

In all cases, even where the credibility of a source has been impeccable, an earnest attempt has been made to trace each citation to its original source and in almost all cases has been successful. While seemingly a scholarly crotchet, such a procedure has on several occasions revealed that some quotations were incorrectly or incompletely copied from their sources or that equally incisive material lay close by. Wherever this procedure has failed, the comment "Cited by...." or "Summarized by...," followed by a name appears in the List of Sources.

Much of the scholarly work takes place in an undergrowth of references. Many citations were in fact doubly or trebly embedded: thus, Ballard citing Holmes citing Levy, Steiner citing Florio citing Bruno, or Ballard commenting on Tytler by citing Newmark and Widmer. Passages from St. Jerome, which came from Labourt's bilingual French and Latin edition, were first encountered in Ballard's French citations. Two passages by Luther cited by Rener in German and one cited by Ballard in French eventually led to Luther's full text in his *Sendbrief vom Dolmetschen* of 1530.

All errors in fact and procedure are the author's responsibility alone, and an attempt will be made to remedy them in future editions, which are somewhat easier to produce and publish in this electronic medium than in print. It is indeed hoped that ATA members and others will suggest further quotations that should have appeared in this first version. Perhaps these will suddenly become plentiful, now that the initial work of discovery has been undertaken. It is sincerely hoped that *A Translation Triptych* will prove of interest to translators, interpreters, terminologists, and all those concerned with language both within the ATA and beyond it. Perhaps this presentation can also prove of use to the various translator- and interpreter-training programs both here and abroad.

1 Arrowsmith, William and Shattuck, Roger, editors. *The Craft and Context of Translation: A Critical Symposium*. Garden City, N.Y.: Anchor Books, 1964. Preface, p. xiii.

2 Bloomfield, Leonard. *Secondary and Tertiary Responses to Language*. In *Language* 20: 45-55 (1944) and in C.F. Hockett (ed.). *A Leonard Bloomfield Anthology*. Chicago: University of Chicago Press, 1987. For the discussion mentioned here: Gross, Alex: *Limitations of Computers as Translation Tools*. In Newton, John (ed.). *Computers in Translation: A Practical Appraisal*. London: Routledge, 1992. pp.96-130.

3 Steiner, George. *After Babel: Aspects of Language and Translation*. Oxford: Oxford University Press, 1975. p. 238.

4 MacFarlane, John. *Modes of Translation*. In *Durham University Journal*, 45: 92-93, 1953.

5 Miles, Josephine. *Poetry and Change*. Berkeley: University of California Press, 1974. p. 200.

LIFE IS A TRANSLATION—HOW TO RECOGNIZE THE FOOTPRINTS OF LIFE WOVEN INTO THE SIMPLEST TEXTS AND EXAMINE WORDS FOR THEIR "CULTURAL CHARGES"

Colette R. Kent
American Translators Association

Abstract: Conceiving language as a tool to be used in life situations overlooks the fact that "language is the stuff of life". I will step gingerly in those territories where language and life are intricately entwined like plant life and water in a Florida swamp. I will touch on some quasi-philosophical concepts such as Bruce Chatwin's in "The Song Lines" and Noam Chomsky's biological connections. I will investigate why certain words and expressions (not false friends and not especially obscure) will explode after insertion in the text. The reason for these "cultural bombs" lies in different "cultural charges" in the two languages. One may find such ordinary words lying in the falsely secure path of a commercial, technical or descriptive translation - they are "land mines". In a conversation, they are sometimes offensive or they distort the argument. My purpose is to define the different cultural buildups that have eventually crystallized into - for instance - "Fame" in English, and "Renommée", or "Gloire" in French. Lifestyles, cultures, commercial methods, the elements of life as it is lived in different countries, will feed and map the language of those countries.

What I want to explore here is not the well known problem of "false friends", but rather "fickle friends". Whether they are cognates or not, these words, given as translations, can be trusted part of the time, but they may also turn around and suddenly "stab you in the back" or poison your carefully elaborated work of translation. I would like to support my case with a quotation from T.S. Eliot: "Words strain, crack and sometimes break, under the burden, under the tension, slip, slide... decay with imprecision..." I have lived astride two cultures long enough to know that ordinary words, mundane words, hold two different keys for two minds rooted in different pasts. Let us look at these words, let us see in what manner they have been "loaded" with a cultural charge or, if one compares them to delicate crystals, given a certain colour or resonance.

Take the word "malice". Hardly a "false friend" if you look up in the Harrap's. All the expressions built around "malice" are pretty much the same in the two languages. And yet, the French "malice" is ten times lighter, more innocent, than the English. The French meaning is bent, "tuned" one might say in the direction of "malicieux" which means playful, while the English version owes a lot to the legal connotation of malice. If I tell my neighbor that his kid is making our dog bark, but "je n'y vois rien de malicieux" I just mean the kid is not trying to tease the dog. "I don't see it as malicious", retranslates as "délictueux", which means somehow that the kid might be breaking the law, entering the property against our will, etc...Enough to start a neighborhood war.

Another loaded word is "pure". Myriads of connotations are summoned in people's minds. I am sure it is used in psychologists' tests. Some people see moral implications, some people see a straight jacket of rules (for instance the old advertisement for Ivory soap - 99.9% pure). But I realized -- in a situation in Ethiopia where our water was supplied by the military post -- that "pure" meant one thing for the military health authorities and another thing for me. "Pure" for them meant enough chlorine to bring down the count of coliforms and eliminate nitrites. "Eau pure" for me meant free from anything that didn't belong, untouched, unadulterated, unblemished, as close to "pristine" as possible. The baby's health was at stake. But as I see it now, my "line" of connotations for "pure" was very French. And that same notion, I realized later, was better expressed in English by "pristine". Look at preserves in a supermarket. The label "Pur fruit, pur sucre" in a French épicerie always meant to me there was no pectin or citric acid (not that there is anything wrong with these ingredients). Here. "pure preserve" means really "preserved according to the rules". What is the "line" of connotations that we may trace back in the English "pure"? Maybe a little bit moralistic, may be centered on the quality of intention. What is the French line of connotations? I can think of "perfect", "absolute", "without mixture, or compromise", and it goes back very far in history. Starting from "Les Cathares", the religious sect who derived their name from the Greek and called themselves in French "Les Purs" or "Les parfaits", going through the political attitude baptized "pure et dure" (without any compromise whatsoever), we find this association of "pur" with "untouched", "unattenuated", "all of one substance", "perfect" in literature as well as in traditional speech. Let us look for instance at a poem of Paul Valéry : "Quel pur travail de fins éclairs consume -maints diamants d'imperceptible écume", which is well rendered in the fine translation of Graham Dunstan Martin by "How finely worked out" - here we have perfection; and in "La Jeune Parque", "aux silences suivis d'actes purs" (which means acts that are just acts, free of all other purposes). Translated by "pure acts" it would mean "acts of purity" which would destroy the poet's intent. I remember some interlocutor in a cocktail party who thought that "Critique of pure reason" was the ethical part of Kant's work. In his mind it meant "reason applied to what is pure", while the French translation "Critique de la raison pure" (the workings of reason by itself) is a somewhat closer rendition of "Kritik der Urteilskraft. On a completely different level "Pure laine" or "Reinwollen" will become "All wool" in English.

On the contrary "Success" seems to be more meaningful and loaded in English, and more mundane in French. There is a myth behind the American "Success" ("Réussite" comes closer to it than "succès"). For full effect the French translation should carry extra emphasis. "La pièce fut un triomphe". The word "succès" in French has been worn out, emptied by its use in commerce and everyday chatter (many small stores called "Au Succès", some type of pastry called "des succès", "un succès de librairie" to mean a best seller). My mother would say "ma tarte est un succès" meaning that people asked for seconds. What does "Success" represent in America? Here is a quotation from the New York Times Book Review. "Success is the great American secular religion, a 'drama of economic salvation' ..... The gospel of success promises salvation (wealth) through faith (ambition, perseverance) and good works (industry, frugality)." I would like to mention that in this sentence, the words "secular", "faith", "good works" "ambition" are all so culturally loaded that it would take a small book to explain the shifting clouds of meaning they carry with them in the two cultures. How can "success" be transformed into a bomb. Just tell your husband that his brother (a lawyer) has no success (meaning his court cases progress slowly) because he does not stick to "pure" practice (meaning he also teaches law).

We can visualize words as living organisms, plants maybe. They develop according to their roots, and their genetic make up, and then they can be bent in one direction by the sun, dwarfed or made to grow by different atmospheric conditions, affected by parasites that give them a different coloration. If we put a word at the center of our graphic representation, we must give it a very long root which is the history of the word. Then the word grows and sometimes divides itself. It can be appropriated by a discipline (law, science, sociology, etc...) and such an appropriation will be reflected in the subsequent meanings of the word as we have seen with "malice". The most recent use of a given word will give it a certain "coloration". The socio-economic group that uses the word most will suddenly saddle it with added meaning.

We could mention "words that have a past" (like a woman or a man with a past). Such a word is "claim" a real Don Giovanni of a word, with a thousand and three intrigues (i.e. connections) to put on the list. Granted a long column in the Harraps, a whole page in my legal dictionary for the various meanings -- réclamation, sinistre, revendication, créance, demande, droit, concession (de terrain), proclamation, prétention, titre, recours contre -- this word shows it has been hobnobbing with history, administration, civil law, patent law, insurance, advertising, land surveying, journalism, who knows.... Be careful that the association of this word with one category does not corrupt your translation through an unfortunate insertion.

"A claim to fame" is fairly easy to translate "Un titre de gloire", but it means handling two extremely unstable chemical components. "Fame" is loaded. It corresponds to three or four French words, each of which is not a "true" translation because "Fame" in English carries both the old and the new, the popular and the high class. One can speak of the fame of Marilyn Monroe, and the fame of Albert Einstein. In French, it would be "célébrité" for the former and "gloire" or "renommée" for the latter. "Gloire" carries an antique aura. I am told that "fama" is the same in Italian. I am told that one would speak of the "reputazione" of Fausto Coppi - but "réputation" would not do for Louison Bobet (a French cyclist). "Célébrité" applies here and also "un grand nom du sport".

In the preface to his novel "Glory", Nabokov explains why he chose this word over "Exploit", a word closer to the Russian "Podvig". This is what the English "glory" evokes for him: high adventure and disinterested achievement. It appears here that the word glory is bent in a particular direction that only borders on the French "gloire". We can say that "fame", "glory", "celebrita" and "memoria" in Italian, "podvig" in Russian, "gloire", "célébrité" et "renommée" in French, all correspond to cultural myths -- only they are not the same myths. Because we spend our lives with words, we give to them as much as they give to us. Words are like dogs: they come to resemble their masters.

Here are some French myths, or should we say archetypes. "Gloire" always crowns somebody with rays like the sun, or with a crown of laurels. "La Renommée" flies in the air with a trumpet. "Célébrité" is transient. Here are some American myths. "Fame" evokes neon lights, Broadway, Hollywood, newspapers headlines, razzmatazz. And then there are myths that are purely American, and it is hard for a translator to do them justice. Wilderness is one (between the French "nature" redolent of Ronsard's nymphs and Rousseau's benign paths and the American wilderness, there is an ocean wide separation); another one is fulfillment which gathers elements of "vocation", "satisfaction", "accomplissement" and behind which whole generations of American gurus and self-help books can be detected.

In Bruce Chatwin's "The Song Lines" , the "Songs" are a cultural property inherited by the members of a tribe. For them life, in the form of compulsory paths to be followed by the aborigines in a sort of pilgrimage (but the route is also a trade route), has to be "enacted" if this is the word, in the singing of the song cycle of their ancestors. There was a form of inheritance that made the members "song owners". They owned a stretch of the Ancestors' song and the stretch of country over which the song passed. In Bruce Chatwin's words "One after the other, each owner would sing his stretch of the Ancestor's footprints. Always in the correct sequence".

What a splendid metaphor capable of synthetizing land, life and words. And the myth behind it is that the Ancestors "sang the world into existence". They sang along the way and named each thing, and that was the creation of the thing. And the Ancestors also left a trail of "life cells" or "spirit children". These spirit children can be transmitted to a pregnant woman, when she walks on the trail. All the myths reinforce the idea that reality lives "in association with words". At the same time, the words do not have to be the same since there are 200 aboriginal languages in Australia. What emerges is a picture of a "path" (which one might call a way) connecting to words (which may change along the way) and a "tune" (the tune always stays the same). On each path the song has been handed down by an ancestor. That is a valid symbol of cultural tradition. The words change according to the territory one crosses, but the tune stays the same (a symbol of translation).

For the aborigines, language is so physical that it has to be walked. And the walk is so spiritual that it has to be sung. These are the movements, the ritual ceremonies that make language "the stuff of life". The child makes the world in which he lives by naming it. So did the ancestors. And that, according to Bruce Chatwin, is "true poetry" (and Paul Valéry would have approved!).

In a different, but parallel field, Chomsky speaks of a biological "hard wiring" (shall we say paths?) and associates this concept to his tree structures (his diagrams or mappings of deep structures). The arrangements of meaning in the human mind obey certain laws. What do the aborigines say? One has to sing the verses in the proper order. "To sing a verse out of order was a crime. Usually, it meant the death penalty". "I see", says the anthropologist in a dialogue of the "Song Lines", "It's the musical equivalent of an earthquake." "Worse! It would be to uncreate the Creation."

So my argument goes like this. We go through life in the company of words. And the words are living things. Handled by us individually or in groups (military circles, advertising circles, scientists, ordinary Joes) the words will crack, bend, slip and also expand and acquire forces of suggestion. They should be handled with respect and in their natural environment. Leaving the handling of the meaning of words to machines is a sin. And thinking like a machine is a sin. Anybody or any thing that has a knee jerk reaction to words (I said "knee jerk", I did not say "fast" -- I have the greatest respect for interpreters) will let the meaning escape, drop the couplet of the "Song Line". And we know what that means.

FIGHT THE WINDBAGS!

Arthur Lindsay
DiplArch • FRIBA • SIA • ACIArb • MIL • MITI • BDÜ • ÖÜDV
Universitas

Keywords: Abbreviations, Economy of Expression, Polysyllabics, Prolixity, Technical Illiteracy.

Abstract: Epidemic logorrhœa and verbal flatulence. Acronymic acrimony. False plurals. A fashion of false foreign friends. Other forms of intellectual snobbery. Political correctness gone mad. A plague of — and on — hypersuperlatives. Redundant qualifiers, prepositions, and adjectives. Illiteracy in high places. Lack of clarity, grammar, and sense; the duty to fight perpetrators of linguistic crimes. Words, our tools of trade. Call out the cavalry!

1 EPIDEMIC LOGORRHŒA AND VERBAL FLATULENCE

Since the last time you were kind enough to listen to me nine years ago in Cleveland, windbags have proliferated. No statistics are available, but personal observation suggests that they now account for over 95% of the industrial world's population.

Probably the one passage we all remember in *Don Quixote* is the fight with the windmills. But for us linguists there is nothing *less* quixotic than fighting the world's ***windbags.*** We trip over them everywhere. They are particularly prolific among bureaucrats, lawyers, politicians, preachers, and teachers. But worst of all are the media people that daily invade our privacy with the spoken and the printed word.

The less substance there is in what they have to say, the greater the temptation to resort to meaningless bombast, verbal flatulence, logorrhœa. For example, as I watched the theatrical posturing of lawyers in the Simpson case, it seemed to me that they have struck from their vocabulary the simple word *before* and replaced it by the longer term *prior to.* This silly speech habit is a form of intellectual arrogance, and its constant repetition is a bad influence on all the less critical ears that listen to such proceedings. Or take the expression *with regard to* — at the beginning of the case the learned judge even pronounced this *with regards to* —: usually, when we use this term, *about* — or at worst, *as regards* — would be shorter and do just as well. But then, time is money: the lawyers on the case are obviously starving and need the cash.

As proof of how insidiously pervasive the habits of windy speech have become even among those alert to their dangers, who use language as their daily tool of trade, forgive an outsider if for a moment he takes issue with the theme of this conference, *Connections.* Wouldn't *Links* have been just as good?

2 ACRONYMIC ACRIMONY

Acronyms have come to stay. If heaven knows why, it hasn't yet let us into the secret.

The use of abbreviations is clearly a major growth industry. Maybe it's simply a kind of antidote to the growth of superfluous words. And I suspect that, because we are rather ashamed of our misdeeds and want to hide what we mean, we now call abbreviations *acronyms*. This form of intellectual snobbery has two purposes: to show that we had the benefit of higher education — indeed, a classical one — and that we are members of the exclusive club who know both what an acronym is and what the string of letters we happen to be looking at means.

I enjoy deflating that kind of arrogance. Even the least technically gifted among us knows that an LED is a light-emitting diode, and that one can use these electronic components in a display. It is thus perfectly proper to use the term *LED display*. We also know that the standard abbreviation for a liquid-crystal display is *LCD*. Look out for the next time some windy texter or media conveyor utters the pleonasm *LCD display*.

Most of us have heard of GPS, the Global Positioning System that makes use of satellite signals to compute positional and navigation fixes in space and on the ground. I know an international manufacturer of ground-based GPS receivers that insists on calling its latest product the *GPS System XYZ*. This proves that the people who work there and are responsible for brand names have chosen to forget — or have become ignorant of — what GPS stand for.

A recent TV news program supplied another such nonsense: *HIV virus*, i.e. a *human immune-deficiency virus virus*. Another piece of such windy garbage was a reference to *CET time* (= Central European Time *time*).

In two recent copies of *Time* (19 June and 17 July 1995) I stumbled on yet another illiterate habit now proliferating in the printed media: to show abbreviations such as NATO in lower-case. It would be bad enough to refer to the North Atlantic Treaty Organization as *Nato* and the United Nations Organization as *Uno* (correctly: *NATO, UNO*). In this case they had obviously run out of all capital letters!

3 FALSE PLURALS

The word for a given fact, a piece of information, is *datum*. As you know, this neuter-gender noun comes to us from the past participle of the Latin verb *to give, dare*. Its plural form is *data*.

All that had to happen was for some illiterate software engineer, an ignorant writer of user manuals, or a media genius to turn the neuter-gender plural into a feminine singular. Thus we now continually read such nonsense as *this data*, instead of *these*. I hardly need add that I recently stumbled on the false plural *datas* formed of this plural noun. What's the betting that *media* will be the next candidate for this pervasive, galloping illiteracy?

What makes me think so? The stamp we need in our passports to visit foreign lands is ***a** visa*. But that word is in fact the plural form of another Latin neuter-gender noun, *visum*, from the verb *videre*, to see. It stands for something seen or inspected. And we used to think that a Latin-based higher education was a prerequisite to entry for the diplomatic service!

4 A FASHION OF FALSE FOREIGN FRIENDS

A spider wraps — envelops — its prey in sticky gossamer threads that it secretes for the purpose. Has anyone ever heard of a spider *un*veloping its prey? Of course not. Not yet, at least!

I believe that most of us who still occasionally send a letter by ordinary mail, rather than by fax or modem, put the sheets of paper into an envelope. But I have lately noticed the spread of a highly infective French micro-organism, the ***unvelope*** virus. Though both Webster and Shorter Oxford allow this pronunciation, they discourage it.

But there's worse to come: since the United Nations, in what we used to know as Yugoslavia, created so-called safe areas in Bihac, Gorazde, Banya Luka, and so on, these zones have also sometimes been referred to as *enclaves*. The silly habit of calling them ***unclaves*** has spread unchecked in recent newscasts.

I await with trembling the first appearance of analogous misconstructions such as *unvironment, unclosure, and uncyclopedia.*

5 OTHER FORMS OF INTELLECTUAL SNOBBERY

In the beginning was the word. Biblical scholars now generally agree that the beginning of St John's gospel is the story of creation retold on another plane.

The human ability to use words *is* a decisive intellectual act of creation. Watch a child as it acquires language and see how it creates the physical world around it by the power of words. An undefined *thing* becomes a *glass*, the *glass* becomes a *glass of water*, and the *glass of water* in turn becomes a *glass of cold water*.

In general terms, the purpose of language is to define, explain, reveal the world to ourselves and to others. In this view, therefore, the use of words to falsify, mislead, obscure, to hide or misrepresent the truth is a form of intellectual snobbery and a crime against our common humanity.

Such snobbery comes in various guises. In his famous *Dictionary* published in 1755, Samuel Johnson drew attention to and roundly condemned one of these, the use of words of Latin, Greek, Italian, or French parentage instead of a simpler, more direct term of Anglo-Saxon origin. Another all-too-common habit is to use long words where shorter ones would do as well or better. Long, involved sentences often accompany and exacerbate this pernicious habit.

Prolixity comes in many guises. It always stems from careless writers and speakers who try to impress the reader or listener by their erudition. To me they merely succeed in displaying their ignorance. Here are a few specimens I picked up in recent weeks: the *angulated* flight deck of an aircraft carrier, *angled* would have done; this was inevitably followed by the cognate neologism *angulation,* instead of *angle*; *could be possible*, instead of *would* or *may be possible*; *linear area* from someone who clearly does not know that a line has only one dimension, length, and an area has two, length and width; *appear to be* rather than *seem*; *in the interim* instead of *meanwhile*; *circular ring*; *in close proximity* where *very near* or *very close* would be shorter and better; *participate* instead of *take part*; *requires (= needs)*; *vanquish (= beat)* an opponent in sport; *ascertain (= find out)*; *following (= after, next)*; *half-decade* (the latest Time magazine gobbledegook for *five years*); the ever recurring *approximately* for *about; conversed with (= talked to, spoke to)*. And why use *however* when *but* is usually enough?

The German preposition *durch* means *through* in a physical, spatial, and cognate sense. It also means *by, by means of.* Why is *through* now becoming a standard cliché in English for *by*?

Other recent examples of such windy arrogance are: *stricken* for *struck (out* or *off)* (I may be stricken with a disease, but a statement cannot be stricken, i.e. deleted), *decedent* instead of *deceased* or *dead, archetypical (= archetypal), two individuals (= two people), two-hour hiatus (two-hour break)*, and so on and on and on.

But don't mistake my meaning. I am not advocating a purification campaign like the absurd one perpetrated in the 1930s on the German language by Hitler's propaganda chief Goebbels. Words such as *Personenauto* and *Lastauto* became taboo and were replaced by the windy monsters *Personenkraftwagen* and *Lastkraftwagen*. Inevitably, these then became abbreviated to *PKW* and *LKW*. *Garage* became *Kraftwagenabstellraum* or some other such polysyllabic nonsense. Yet, mysteriously, their motorways escaped this Germanizing mania and have always been known as *Autobahn*.

Nor do I believe in an official language watchdog such as the *Académie française* that invents such etymological nonsense as *coquetel* as the proper spelling for a *cocktail*. But I suggest that we, the language professionals, to whom words are tools of trade and who depend on words to be reasonably unambiguous, must be ever watchful against the abuse of language, fight it wherever we meet it, and use words only to mean what they are supposed to mean.

6 POLITICAL CORRECTNESS GONE MAD

I was only half surprised to read a recent report that a new Random House dictionary proposes in all seriousness the use of such unetymological monstrosities as *herstory* and *theirstory* that its editor apparently thinks is the politically correct way to replace the word *history*. Allow me a few moments to speculate on other possible ramifications of this new, but surely mistaken approach to the English language.

As an amateur historian, I know that since the beginning of time the world has been a matriarchy in which men have received the semblance of power, but where women have always kept the reality of it firmly in their hands. All my life I have fought for equality of the sexes — I believe *gender equality* is now the preferred, politically correct term for this — though I have always known that in my lifetime we men will never be treated as women's equals. I have my wife's explicit permission to tell you what I'm going to say. I therefore plead with all who are sensitive about this sort of thing to forgive an ignorant foreigner in advance his eccentricities and offences against political correctness. So, let me speculate:

If we take this example to its logical, if absurd conclusion, we will in future no longer call a sibilant sound a *hiss*, but a *hers*; theirstamines will replace histamines; and histrionics must, of course, become herstrionics. And what of histogram, hispid, and Hispanic? Must we also ban these words and change them at a dictionary editor's whim to herstogram, herspid, and perhaps Theirpanic? I hope not! Let's be sensible and stick to conventional etymological rules.

Incidentally, for those who don't know what a hispid is: it's a leaf-beetle and, in a botanical context, something covered by short stiff hairs or bristles, a kind of natural crew-cut.

7 A PLAGUE OF — AND ON — HYPERSUPERLATIVES

At school most of us learned about comparatives and superlatives. To form these in English, we insert the determiners *more* and *most*, and for most of the exceptions we append the endings *-er* and *-est*. With these we make up comparatives such as *more often, more severe, heavier, stronger*; and so on, and their superlatives *most often, most severe, heaviest, strongest*.

Optimus, optima, optimum, respectively, are the masculine, feminine, and neuter nominative forms of the Latin superlative *best*. We derive the verb *optimize*, the adjective *optimal*, and the nouns *optimism, optimist*, and *optimum* directly from it. Imagine my surprise to see, in a recent text for translation, the phrase *a more optimal solution would be to....*. The author might have avoided this solecism by sticking to the plain adjective *good* and its ordinary comparative *better* or its superlative *best,* instead of using the pretentious *optimal* and then creating an absurdity by forming a comparative of it.

We often come across other examples of such ignorant comparatives and superlatives — I call them hypersuperlatives — that sloppy writers use with absolute terms. Here are some random samples from my collection in the last few months from watching that intrusive instrument of the disinformation highway, television: *absolutely essential, especially essential, maximum top price, absolutely lowest rock-bottom bid* (how's that for a double hypersuperlative?), *higher maximization of profits, very minimal, absolutely difficult, not very crucial, a very unique trial,* and similar forms of windy nonsense. Other absolutes we cannot qualify further by adding *more, most, very, absolutely, extremely, perfectly* and so on are *excellent, ideal, perfect, superb, supreme, universal,* and so forth.

8 REDUNDANT QUALIFIERS, PREPOSITIONS, AND ADJECTIVES

A windy solecism one encounters all too often is a qualifying adjective such as *all* or *whole* in conjunction with the preposition *throughout,* which already means *through and through, for the whole of*; for example: throughout the day. To write *throughout the whole day* adds nothing but wind. Other examples heard or read in recent weeks: *all the way through the entire..., throughout the entire flight, a large abundance,*

Then there are redundant prepositions — *up, down, in, out* — tacked on to verbs. You can switch a machine on and off, and you can shut it down, but you can't open down a factory or a market. Then why do we constantly hear of opening them *up*? What difference is there between opening a factory and entering a market, and opening either of them *up*? What's the difference between slowing your speech and slowing it *down*? Between entering a house and entering *in* — or *into* — it?

One of the prize specimens in my collection of nasties is *I saw this observation* when the person concerned simply *made* it. Other pet uglies are: *period of time* — *period* alone would do, because it refers to a continuous space of time; *protocol* for *report*; *one time, two times* instead of *once, twice*; *in the event that* where *if* would do. And I leap out of my skin several times a day when some windbag tells me that he or she is *able to* do this or that when they simply *can* do those things.

Some others in my most recent trawl: *manoeuvre a movement, software program* (either *software* or *program,* but not both!), *actively consider* (what happens when you consider something passively?), *time and time again* (*time and again* means the same), *inside contents* (ever heard of *outside contents*?), *environmental surroundings, at an earlier time* (*earlier* would say no less!), *during the course of* (why not *during*, or ***in** the course of*?), and *very small little*.

Finally, there are redundant adjectives. For example, *in **actual** fact* is simply a windy way of saying *in fact*. And even that is a term we can usually leave out.

9 ILLITERACY IN HIGH PLACES

Illiteracy takes several forms: the inability to read or write, the wrong use of words, and the inability to spell correctly. The first is usually curable by diligent study and competent teaching. The second is a matter of bad habit, curable by strict self-discipline. The third demands a willingness to doubt the correctness of what you have written and to look it up, memory training to help you remember what you have found, and the self-discipline to apply it. The first is not our concern here and now. I have already talked about some aspects of the second. We will therefore briefly consider only the third of these.

For far too long most schools in the countries that are traditionally part of what, for want of a better term, we refer to as Anglo-Saxon culture have neglected language training. Teachers have concentrated on promoting in the student what they fashionably called *self-expression*. What mattered was that you said whatever you had on your mind, no matter how you said it or spelled it. This form of tunnel vision has resulted in a generation that has too much to say, does not listen, and simply can't spell. The process has become all but irreversible, because we now have a generation of teachers who talk too much and also cannot spell!

An example: David, a friend of mine from Iowa, teaches English at a Swiss secondary school; what I think you would call a *senior college* in this country. When I was living in Switzerland, I was a member of the governing body of a local adult-education association. One of my duties was to edit and publish the annual program. It was our custom to ask those who were to run courses to prepare a synopsis of the contents for publication. One year David offered to do a course about the USA. Because he would give his course in English, we decided to print his synopsis in English, to demonstrate the level of language skills that the course would require in the students. After several reminders David at last supplied his. To my surprise I found it full of misspellings, ungrammatical and incomplete clauses, and wrong punctuation. When I tackled him about this, his reply — "it's only a rough draft" — completely floored me. I had always believed that correct spelling, grammatical construction, and proper punctuation become second nature, regardless of whether you're preparing a draft or a finished piece of writing, all the more so if you happen to be a professor of English!

10 LACK OF CLARITY, GRAMMAR, AND SENSE; THE DUTY TO FIGHT PERPETRATORS OF LINGUISTIC CRIMES

Most people would agree that a living language must continually change and grow. But change and growth cannot go totally unchecked. A vine left to nature cannot produce the best grapes for the table or for making wine; a trained horticulturist must prune to make it produce the best fruit. Similarly, we who are skilled word gardeners must prune our language, rather than leave it to natural processes alone by exposing it to the damage done by the careless and the ignorant.

Earlier in this paper I suggested that as language professionals we have a duty to guard against the perversion of language. There are three distinct kinds of such abuse: of clarity, grammar, and sense. Usually they go together.

I also said that I did not believe in language-purification campaigns like that in Nazi Germany, or in language watchdogs similar to that in France. What other options are available?

Recently, a commission of German, Austrian, and Swiss academics hired as consultants by the publishers of Duden and Wahrig, the two standard reference dictionaries of the German language, with about the same status as Webster's and the Oxford English Dictionary, recommended a number of spelling and grammatical changes in German usage to take effect early next century. They decided to recommend changes that reduce the language to its lowest common denominator and accept the most common spelling mistakes in German as "natural". These proposals show that the commission's members are cowards, because it simply is not good enough for teachers to admit defeat and accept that they have proved incapable of teaching people to read and write properly in their native tongue. I believe that in such a situation their proper task is to rethink what and how they teach!

As language professionals and daily users of words as our tools of trade, we can make a major contribution by applying constant pressure on teachers of English, and on all the abusers and perverters of language who cross our path. By the way, my daughter happens to be a teacher of English.

When I read a book, a periodical, an advertisement, I make a note of the malapropisms, mistakes, redundancies, and grammatical or linguistic solecisms that I find, and bring them to the attention of the perpetrators, whoever they are: author, editor, advertising agency, or the advertising agency's client. Some take offence, others are grateful. This is a useful form of quality control that helps maintain standards. If enough of us show that we care enough about our language, the offenders will in time take greater care of how they use words. And let's not forget the politicians, who along with the media people share the doubtful honour of being the worst perverters of language of them all!

11 CALL OUT THE CAVALRY!

Words are at the root of all human genius. Language is a distinctively human skill, a major tool of human creativity, the mirror and yardstick of civilization. Primitive societies can exist by oral communication. But without the written word, no complex society can exist. Without literacy, language is a clumsy, primitive tool that cannot communicate beyond its narrow bounds of space and time.

If education at all levels, from grade school to university, fails to teach linguistic skills, literacy, a proper understanding of words and what they mean, and does not give language the place it properly deserves as the vehicle of culture and the means of communication, our civilization cannot survive. In that case we discard millennia of human creativity that the ability to transmit our thoughts in writing to other members of the human race beyond space and time has given us.

It's high time to call out the cavalry. That means us!

SOURCES OF DISCREPANCY IN TRANSLATING

William H. Snyder
SUNY Binghamton

KEYWORDS: Translation Theory, Translation Error

ABSTRACT: There are legitimate reasons for disagreement between translations and translators, but where differences might arise as the result of error or translator bias there are steps which the translator normally takes to recognize and avoid such "real" error. Even when the translator exercises caution, however, and makes every effort to avoid mistakes, there still seems to be no way of guaranteeing totally error-free translation. The translator works from an intuitive knowledge of language and thus the understanding of what constitutes this knowledge is essential to understanding the nature of language itself. Beyond our strong desire to provide the best possible translation, however, we find that error can be particularly important to us in the information it provides about the art of translation and about language itself, Ultimately knowing what makes up the intuitive knowledge of language is also essential in creating more accurate machine aided translation.

INTRODUCTION

What professional translator would be indifferent to the possibility of error in his translation? Virtually every translator is haunted by the concern that error in some form will creep into a translation just waiting to emerge at a later time and embarrass the translator or, what is an even greater concern, cause actual harm. We all observe the most obvious ways of avoiding error; we proof-read our final product searching for errors in spelling and punctuation, for readability and accuracy, we check all the problem places in the source text and we carefully compare the final text with its source. Most translation work is routine and these standard precautions eliminate the most likely sources of error. Certainly, at this point we should be finished with our work. But in work which is not strictly routine, is there anything which we might add or something we might have done differently to insure a more accurate final product? And: what can we discover about language and translation from the errors themselves?

In my own personal experience I have translated, taught translating to others, and theorized about the nature of language in my linguistic research. My interest in translation error has its origin naturally enough in the teaching of translation where the errors commonly made by students in their work has become a focal point, but in the course of correcting student work - which means comparing student translations with my own - I have also discovered that I myself am not immune to error. This has also been a great teacher of humility and certainly has encouraged modesty when it comes to my own translation work. Correcting translations has also caused me to become more preoccupied with translation error in general and this in turn has led me to try to identify different types of error in the hope of ultimately finding some means of eliminating the sources of error and of teaching others how to translate without error.

Early this century the physicist Robert Milikan first demonstrated that the electron had its own unique electrical charge and in his famous 'oil drop' experiment succeeded in measuring this charge. Within just a few years of Milikan's discovery, however, another researcher using a different method of measuring the electron's electrical charge came up with significantly different results. Reviewing his work Milikan found that he had made an incorrect assumption about the way the oil drop passed through the surrounding medium. In the end, correcting his error had led to yet a second discovery and also to a better understanding about the relative size of the air molecules. His error - when corrected - led to an increased understanding of the physical nature of our world. In language study errors in speech - slips of the tongue, for example - have been a rich and invaluable source of information on the way in which the human brain processes language. Language errors in translating, however, have not been fully evaluated for the information they can yield about translation or language, but this source of error, too, can provide profound insight into the way in which language works in the question of how we understand language, how we communicate with each other, and what knowledge of language actually means.

Error in translation is essentially different from the sort of error which occurs in scientific research where the researcher creates the environment and defines all parameters. Errors in scientific research result from erroneous assumptions and occasionally from blunders in laboratory techniques. The translator does not define the parameters of the source text, thus translation error comes in a variety of forms, most quite different from those found in the physical sciences: first, of course, there is simple ignorance and close to this simple bungling. The student who several years ago in translating a piece of German art criticism turned *den Fährmann Charon* (the ferryman Charon - with the German accusative article) into a *Mr. Faehrman* - even omitting the incomprehensible *Charon* - and succeeded in making several basic errors with this one simple phrase. *Fährmann* does mean, after all, *ferryman* and is not a personal name. The mistranslation is in part the consequence of basic ignorance of a part of the lexicon of the language; it is an error which might have been caught, however, had the translator been more curious about that next word in the sentence. After all, who is *Charon*? one might ask. This error arises from an ignorance of the cultural treasure of both the source and target languages, a bit of ignorance which has become all too frequent in recent years, and lies at the bottom of many translation blunders. The particular student was otherwise a very competent translator and this particular potential source of error lies otherwise hidden - but along with how many others?.

A different sort of error is more characteristic of a non-native speaker translating into his second language. The quaint little toy bird made in Taiwan boasted on the label of its package that "it move, it dance, it sing." And then there is the hotel in Greece which has translated its name into English - obviously in the hope of encouraging English speaking travelers - but its English name "The John" gives, possibly, the wrong sort of encouragement. The Taiwan example shows an inadequate knowledge of the basics of the grammar of the target language, the Greek translation displays an embarrassing unfamiliarity with the culture and language usage in the target language.

One particularly memorable student introduced a unique type of error I have never encountered anywhere else. His knowledge of the source language was far less adequate than his academic credentials indicated but he did not let himself be intimidated by his truly modest linguistic talents. Unlike many students of translation - and some translators as well - he had seized on the fact that the translation should produce acceptable, understandable sentences in the target language, but he went about his work by picking out those few words which he recognized in each sentence of his source - apparently in the belief that looking up unfamiliar words in a dictionary is forbidden. He would then construct from these few words a proper sentence in the target language. The result is not easily described. Each individual sentence presented a tidy little package in the target language but the package had nothing to do with the sentence before it or with the sentence that came after. I also quickly discovered that error of this magnitude can not even be corrected.

Of course, the most frequent and foolish error in understanding the art of translation is the naive assumption that translation simply means substituting a word in the target language for each one in the source. Translators seem to be the only people who understand that this does not work. But at this point a curious sort of error arises out of a fundamental flaw in the process of translating itself. It looks like this: in order to translate correctly we must have at our command an extensive knowledge of the lexicon of both the source and target languages but our understanding of the concept of *word* itself causes serious problems. Consider for a moment that meaning may be expressed in what we refer to as idiomatic expressions by which we mean that the meaning of the words in an utterance taken collectively does not give the actual meaning of the utterance. Thus "He lost his head over the beautiful Tyka." does not really mean that he *lost his head* nor does "His ideas are very much off the wall." does not mean *off the wall* and "es gibt" does not mean *it gives* nor does "il y a" mean *it here has.* We also recognize that "Forget-me-not" is one word and "Korkenzieher" (cork screw, literallly 'cork puller') is just one word as well, but in some languages what we would think of as a sentence can consist of just one - long - word containing the subject verb and objects. Our definition of *word* is a flawed definition; its origins lie in prehistoric times long before the existence of the English language; recent elaborations introducing concepts such as morpheme and phoneme do not remove the inherent misconception, they only complicate it. Words consist of a succession of sounds to which meaning is attached. Sentences are not essentially different from words; they are merely more elaborately constructed words - they are compound words with clearly defined construction. In actual speech, sentences, like words, also flow uninterrupted one into the next and like words sentences can only be understood in the total context in which they occur. In terms of translation this clearly indicates that we must have read the complete text in order to understand the total meaning before we begin with translation. All translators are familiar with this important guideline but most of us - perhaps all - tend to ignore the problem in our haste to get the job done.

In a research article by V. L. Allen and M. L. Atkinson (Detection and perception of deception, in Th. Sebeok & R. Rosenthal eds. The Clever Hans Phenomenon, New York: New York Academy of Sciences, 1981), the authors ask subjects in their study to examine a film of a boy who was suspected of lying about cheating. Some of the subjects were led to believe that the boy was indeed lying while the others were simply asked to look for behavioral clues that would betray attempts at deception. The researchers found that the subjects who "knew" the boy to be lying analyzed his behavior on a much finer level than those who were simply looking for clues to possible deception. The knowledge that they were viewing deceptive behavior actually elicited the detection of twice as many nonverbal gestures indicating deception. The study clearly showed that the bias which the researchers had introduced had led to an excessive analysis of small details in which every detectable behavioral gesture was interpreted as revealing deceptive behavior. These details had in fact no real significance in detecting lying behavior on the part of the actor who played the deceptive child on film.

In translating a comparable error is introduced when the text is read closely but without a prior understanding of its full meaning. This can happen when the translator has not read the text thoroughly beforehand or in his focus on detail he has lost sight of the overall picture or, as is often the case, the translator has not understood the text. When the source text is translated word by word without attention to the full text, the many possible meanings of each individual word can, in the absence of the limitation imposed by knowledge of the full text, only lead to an over-analysis of the text and to confusion the final translation. A study by George A. Miller and Walter G. Charles on "Contextual Correlates of Semantic Similarity" (in: Language and Cognitive Processes v.6, n.1, pp. 1-28, 1991) compares semantic and contextual similarity for pairs of nouns that vary from high to low and finds an inverse relationship between similarity of meaning and the discriminability of contexts. It is significant for us that the similarity of context is the deciding factor in determining the meaning of words in source and target languages.

The meaning of the word is defined and delimited by the sentence in which it occurs, the meaning of the sentence by the next larger context. This progressive encapsulation of meaning can expand to include an entire book or it may be confined to a very few words within some immediate external context. The significant role of context in determining meaning has, however, received until recently very little attention from linguists and philosophers who deal with language phenomena (cf. Roberts, Lawrence. *How Reference Works*. Albany : State University of New York Press. 1993.; Levinson, Stephen. *Pragmatics*. Cambridge & New York: Cambridge University; W. H. Snyder. Aspects of Context in Translating. Vistas. Proceedings of the American Translators Association. Medford, NJ.: Learned Information, 1994). By way of illustration of the significance of context we can draw on an example from an Icelandic saga which tells of the adventures of a pious young Christian - Authun. The hero uses all of his money to buy a polar bear which he then plans to take to Denmark where he hopes to make a present of his treasure to the king. He eventually succeeds in getting to Denmark where one of the king's stewards demands that Authun give him a half share in his bear in exchange for food for the now penniless Authun and his bear. But when Authun comes before the king he reveals the steward's duplicity: "... keypti ek þat meþ allri eigu minni, og nú er þó á orðit mikit fyrir mér, ek á nú hálft eitt dyrit, ok segir konungi síðan, hvérsu farit hafþi meþ þeim Áka, ármanni hans." ("... I bought *that* with all my possessions and now though has happened something big to me, I have now half of one animal, and (he) tells (the) king then how it had gone with that Aki, his steward.") This is translated by Hermann Pálsson: "... when I came to see this man Aki my last penny had been spent; I was dying of hunger, and so was the bear. Now the gift's been spoiled. Aki would not help me or the bear unless he got a half-share in it." Knowing some of the tale we can understand partly what the original line means but the sentence standing alone would be meaningless when translated word for word and without knowledge of what had gone before. Pálsson has gone beyond this simple translation and introduced into the translation material from the preceding lines in order to clarify the meaning of the pronoun 'that.' He also adds to the translation in order to explain the 'how it had gone' in this particular line while at the same time adjusting the syntax (the original "I bought ... he tells") to fit contemporary English. This episode occurs within a longer narrative; its meaning must be gained from the immediate context but it is also essential to the meaning of this brief passage that we understand all that has gone before it.

The word for word translation may seem at first glance to be a trivial source of error, something which would rarely occur, but in practice it is an error which I have encountered frequently - and not just in student work, but even in highly regarded translations of literary works. It is an error which is particularly prominent in machine aided translation where the computer program can not and does not understand the source text. I am reminded here of one of our colleagues who found a machine translation in which the English word 'patient' as in 'hospital patient' had been translated into German as 'der Geduldige', i.e., 'one who is patient, forbearing' instead of German 'Patient.' The translation program here takes the more frequent meaning of the English word rather than the meaning of the word in its specific context. We tend to find such errors amusing but, as professional translators, we are also irritated by and concerned about mistranslation and we certainly do not want to see significant error where there is actual potential for harm to others as a consequence.

Njozi, Hamza Mustafa (Text and Context in Translation: The Phenomenological Approach. 1991) has analyzed the phenomenological approach to text translation and the deconstructive approach and, contrary to the assumption of the post structuralists that it is impossible to transfer pure meaning from one language to another, has concluded instead that exhaustive translation is possible though difficult. Unfortunately for his argument and for us as translators, both sides of this discussion are essentially in error. The one assumes that the inherent plurality of meaning (of words) makes precise determination of meaning impossible, i.e., the translator is translating word for word without regard to context - which, as we know, would be foolish. The phenomenological approach takes the more traditional concepts of translation, proceeding from the conviction that it is possible to make an independent, objective judgment about reality, i.e., that the translator understands precisely what the author means. This latter argument thus assumes that the translator and the author necessarily have the same total knowledge of reality - including language - which is patently impossible. The translator can and does only approximate in the target language the meaning of the original text as he understands it. Recent discussions of context and meaning mentioned above have revealed some of the error of this assumption.

Some years ago the linguist John Lyons suggested that "Semantics is concerned with accounting for the degree of uniformity in the 'use' of language which makes normal communication possible." It seems clear to me, at least, that in all discussions of the way in which meaning is expressed in language - and this is after all what translators deal with - there is an implicit assumption that we all understand each other with few if any mistakes. But these discussions themselves show profound differences even in what philosophers and linguists believe is meant by 'meaning.' In reality we do misunderstand each other quite often, but in day to day communication we quickly correct these misunderstandings and they go virtually unnoticed. In those situations where the concomitant correction is absent - in translating the written text, for example - the burden of correcting misunderstandings lies with the translator. In effect we are trying to correct misinterpretations where we may not even be aware of their existence. It seems an impossible task and we should wonder that translators are generally as reliable in their task as they indeed are, but this is a thought which should inspire great humility in us.

We do everything we can to avoid error in translation. Well developed editing skills help us catch our 'clumsy' errors - wrong word, bad spelling, bad punctuation, etc. Error arising from ignorance of the cultural context or the technical area knowledge of the source text we avoid by knowing as much as possible about the culture or the specific subject matter. Beyond this we are at the mercy of the limitations of our intuitive powers. We will always make errors because we have simply not understood our source; this is a real human limitation is not readily overcome. In the machine aided translation, however, we have serious difficulties in building into the program that intuitive understanding of the language which allows us to know as much as we do of the meaning of words and sentences within the totality of the a source text. While I do not think this an impossible task, I have yet to see any translation programs which can deal well with this specific difficulty. At best, however, the machine translation will be no better then human translation, i.e., at best it too will still make the same sort of errors that we human translators make.

REFERENCES

1 Allen, V. L. and Atkinson, M. L. . Detection and perception of deception, in Th. Sebeok & R. Rosenthal eds. *The Clever Hans Phenomenon*, New York: New York Academy of Sciences, 1981

2 Levinson, Stephen. *Pragmatics*. Cambridge & new Yord: Cambridge University Press. 1984.

3 Lyons, John. *Introduction to Theoretical Linguistics*. Cambridge : At the University Press, 1968.

4 Miller, George A. Charles, Walter G. Contextual Correlates of Semantic Similarity. *Language and Cognitive Processes* v.6 n.1, pp. 1-28, 1991

5 Njozi, Hamza Mustafa. *Text and Context in Translation: The Phenomenological Approach*. 1991.

6 Roberts, Lawrence. *How Reference Works*. Albany : State University of New York Press. 1993.

7 W. H. Snyder. Aspects of Context in Translating. *Vistas. Proceedings of the American Translators Association*. Medford, NJ.: Learned Information, 1994.

Connections

German

"COARSE, LOUD, AND EXTREME"? LUTHER'S BIBLE REVISIONS COMPARED WITH THE GREEK ORIGINAL

Geoffrey S. Koby
Kent State University

Keywords: Luther, Bible Translation, Early New High German, Revision

Abstract: Between first publication of Martin Luther's German Bible translation in 1522 and the last edition before his death of 1545, Luther and his team of Bible scholars made thousands of revisions. Some have seen Luther's earlier language as vulgar and coarse, especially in the Gospels, and have suggested that one major principle of revision was moderation of his language over time. This paper compares the 1522 and 1545 versions with the Greek original, showing that Luther's revisions fall into three main categories: revisions making the translation adhere more closely to the original, those that make it more independent, and those which are primarily stylistic changes not affecting the distance from the original. Luther's primary concerns are adherence to the original and comprehensibility in German. The two sometimes conflict, and when they do, the original always takes precedence. Other identifiable translation principles are aesthetic/stylistic considerations, revisions to approximate a more popular style, correction of logical/factual errors, transference/avoidance of foreign words, quality of the German phrase, approaches to the emerging German standard language, and textual/grammatical revisions. All types are presented, with examples, and summarized. The evidence casts Luther in a positive light: as a translator deeply concerned with the accuracy and readability of his translation.

1. INTRODUCTION[1]

When Martin Luther started translating the Bible from Greek into Early New High German in 1521, it is probably fair to imagine that he did not expect to still be working on revisions twenty-four years later. The first edition of his translation of the New Testament was published in September of 1522, and he read proof on the 1545 edition of the Bible, the last version published before he died in 1546. Luther's revision work began immediately after the printing of the September Bible in 1522, as Wolf points out: "In the space of a few weeks, Luther made more than 500 changes in its [the Bible's] language, which were included in the printing of the December Testament" (Ref. 21).[2] In principle, every word of the translation was subject to comparison and change: "Luther realizes that he is not always immediately successful in grasping the content of the original text and then reproducing this content adequately in the target language. The translator never stops learning. On the one hand, he gains ever-better command of the foreign language by constant effort, and on the other hand, he constantly acquires new insights into his own language. Such efforts are the translator's burden. In the most extreme case, as with Luther, it means a lifetime of revision" (Ref. 10).[3] Some scholars have seen Luther's lexical choices in the 1522 version as radical and provocative: "coarse, loud, and extreme" (Ref. 6) and the Gospels are emphasized as being especially coarse (Ref. 19). In this paper, however, comparative data will show that Luther's changes are primarily motivated either by the source text or by comprehensibility considerations, and his original translation is not as radical as some have seen it. Specific examples which have been used to argue for the radical interpretation will be reconsidered in light of the Greek text.

We should not imagine, however, that Luther revised the Bible alone. The work was too important to be left in the hands of a single individual. Luther's contemporary opponents, such as the Catholic Bible translators Emser and Witzel, had already expressed the need for a team of translators, and over the course of his life, Luther formed a group of learned friends and colleagues with whom he spent many hours discussing the correct translation of passages which he did not succeed in translating smoothly the first time around (Ref. 10).

The Gospel of Matthew was subjected to a close reading in facsimile editions of both the 1522 New Testament (hereinafter: 1522) and the 1545 Bible (1545), in order to identify major lexical changes. A lexical change is defined as any revision which affects the meaning, including changes as small as a verbal prefix or as large as recasting an entire phrase. A gospel, rather than some other portion of the Bible, was chosen both for its potential radical language and for its combination of historical and theological content. In this way, word choices could be found in sufficient numbers to find underlying regularities. 185 major lexical changes were identified and compared with the Greek source text, then assigned to three major classes: I. changes which bring the translation closer to the original Greek meaning, II. changes which diverge from a close rendering of the source text, for comprehension or esthetic reasons, and III. changes which are neutral with regard to the source, originating from target language (German) considerations.

The principles of revision fall into ten categories, or types, which are assigned to the three major classes. Class I includes only one category, original Greek meaning (Type O). Class II, divergences, includes factors of understandability (U), esthetics (E), popular language (P), and logic (L), as well as the elimination of foreign loan words (F), while Class III, source neutral changes, covers changes which improve the quality of the translation as a text in German (Q), as well as textual revisions (T) such as changes from an adjective to a verb, new words which are more consonant with the emerging modern standard German language (S), and error correction/carelessness (C). Each change was assigned to one or more categories, in order to account for all of the possible motivations behind each change. No obvious occurrences of error correction/carelessness were found. A summary of the categorical distribution of the forms is found in Table 1.

Class	Type	Number (n=185)	Percent
I	O	77	42%
II	U	56	30%
	E	22	12%
	P	14	8%
	L	12	6%
	F	5	3%
III	Q	29	16%
	S	21	11%
	T	7	4%

Table 1 - Summary of Class and Type Distribution

In order to determine the extent and impact of the revisions, three historical reference works were consulted: Grimm's *Deutsches Wörterbuch* (Ref. 11), Phillipp Dietz' *Wörterbuch zu Dr. Martin Luthers deutsche Schriften* (Ref. 8), which was only completed from *A* through *Hals*, where possible, and Franke's *Grundzüge der Schriftsprache Luthers* (Ref. 9), which, though not as complete as a dictionary, contains many useful definitions, particularly of words used by Luther differently than used today. The recently published first volume of the *Frühneuhochdeutsches Wörterbuch* (Ref. 2) was consulted for *anschnauben* and *anfahren* (Matth. 19:13). The principles of translation which others have determined are examined in the discussion of each type. Definition translations are my own.

2. CLASS I, TYPE O: ORIGINAL GREEK MEANING

As in any translation, the original text is the primary authority. In Biblical translation, especially in Luther's day, the Greek source has unusual authority as one of the three "holy languages" along with Latin and Hebrew. Since Luther translates with the intention of preserving all of the meaning of the original, the Greek source text is his guide, even when German style might suffer. In his own reflections, written in the interesting mixture of Latin and German which he apparently preferred (and possibly thought in?), Luther states that he sometimes had to reject a word, even though he would have rather have lost a "red guilder," because, as he wrote, "as well chosen as a word sometimes might be, it still doesn't always match the sense" (Ref. 10).[4] Of the 185 major lexical changes in Matthew between 1522 and 1545, 77 (42%) of the changes are Class I, making the German translation more closely consonant with the original Greek text, while 41 (22%) make the German diverge. Some examples of Type O changes:

Vs.	Luther - 1522	Luther - 1545	Greek NT	Literal Greek
17:5	gehorchen yhm	Den solt jr hören	αὐτοῦ ἀκούετε	him hear ye
13:57	daheym vnd bey den seynen	in seinem Vaterland vnd in seinem Hause	ἐν τῇ πατρίδι αὐτοῦ καὶ ἐν τῇ οἰκίᾳ αὐτοῦ	in his [own] country and in his [own] house
6:28	nehen	spinnen	νήθει·	do they spin

Comparing these German examples with the Greek clearly shows that Luther chose to bring the translation closer to the original Greek meaning in the later revision. In 17:5, "hearing" is not *gehorchen*, obeying, but *hören*, in 13:57, *daheym* is 'at home,' which misses the meaning of 'own country,' and in 6:28, there are technical differences between *nehen* 'sewing' and *spinnen* 'spinning.' 48 revisions (26%) have Type O as the sole reason for the change. In 3:3 for example, the 1545 version is grammatically closer to the Greek, as Juntune points out: "Ma. 3:3 is probably best analyzed as a more accurate or literal translation, as in Greek βοῶντος does not grammatically modify 'voice', but is a genitive masculine participle. The 1522 translation may reflect the Hebrew of Isaiah, from which this is translated into Greek" (Ref. 13):

Matthew 3:3

(1a) 1522: Es ist eyn *ruffende stymme* ynn der wusten/
'there is a *calling voice* in the desert/'

(1b) 1545: Es ist eine *stimme eines Predigers* in der wüsten/
'there is a *voice of a preacher* in the desert/'

(1c) Greek: Φωνὴ βοῶντος 'voice of one crying'

(1d) RSV The voice of one crying in the wilderness [5]

Eight Type O examples show omissions in 1522 which were inserted at a later time. Given the conditions under which Luther translated the *Septembertestament* (alone, in hiding in the Wartburg, and without access to reference works or other Biblical scholars), there are two possible reasons for such omissions: the most probable is that the Greek text from which Luther worked did not contain these phrases, while it is also possible that some of them represent simple errors on Luther's part.

The other Type O changes, while still motivated primarily by the Greek text, also show other possible reasons for the change. Of these, most improve the quality of the German text. Similarly, the most numerous group of counter-Type O (!o) examples are motivated by understandability reasons, along with several others. These types will be discussed under their respective headings.

3. CLASS II, TYPE U: UNDERSTANDABILITY

"The changes which Luther makes to his Bible translation can for the most part be divided into the following categories: they are either required by the source text, or serve to make the text more understandable, or are motivated by considerations of language aesthetics" (Ref. 21).[6] This principle, though mentioned only by Wolf, is the primary motivator for changes which take the German text away from a close rendering of the Greek text. 56 (30%) changes are made for this reason. No changes were found which would make the text *less* understandable. 22 (12%) changes are made counter to the Greek text, while five (2%) are simultaneously closer to the Greek and more understandable. 11 (6%), which improve understandability, are also motivated by the quality of the German, that is, the principles work in tandem. An example of Type U will serve to illustrate this type:

Matthew 1:18

(2a) 1522:	Als Maria seyne muter dem Joseph vertrawet war/*ehe sie mit eynander zu hausz sassen/*
	'When Mary his mother was betrothed to Joseph/*before they sat together at home/*
(2b) 1545:	Als Maria seine Mutter dem Joseph vertrawet war/*ehe er sie heim holet/*
	'When Mary his mother was betrothed to Joseph/*before he fetched her home'*
(2c) Greek	πρὶν ἢ συνελθεῖν αὐτοὺς
	'before came together they *(to cohabit matrimonially)'*
(2d) RSV	When his mother Mary had been betrothed to Joseph, before they came together

Here, Luther strives to translate the concept of marital cohabitation, which does not occur as a convenient verb or collocation in German. The 1522 version could mislead the German reader by emphasizing mutual co-residence, while the 1545 text implies the consummation of the marriage.

Another of the Type U examples must be considered as well, one of only two in the entire gospel which support the argument that Luther was moderating radical or coarse language (the other will be discussed under Type E):

Matthew 27:27

(3a) 1522:	vnd samleten vbir yhn die gantze *rotte*	'and the whole *mob* gathered around him'
(3b) 1545:	vnd samleten vber jn die gantze *Schar*	'and the whole *troop* gathered around him'
(3c) Greek	σπεῖραν	'band of soldiers'
(3d) RSV	and they gathered the whole battalion before him	

Greek σπεῖραν is neutral, simply a 'band of soldiers', and in the 16th Century *Rotte* could also simply mean 'a group (especially of soldiers).' It is in this sense that it is used here. Grimm (Ref. 11) mentions that, even in the 19th Century, *Rotte* retained the specialized military meaning, but also (and among lay people, usually) meant 'evil group, mob'. Luther uses the pejorative sense elsewhere in the Bible, in the sense of 'Sekte, Partei', but all of these verses imply a negative attitude towards the mentioned group (Ref. 9, 13). Furthermore, according to Grimm, Luther's use of this word outside the Bible referred specifically to the non-Lutheran groups in opposition to him, and in this sense had a distinctly pejorative meaning. *Schar,* though originally a military term for a certain number of soldiers, and later for any group of soldiers (14:2170), usually means 'any large group of people' (14:2172). The change of *Rotte* to *Schar* is almost neutral with respect to the Greek text, possibly somewhat less precise in a purely military-technical sense. However, given Luther's target audience and his use of the word elsewhere, the change removes the pejorative overtone from the text, and in this sense is indeed more moderate.

4. CLASS II, TYPE E: ESTHETIC CONSIDERATIONS

Many have seen esthetics as one of the guiding principles for Luther's work (Ref. 3, 7). Others see Luther's early language as representative of his role as a radical reformer (Ref. 19), and this opinion is still current in some circles: "The language of his early translation of the New Testament of September 1522 contains a number of harsh and even radical words and phrases which were changed or dropped in his later translations. The language of this September Bible is not only popular and plain, but coarse, loud, and extreme" (Ref. 6). These opinions can be refined to what is essentially a single principle; as Baeumer puts it: "From [1525] on, Luther's Bible language changed from the coarse and extreme to a refined and standard vocabulary and a more cultivated style" (Ref. 6).

For the Gospel of Matthew, three aspects of esthetic changes must be noted: first, they cannot be the primary motivation, since only 22 (12%) of the 185 changes are made for any esthetic reason; second, four changes are made counter to esthetic principles; and third, it is rare to find a change made solely for esthetic reasons. Only seven (4%) are purely Type E (Table 3), and only one goes directly counter to the original Greek. One of the seven has been seen as a moderating change:

Matthew 19:13

(4a) 1522:	die iunger aber *schnaubten* sie *an.*	'but the disciples berated them'
(4b) 1545:	Die Jünger aber *furen* sie *an.*	'but the disciples rebuked them'
(4c) Greek:	ἐπετίμησαν	'rebuked'
(4d) RSV:	The disciples rebuked the people;	

For *anschnauben*, originally the snorting of horses (1:447), Grimm lists *anfahren*, 'to speak with strong words, attack with words' (1:322). Both words carry a strong negative connotation in modern German. Luther uses *anschnauben* in the Old Testament in 1545 in an adversary situation (1. Samuel 25:14). Dietz defines *anfahren* as 'to speak to someone with strong words,' and *anschnauben* as '**strongly** *anfahren,* sneeringly speak to' (emphasis mine). In the *Frühneuhochdeutsches Wörterbuch*, these words are near synonyms, but *anschnauben* includes the synonyms *anherrschen* 'bark at' and *anschnauzen* 'snarl at.' The harsher tone of *anherrschen* and *anschnauzen* suggests that it is this additional, negative character of *anschnauben* which may have led Luther to change it to the less aggressive *anfahren*.

The four counter-esthetic changes, considered only from the point of view of the German text, could be seen as making the text stronger, more explicit, and therefore less esthetic or refined. However, three of the four adhere more closely to the original, showing that the original meaning can take precedence over esthetic considerations. (Note, however, that there are also four esthetic changes which more closely follow the original.) The fourth, 17:14, not only strengthens the tone of the passage, but at the same time directly counters the original:

Matthew 17:14

(5a) 1522: tratt tzu yhm eyn mensch/vnnd *beuget die knie gegen yhm*
'a man stepped up to him/and *bent the knees towards him*'

(5b) 1545: trat zu jm ein Mensch/vnd *fiel jm zu füssen*
'a man stepped up to him/and *fell at his feet'*
(5c) Greek: γονυπετῶν αὐτῷ 'kneeling down to him'
(5d) RSV: a man came up to him and kneeling before him said,

In 1522, Luther uses *beugen* where modern German would expect *biegen*, particularly in *das knie beugen* 'bending the knee.' *Knie* appears with both verbs, often in the context of showing servitude, but also of kneeling in prayer. The numerous citations of Luther's usage in Grimm 1:1743 show this phrase primarily in the sense of showing allegiance, respect, or servitude, as in 2 Kings 1:13, or in the unchanged phrase in Matthew 27:29, where the Roman soldiers *beugeten die knie fur im* 'bent the knees before him' in mock homage. Dietz agrees that *beugen* occurs frequently, particularly for 'physical *biegen*, flectere'. The earlier version also reflects the original Greek use of γονυπετέω from γόνυ 'knee' (Ref. 13).

For the 1545 phrase, Grimm lists *ich falle zu füszen* 'I fall at your feet' implying 'I am an obedient servant,' as well as the verse in question, similarly in Mark 5:22. The implication of "zu füszen" is often subordination, as shown by Grimm's numerous examples under *fusz*, but includes the specific meaning of kneeling for deferential, humble requests, as in 1 Sam. 25:24, or Mark 5:22. In the marginal gloss to 2 Kings 2:3, Luther says "to be at the feet means to be a disciple and subordinate," but Dietz also notes that the act occurs "as a sign of praise or worship." It would seem, then, that Luther's change in this case was primarily motivated by understandability, even though it is counter to both esthetic principles and the Greek original. In this case, instead of showing homage, or praying, the person in question is kneeling to make a request, and this meaning is clearer in the later translation.

5. CLASS II, TYPE P: POPULAR LANGUAGE, TYPE F: FOREIGN WORDS, TYPE L: LOGIC

Vs.	Luther - 1522	Luther - 1545	Greek NT	Literal Greek	Type
11:25	verstendigen	Klugen	σοφῶν	wise	p
5:44	benedeyt die euch maledeyē	Segenet die euch fluchen	εὐλογεῖτε τοὺς καταρωμένους ὑμᾶς	bless those who curse you	f
27:5	erwurget	erhenget	ἀπήγξατο	hanged himself	l

These three categories represent the less significant principles of Luther's revisions in this class. Only 14 (8%) of the change are of Type P, which Baeumer (Ref. 6) saw as popular and plain, Arndt (Ref. 3) called a style close to everyday speech (above) and where Schirokauer notes, "Luther ... varies word forms and syntax, changes the word order, usually according to the requirements of spoken language" (Ref. 19).[7] Six of this group also have other reasons for the change, while eight run counter to other principles in approaching a more popular style.

Both Wolf (Ref. 21) and Wodňanská refer to the second type, foreign loan words: "Luther seeks to be close to the people, and avoids foreign elements: he later replaces foreign words such as *benedeien* and *Firmament* with *segnen* and *Himmel*" (Ref. 20).[8] Five examples were found, all involving *benedeyen* and *maledeyen*. However, there are also four counterexamples, as Becker points out (Ref. 7). In 154, Luther returns to the original Greek by replacing the German words *feyertagen, gulden,* and *meyster* (twice), with *Sabbath, Stater* (a Roman coin), and *Rabbi*.

Finally, in Class II, simple logic (as a subset of general understandability) is a motivating principle in a few changes. These 12 occurrences (6%) sometimes take the text away from the Greek, as in:

Matthew 9:15

(6a) 1522:	des breutgamsz kinder	'the bridegroom's children'
(6b) 1545:	die Hochzeitsleute	'the wedding people'
(6c) Greek:	οἱ ὑοὶ τοῦ ςυμφῶνος	'the sons of the bridechamber'
(6d) RSV:	the wedding guests	

Here, the literal Greek is so distant from the way this meaning would be expressed in German as to be incomprehensible to the target audience, since the linguistic logic of the 1522 version states that the bridegroom already has children. This would be illogical to the German speaker. Logic is also at work in the three occurrences (26:49, 27:29, 28:9) of the Greek greeting Χαίρε(τε) 'Hail'. In 1522, these phrases, *gott grus dich* and *got grusse euch*, include the word *got(t)*, as is still customary in Modern (Southern) German *Grüß Gott* 'God's greeting'. The word is not implied in the Greek, and in 27:29, Roman soldiers mocking Jesus are hardly likely to have used a religious greeting. Consistency in translation would have required the change of the other two occurrences as well.

6. CLASS III, TYPE Q: QUALITY OF THE GERMAN TEXT; TYPE S: STANDARDIZING INFLUENCES; TYPE T: TEXTUAL (GRAMMATICAL) MODIFICATIONS

Vs.	Luther - 1522	Luther - 1545	Greek NT	Literal Greek	Type
5:42	der von dyr borgen will	der dir abborgen wil	τῷ αἰτοῦντί σε	To him who asks of thee	q
14:26	eyn spugnisz	ein Gespenst	'Ότι φάντασμά	An apparition	s
18:35	vō hertzen	von ewrem hertzen	ἀπὸ τῶν καρδιῶν ὑμῶν	from your hearts	t

Naturalness in German is the third class of revisionary principle which Luther adheres to. This means that he is always seeking the best expression of a thought in German, not merely an equivalent one. Wodňanská points out that nouns taken over by Luther in 1522 from the Greek or Latin text are later expressed as verbs (Ref. 20). There are 29 changes (16%) which affect the quality of the German text (Type Q). Six reflect Greek γέγραπται 'it has been written', a legal expression changed from *ist geschrieben* to *steht geschrieben*, denoting its more permanent nature. Most of the Type Q expressions are also in accordance with other factors, particularly the original text. This is hardly surprising, since improvements in the quality of the text which counter the original are rare. Of the five which occur in Matthew, two are for understandability reasons. The remaining three, though counter to the literal meaning of the Greek text, lose very little meaning.

21 changes (11%) reflect standardizing tendencies (Type S). Some see Luther's effort towards eliminating localized or dialect forms and the care he took in working with the Bible as a primary driving force: Luther attempted to write a more widely acceptable form of German, "by suppressing purely East Middle German special forms more and more as time went on" (Ref. 5).[9]

> In the tracts and pamphlets of the first five years, his writing was hardly carefully considered - and was thus far more idiosyncratic than, for example, the language form of an orderly chancery of the time, with professional officials. Even the

language of the September Testament of 1522 was marked by far more care. And yet, already in the December Testament of the same year, almost every line was changed, and almost always to forms which are identical to Modern Standard German or comes closer to it (Ref. 4).[10]

Last of the types of change is that of textual or grammatical revision (Type T). Becker also cites purely textual principles accounting for some of the differences between 1522 and 1546: Nouns become verbs in 1546, nouns become adverbs in 1546, concrete nouns become abstract, repetitions of content or words are eliminated, and words implied by context in 1522 are made explicit in 1546 (Ref. 7). Because these principles are only comparative between the German versions, they are of little use. Most of the changes Becker cites are actually occurrences of other types, so that just seven (4%) can be attributed to this type, and only one (18:35) is apparently solely motivated by a textual principle. At the same time, six examples were found which counter these textual principles. Type T changes, then, are best explained by reference to other principles.

7. CONCLUSION

In all of the different types of revisions discussed above, there are significant numbers of examples which overlap into the other classes. Only Type O and Type U show large numbers of changes which are solely motivated by these two principles, and they also play significant roles in most of the other, less important, subtypes. It is thus clear that the most significant principle of Luther's revision of his Bible translation is his attempt to adhere more closely to the original Greek text (Class I). Where the revisions are contrary to the Greek text (Class II), the primary reason is to resolve misunderstandings arising from a too literal translation, while esthetic principles are secondary. The other types in Class II, popular language, logic, and foreign loan word changes, are subtypes of the larger principle of understandability. The Class III changes, neutral with regard to the Greek text (quality, standardization of language, textual/grammatical revisions), are clearly only minor principles of revision.

The traditional view of Luther's Biblical lexical choices in 1522 as being "coarse, loud and extreme," vulgar, and radical, especially in the Gospels, is insufficient to explain the vast numbers of revisions which were seen even in only one small section of the Bible translation. In the changes considered here, the majority do not affect the tone. It would seem that Luther's primary concerns were twofold: first, to acknowledge the primacy of the Greek text, which stood as a constant corrector of his attempts to render the Bible in German, and second, clarity and exactness in his German. Although Luther's other writings do exhibit a tendency to condemn his opponents in the strongest possible terms, radicalism seems not to have been a guiding principle in the Luther's translation of the New Testament. Thus we should see Luther as a dedicated Biblical translator and revisor, who constantly strove to achieve the most complete, exact, and German translation possible while remaining true to the Greek original.

NOTES

1. A much longer discussion of this topic, including all of the relevant German and Greek data, will be appearing in the *American Journal of Germanic Linguistics and Literatures.*

2. "Innerhalb weniger Wochen nimmt L. daran mehr als 500 sprachliche Änderungen vor, die in den Druck des Dezembertestaments (1522) eingehen."

3. "Luther [macht] die Erfahrung, daß es nicht immer auf Anhieb gelingt, den Inhalt der Vorlage zu erfassen und diesen Gehalt dann in der Zielsprache mit adäquaten Mitteln wiederzugeben. Der

Übersetzer lernt niemals aus. Aufgrund ständigen Bemühens beherrscht er einerseits die Fremdsprache immer besser, andererseits gewinnt er auch stets neue Einsichten in seine eigene Sprache. Solches Bemühen macht die Last des Übersetzers aus. Im äußersten Fall bedeutet es, wie bei Luther, lebenslange Revisionsarbeit."

4. "Ich mus offt ein gut wort in translatione verwerffen, das ich lieber ein rothen fl. verlirn wolt quam illud verbum, quia quantumcunque nonnunquam elegans est aliquod verbum, tamen quandoque non convenit ad sententiam."

5. All verse references are to the Gospel of Matthew and will be cited as Chapter:Verse. Examples include the Revised Standard Version text for reference.

6. "Die Veränderungen, die L. an seiner Bibelübersetzung vornimmt, lassen sich großenteils in folgende Katagorien gliedern: sie sind entweder vom Urtext her bedingt oder dienen der besseren Verständlichkeit oder beruhen auf sprachästhetischen Erwägungen."

7. "Luther ... variiert Wortformen und Satzbau, ändert die Wortstellung, meist nach den Erfordernissen des Gesprochenen..."

8. "Luther sucht die Volksnähe und vermeidet Fremdes: Fremdwörter wie *benedeien* und *Firmament* in seinen älteren Bibelausgaben ersetzte er später durch *segnen* und *Himmel*."

9. "indem er rein ostmitteldeutsche Sonderformen mit der Zeit mehr und mehr zurückdrängte"

10. "In den traktaten und kampfschriften der ersten fünf jahre war seine schreibe kaum reflektiert - deshalb weit willkürlicher als z.b. die sprachform damals in einer geordneten kanzlei mit professionellen beamten. Schon die sprache des septembertestaments 1522 war von weit grösserer sorgfalt geprägt. Und doch wurde bereits im dezember testament desselben jahres fast in jeder zeile geändert und beinah immer zu formen, die mit dem neuhochdeutschen identisch sind oder ihnen näher kommen."

REFERENCES

1. ---. N.D. *The Greek-English New Testament, Being the Original Greek Text with a Literal Interlinear Translation*. Philadelphia: McKay.

2. Anderson, Robert R., Ulrich Goebel, Oskar Reichmann. 1986- . *Frühneuhochdeutsches Wörterbuch*. Band 1. Berlin/New York: de Gruyter.

3. Arndt, Erwin. 1962. *Luthers deutsches Sprachschaffen. Ein Kapitel aus der Vorgeschichte der deutschen Nationalsprache und ihrer Ausdrucksformen*. Berlin: Akademie-Verlag.

4. Arndt, Erwin. 1984. "Luthers Bibelübersetzung--eine revolutionäre Tat." IN: Schildt 1984a. Vol. 1. Pp. 59-76.

5. Bach, Heinrich. 1984. "Wo liegt die entscheidende Wirkung der 'Luthersprache' in der Entwicklung der deutschen Standardsprache?" IN: Schildt 1984a. Vol. 1. Pp. 96-107.

6. Baeumer, Max L. 1984. "Luther and the Rise of the German Literary Language: A Critical Reassessment." IN: *The Emergence of National Languages*. Ed. Also Scaglione. Ravenna: Longo Editore, 95-117.

7. Becker, Ilse. 1935. *Luthers Evangelienübersetzung von 1522 und 1546*. Phil. Diss. Köln.

8. Dietz, Phillipp. 1870. *Wörterbuch zu Dr. Martin Luthers deutschen Schriften: A-Hals*. Leipzig: Vogel.

9. Franke, Carl. 1914. *Grundzüge der Schriftsprache Luthers. Zweiter Teil: Wortlehre*. Halle: Verlag der Buchhandlung des Waisenhauses.

10. Gelhaus, Hermann. 1989. *Der Streit um Luthers Bibelverdeutschung im 16. und 17. Jahrhundert. Mit der Identifizierung Friedrich Traubs*. Tübingen: Niemeyer.

11. Grimm, Jacob and Wilhelm Grimm. 1854-1971. *Deutsches Wörterbuch*. 33 Vols. Nachdruck. München: Deutscher Taschenbuch Verlag 1984.

12. Hahn, Sönke. 1973. *Luthers Übersetzungsweise im Septembertestament von 1522; Untersuchungen zu Luthers Übersetzung des Römerbriefs im Vergleich mit Übersetzungen vor ihm*. Hamburger Philologische Studien 29. Hamburg: Buske.

13. Juntune, Thomas. 1993. Personal communication.

14. Luther, Martin, Trans. 1522. *Das Newe Testament Deůtzsch*. Wittenberg--Rpt. of original edition in the Universitäts- und Landesbibliothek Halle/Saale. Edition Leipzig. Stuttgart: Deutsche Bibelgesellschaft, 1982.

15. Luther, Martin, Trans. 1545. *Biblia: Das ist + Die gantze Heilige Schrifft/Deudsch/Auffs new zugericht*. Wittenberg--Facsimile edition of the original. Stuttgart: Württembergische Bibelanstalt, 1967.

16. Moulton, Harold K., Ed. 1977. *The Analytical Greek Lexicon Revised*. Grand Rapids, MI: Zondervan.

17. Schildt, Joachim, Ed. 1984a. *Luthers Sprachschaffen: Gesellschaftliche Grundlagen, Geschichtliche Wirkungen. Referate der internationalen sprachwissenschaftlichen Konferenz Eisenach 21.-25. März 1983*. 3 Vols. Linguistische Studien, Reihe A, Arbeitsberichte, 119/1-3. Berlin: Akademie der Wissenschaften.

18. Schildt, Joachim. 1984b. "Zum deutschen Sprachschaffen Martin Luthers--Schwerpunkte und Entwicklungstendenzen der Forschung." *Zeitschrift für Germanistik* 5:210-214.

19. Schirokauer, Arno. 1957. "Frühneuhochdeutsch." *Deutsche Philologie im Aufriß*. Vol. 1. 2nd ed. Ed. Wolfgang Stammler. Berlin: Schmidt. Pp. 898-910.

20. Wodňanská, Annalies. 1984. "Zu einigen Stil- und Übersetzungsproblemen Martin Luthers." IN: Schildt 1984a. Vol. 2. Pp. 35-44.

21. Wolf, Herbert. 1980. *Martin Luther. Eine Einführung in germanistische Luther-Studien*. Sammlung Metzler 193. Stuttgart: Metzler.

...*BUT IT AIN'T GOT THAT SWING:* ASPECTS OF TRANSLATION QUALITY IN TECHNICALLY-ORIENTED ADVERTISEMENTS

Dieter Wältermann
Carnegie Mellon University

Keywords: Advertisements, Advertisement Jargon, Advertisement Decision Process, Advertisement Format, Message Format, Marketing Strategies, Terminology, Terminology Export, High-Technology Products, Terminology Issues, Native-Speaker vs. Non-Native-Speaker Translation Contracts, Terminology Resources, Translation Teaching

Abstract: The United States is one of the leading exporting countries of technology—and here predominantly computer technology, telecommunications, and microelectronics ... three scientfic fields that are generating a tremendous amount of vocabulary every year in form of new terminology. In addition to translating manuals and brochures, American companies realized on an ever-increasing level that their products need to be advertised in another language and another culture in order for them to maintain and even increase their competitive edge. Whenever translations are needed for US products to be sold overseas, American companies can contract the translation work out either to agencies or individuals working in the United States or to agencies or individuals working in the corresponding foreign country. The need and desire for close communication has led the majority of American companies to do the translation work in the U.S. Others, however, chose to have their translation work done by native speaker-translators who reside in the corresponding foreign country. One of the major underlying arguments centered on the assumption that native speaker-translators provide superior translations due to their native speaker status and their immediacy with respect to the foreign culture. This paper takes a close look at the language of advertisements by American companies, as well as translations of advertisements from English into German. It examines the quality of these translations from various linguistic and cultural perspectives, and addresses translation difficulties related to terminology, basic vocabulary, images, and humor as it is used in popular magazines and newspapers.

1. INTRODUCTION

In her 1993 publication *Funktionales Übersetzen*, Christiane Nord writes that "the purpose of an act of translating is defined by the communicative situation for which the product—the translation or the interpretation—is intended" (1993: 9). Clearly, the purpose of any advertisement is to introduce and to sell a given product. While this simplified definition of the advertise-ment concept may help to provide a basic understanding of the major purpose and function of advertising, it falls short of the required and necessary information needed by those entrusted with translating advertisements.

According to Kotler & Armstrong, two of the leading US-American experts in marketing and cross-cultural research related to business issues, advertising is but one of four major promotion tools which make up the promotion mix, or a company's total marketing communications program. The remaining three tools are: personal selling, sales promotion, and public relations (1994: 464-465). Kotler & Armstrong "define advertising as any paid form of nonpersonal presentation and promotion of ideas, goods, or services by an identified sponsor" (487).

In order to gain a better understanding of the underlying principles at work in the process of advertising ideas, goods, or services, and to transfer these principles to the translation process, students of translation as well as translators should receive adequate information and training not only in marketing and its underlying strategies, but also in the general working of marketing and advertising agencies. Knowledge of the major factors influencing today's advertising decisions should represent one of the central points in a translation course introducing students to the skills and knowledge necessary for them to be effective translators of advertisement materials.

Any translation course dealing with advertisements must, by necessity, do more than merely introduce students to language and culture-related issues involved in the translation process. As Nord has already pointed out, it is also necessary to familiarize students with the communicative situations for which the product is intended. In addition, it is absolutely required to present students with an introduction to the field of advertising. Since any translation of an advertisement becomes part of the advertising process, at times even a major component of the initial advertising decision process, knowing the various factors at play in this process will help to facilitate the translator's task. In order to illustrate the position and the role of the translator in this process, the following diagram from Kotler and Armstrong (490) served as a teaching tool in the translation course:

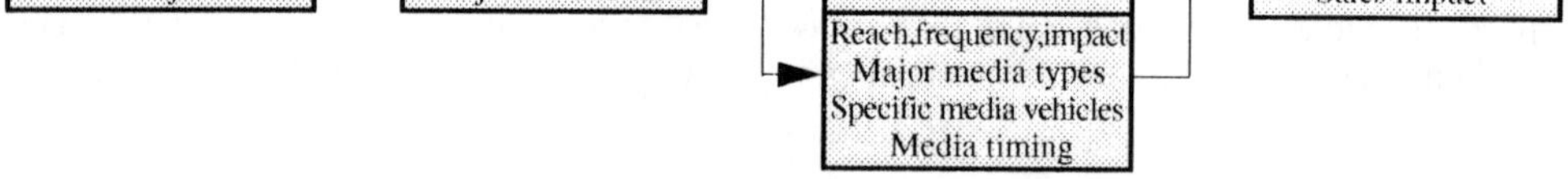

Figure 1. Major advertising decisions without translators

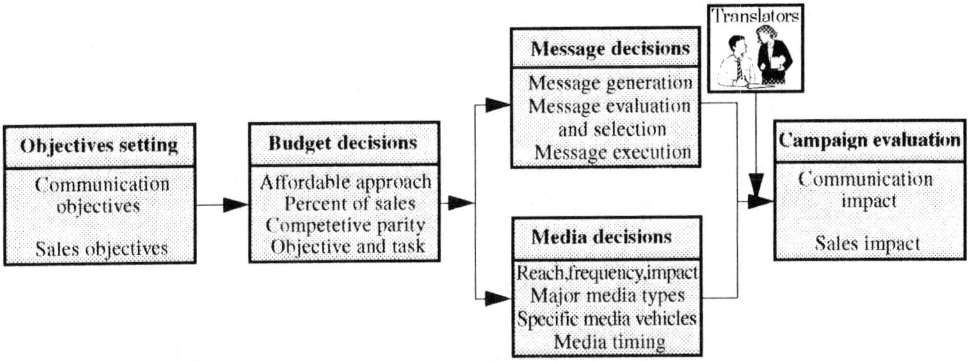

Figure 2. Major advertising decisions with translators

While there may be exceptions, the translator will generally be found at the end of this advertising process. As such, message decisions as well as media decisions will already exist and guide the translator's task. This situation represents the worst-case scenario vis-à-vis the translator's position since her role in internationalization is severely limited. Internationalization is defined as the process in which a product is designed to "meet the requirements of the selected target markets" (Spalink et al., 335) for the purpose of maximizing a company's profit.

Maximizing profit, in turn, is largely dependent not only on facilitating a product's adaptation to the various markets, but also largely dependent on the advertisement strategies employed to introduce the new product to any given market. Therefore, internationalization involves more than simply the preparation of a product—a process often referred to as enabling. If a product is adapted to individual markets or locations, this adaptation is called localization.

In order to teach a practically-oriented translation course, students should be taught these marketing fundamentals, as well as the distinctions between internationalization and localization. They should also be instructed in the differences between product-independent and product-dependent factors within the internationalization process, in order to ensure a company's maximum profit. Internationalization, as laid out by Spalink et al. (336-338). involves the following categories:

Product-Independent Factors:

- **Religious Aspects** — holiday symbols (also applicable to cultural aspects); use of flags and banners (Saudi Arabian flag contains a verse from the Koran -- a bag showing this flag cannot be used for garbage)

- **Cultural Aspects** — different date and time formats; different measuring systems (metric vs. non-metric); currency differences; significance of name usage (*John and Mary* does not correspond to *Johannes und Maria* in Germany or *Juan y Maria* in Spain); landmarks as identifiers (*Hoover Dam* vs. *Brandenburger Tor*)

Product-Dependent Factors:

- **Manufactured Goods** — verifying norms, standards, rules, regulations, and laws
- **Software Products** — modifying symbols, adding multi-byte character sets (primarily for Arabic and Oriental markets); avoid hard-coding of collating sequences
- **Documentation** — expansion of text sources must be accommodated; frames, tables, and texts must not be anchored in original

2. INSTITUTIONAL BACKGROUND

In the fall of 1994, predominantly science-oriented undergraduate and graduate students at Carnegie Mellon University (CMU) in Pittsburgh enrolled in an upper-level scientific-technical translation course. The students came to the translation course after having studied an average of seven semesters of the German language, either at CMU or at another university from which they had transferred. Within the American university system, seven semesters of studying a foreign or second language falls far short of being qualified for translator training. It is, however, possible to tailor translation-based courses to the needs and language levels of undergraduate students by restricting the subject area as well as the linguistic complexities of the course materials and the course's contents.

The average language proficiency of the translation students was rated as Intermediate High on the Oral Proficiency Interview (OPI) scale from the American Council of Teachers of Foreign Languages (ACTFL). This indicated that the students' average knowledge of German would best be employed for translations from German to English only. The materials used in the translation course were, therefore, taken solely from German magazines and journals. At no time were students asked or required to translate from English into German but, instead, work with analyzing German advertisements. The ads chosen for the course were taken predominantly from subject areas with which all students were familiar: computer science, computer technology, and microelectronics. The primary course objectives were as follows:

- introduce basic marketing concepts and advertising processes
- introduce concepts of internationalization and localization
- familiarize students with German advertising
- introduce computer science terminology and terminology sources
- introduce the work of translation agencies and individual translators
- increase sensitivity to the German language and culture
- increase knowledge of the German language and culture

3. LANGUAGE USE IN SCIENTIFIC-TECHNICAL ADVERTISEMENTS

Due to the linguistic and cultural limitations imposed by the students' knowledge of the German language and culture, the beginning of the course concentrated on increasing the students' sensitivity to the German language and culture. The first advertisements used for this training purpose were ads done by German companies advertising German products. By looking at German ads and examining linguistic as well as cultural information presented in those ads, students were made aware of linguistic and cultural subleties, which would otherwise have gone unnoticed. At the same time, the students were able to develop a better sense of appropriate and correct language use before examining translated advertisements from American companies.

A second way of raising the students' sensitivity to the German language and culture consisted of presenting a large variety of advertisements for American products. These translations had been primarily done by translation agencies and freelance translators. As far as translated advertisements are concerned, this step involved an introduction to the work of translation agencies and individual translators working for agencies and bureaus or as freelancers. Familiarizing students with the work of translators also involved a lengthy discussion on where American companies chose to have their documentation and advertisements translated.

Numerous US-American companies have until now assumed that anyone who knows two languages and possesses a dictionary is qualified to translate. At the center of this assumption lie wrong beliefs and a misguided understanding of the translation process as well as the work of a translator. Educating current and future managers and engineers about the intricate details of the translation process can help alleviate this problem. Since the majority of students enrolled in the scientific-technical translation course was majoring in scientific fields, this aspect deserved special attention.

In addition, factors concerning the choice of translation agency, whether home or abroad, and the choice of translators, whether native-speakers of the source language or the receptor language, were discussed. American companies have often "opted to have their translations done in the receptor language country by native-speaker translators" (Wältermann, 45). This decision was frequently justified as follows:

a. native-speaker translators residing in the receptor language country possess a higher level of receptor language skills than native-speaker translators residing in the source language country;

b. native-speaker translators residing in the receptor language country have a better understanding of and access to the markets of the receptor language country; and

c. native-speaker translators residing in the receptor language country possess a higher and much more up-to-date level of technical terminology of the receptor language due to their close proximity to the receptor language and the technical environ of the receptor language. (Wältermann, 45)

While all of these points put forward by American companies may find some justification, it should be obvious that one might also find sufficient arguments to counter any of them. Even point (c) above only holds as long as one does not consider the length of time this translator may have been removed from the technical environment of the target language country. Thus, any company doing business with a translation agency abroad may experience a variety of problems related to restricted communication, restricted access, increased time for product updates, and quality control of the translation agency or the freelance translator.

The majority of classroom discussion on translation quality dealt with advertisements done by translation agencies and freelance translators working in the source language country, namely the United States. This is not to say that those agencies or translators provide inferior products or ads featuring a lower translation quality. An ad placed by the WordPerfect Software GmbH in the March 1993 issue of the German magazine *DOS International* clearly demonstrates that even native-speaker translations done in the receptor language country may lack quality. The text of the advertisement was accompanied by a picture featuring the WordPerfect software package, the *Duden-Grammatik*, and the *Duden-Rechtschreibung*:

> Deutsche Sprache, leichte Sprache
>
> Jetzt gibt es eine Textverarbeitung, die Ihre Texte nicht nur verarbeitet, sondern auch bearbeitet. Denn WordPerfect 5.2 für Windows hat eine neue Grammatikprüfung*, die Ihre Manuskripte in Null Komma nichts auf grammatikalische und stilistische Fehler überprüft.
>
> Und damit Sie alte Texte auch sofort finden, indiziert sie unser neuentwickeltes Quickfinder auf der Festplatte oder im Netzwerk. In Sekundenbruchteilen finden Sie so durch Eingabe eines Stickwortes aus Millionen von Informationen Ihre gewünschte Information oder Ihr Manuskript. (153)

This particular advertisement featured information which could possibly be construed as misleading the reader. The advertisement text speaks of a new grammar checker (*Grammatikprüfung*), and in extremely small print at the bottom of the page the information that the grammar checker featured in this new version of *WordPerfect 5.2 for Windows* does not contain a German grammar checker, but an American-English grammar checker (which customers may exchange for a German version at a later point). In addition, the advertisement text shows a misprint or error which does not serve to promote either a grammar checker or a spell checker: *Stickwortes* should have read *Stichwortes*—not *sticky word*, but *keyword*. In another full-page advertisement in the same journal, WordPerfect speaks of a *komfortable Benutzeroberfläche* (literally: comfortable user interface) and *Dialog-Boxen*, which is generally referred to as *Dialogkästchen*. The term *comfortable* is often misapplied with respect to user interfaces in general, and should have been translated as *benutzerfreundlich* or *anwenderfreundlich*.

The language currently used in advertisements dealing with scientific-technical products, and here specifically with computer products offered by German and non-German companies in popular German computer magazines, often defies common sense and abandons commonly held beliefs about language usage and translation quality. The materials used in the translation course aptly demonstrated the tremendous choice of words at play in designating identical items. While Company A may call its product *Hard Disk* or *Harddisk*, Company B refers to it simply as *Disk*, Company C calls it a *Platte*, and Company D addresses it as a *Festplatte*.

The numerous word choices, namely German and English, and sometimes even a mixture of the two languages which could be referred to as the pseudo-language *Germlish* or *Engman*, and the translations, namely German, appear to be purely random and, in some cases, neither linguistically nor culturally appropriate. Students studying scientific-technical translations are often confused and puzzled, given the large variety of choices facing them. They find themselves caught in the battlefield of linguistic freedom of choice and preciseness of technical language. In order to solve this confusion, one of the first steps must involve a brief comparison of the language found in technical documentation and that found in advertisements featured in popular magazines.

On the one hand, technical writing is supposed to be precise and unambiguous. A closer look at purely technical texts, such as manuals and documentations, will lend credence to this assumption. The arbitrariness found in advertisements does, therefore, violate one of the frequently stated maximes of technical writing: preciseness, clarity, and non-ambiguity with respect to terminology. However, this assumption may only hold as long as the students deal with publications from one company only. Among the many advertisements found in German computer magazines, one company showed consistently superior language and correct use of new terminology. The advertisements by the German company *Daten- und Medienverlag* (DMV) featured various new software products without any language or terminology errors. Advertisements such as those by *DMV* are an excellent terminology source for translators who are constantly looking for new sources with respect to vocabulary and terminology.

Comparing manuals published by Caterpillar Inc. with those from Clark Corporation, two companies manufacturing heavy machinery, or handbooks from IBM Computer Corporation with those from Compaq Computer Corporation, two companies in the microeletronics industry, will reveal differences in language use, style, terminology, and presentation. The differences tend to be rather small with respect to larger companies, since those companies often make a concerted effort to standardize their language and to use standardized terminology if it is available. In addition, larger companies tend to employ translation agencies (or even in-house translation facilities) which tend to have better and easier access to major terminology banks and terminology resources (such as the Internet and its World Wide Web).

4. CONCLUSION

Helping students understand these differences may involve a variety of exercises, such as searching terminology banks in print or electronic format, contacting the International Standards Organization through their WWW home page on the Internet and examining the listing of publications, contacting language information systems which are on-line (such as FLEFO or AT&T Language Line in the United States), and even contacting the corporations in question and addressing terminology questions to the documentation departments of these companies. It should immediately be obvious that the last choice may be rather time-consuming if the name of the responsible person is not known.

During the translation course, students were not very successful in obtaining answers to queries directed at documentation departments of major US companies without having a contact person inside the company. The success rate was much higher when students contacted companies with in-house translation departments. Since terminology-related issues were at the core of this translation course, students spent a majority of their time investigating correct terminology usage and researching terminology sources, either in print or electronic form.

As far as conflicting and even unknown terminology in the fields related to computer science is concerned, the atudents of this translation course appreciated having access to in-house translation departments, such as those found at Siemens, IBM, Hewlett-Packard, and Caterpillar, and access to the publications of the American Translators Association as well as to those of its member chapters. Being aware of the variety of potential resources on terminology and translation in general can help these beginning translators to avoid costly and embarrassing mistakes.

REFERENCES

Kotler, Philip; and Gary Armstrong. *Principles of marketing.* 6th edition. Annotated Instructor's Edition. Englewood Cliffs, NJ: Prentice Hall, 1994.

Nord, Christiane. *Einführung in das funktionale Übersetzen. Am Beispiel von Titeln und Überschriften.* UTB 1734. Tübingen und Basel: Francke Verlag, 1993.

Spalink, Karin; David Kumhyr, and Carla Merrill. "Internationalization and translatability." In: *Vistas.* Proccedings of the 35th Annual Conference of the American Translators Association. Peter W. Krawutschke, ed. Medford, NJ: Learned Information, Inc., 1994. 335-338.

Wältermann, Dieter. "Translation quality: When the natives get rusty." In: *Keystones of Communication.* Proceedings of the 34th Annual Conference of the American Translators Association. Edith F. Losa, ed. Medford, NJ: Learned Information, 1993. 41-48.

WordPerfect 5.2 für Windows. *DOS INTERNATIONAL* 3 (1993): 153.

Connections

Interpretation

INTERPRETER ROLES AND COMMUNICATION DYNAMICS: LESSONS FROM THE SIMPSON TRIAL

Fritz G. Hensey
University of Texas at Austin

Keywords: Interpretation, Rosa López, Simpson Trial, Communications Theory, Communication Triads, Interpreter Roles, Interpreter Codes of Conduct, Court Interpretation.

Abstract: This paper discusses the interpretation of witness Rosa López's testimony at the Simpson trial in February of this year. It applies communication theory to highlight the interpreters' triple role as language broker, transaction mediator and communication facilitator. The focus is on facilitation expressed as overt intervention in furtherance of the proceeding. Wide publicity and continued access to tapes make this material a valuable contribution to greater understanding of court interpreting by language professions and the general public.

1.0 Introduction

The interpreted examination of witness Rosa López at the Simpson trial allowed millions of TV viewers to see and hear court interpreters in action. For four days, viewers watched the work of three skilled court interpreters: Doris Weitz, Alicia Luper and Cecilia Cerda. They observed the three basic modes of court interpretation: simultaneous, consecutive, and sight translation of documents. They saw the interpreter working alone or as part of a team.

The initial proceeding was a hearing to determine whether Ms. López would be allowed to testify as a defense witness and hence out of turn, since at this point the prosecution was still presenting its case. Once she was allowed to testify, out of the presence of the jury, she underwent the normal routine of direct and redirect examination by the defense, cross and re-cross by the prosecution. The target language,

Spanish, was accessible to a large segment of the viewing public. The public interest aroused was reflected in calls and letters to the media, talk shows, and expressions of support or criticism of witness, interpreter, and attorneys. Naturally enough, language professionals have shared this interest and expressed it in publications and at meetings.[1]

2.0 Theories of Communication

Modern theories of communication tend to share a vocabulary taken from the electronic media with terms like sender, receiver, code, message, channel and feedback. Communicative transactions typically involve dyads, meaningful groupings of two parties. One or both parties may be single individuals or groups, but the essential picture is that of a sender or source interacting with a receiver or target. When two parties interact through an interpreter, there arises a communication triad, a meaningful set of three participants.[2]

Some communication scholars believe that triads are more stable communicative units than dyads or larger units.[3] This would seem to be true of interpreted examination of witnesses, because the triad is held together by the strict discipline of legal discourse. Participants are subject to discourse rules as to who asks and who answers, what questions may or may not be asked, and how the interpretation is to be performed.[4]

3. Communication Factors, speech functions and Interpreter Roles

3.1 The informative function: Interpreters as Language Broker, Transaction Mediators, and Communication Facilitators

Linguist and literary scholar Roman Jacobson provides a useful starting point for this study of interpreter roles and communication dynamics. Jacobson elaborated on Karl Bühler's theory of language to develop a set of linguistic factors pertinent to the study of poetry.[5] Translator and translation scholar Peter Newmark discusses these functions as they relate to translation[6]. Table 1 displays Jacobson's six factors in verbal communication, to each of which he assigns a language function. The adjectives in parentheses are those used by Newmark for language functions.

	CONTEXT (Informative)	
ADDRESSER (Expressive)	MESSAGE (Aesthetic)	ADDRESSEE (Vocative)
CHANNEL (Phatic)		CODE (Metalingual)

Table 1. Roman Jacobson's Model of Basic Factors in Verbal Communication combined with Peter Newmark's Model of Language Functions in Translation

Context refers to the linguistic, cultural, and situational content of utterances. In Newmark's terms, utterances oriented toward content are termed informative. To the extent that an interpreter acts as a conduit between speaker and hearer, s/he is at the very least responsible for accurate transmission of the information generated by both parties. Table 1 thus represents the context factor and the corresponding informative function as within the domain of the interpreter in the role of language broker. Figure 1 below symbolizes this role through a model of interpretation in which the interpreter operates as a sort of multi-channel transmitter, analyzing the sender's output and recoding its elements for use by the receiver. [7]

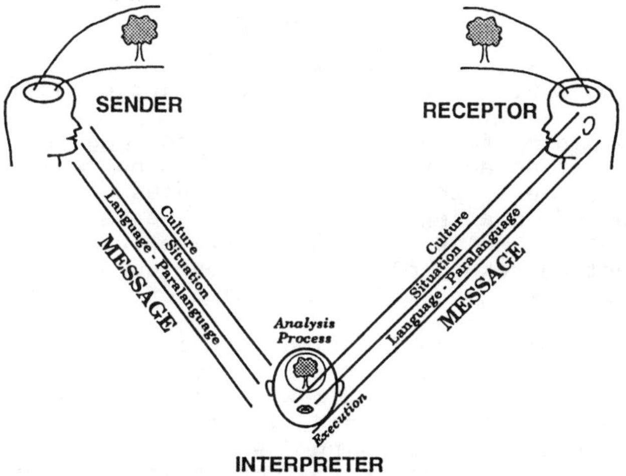

Figure 1. Interpreter as Language Broker

Utterances may be oriented toward the addresser (sender, source ...) of the message, toward its addressee (receiver, target ...), or to toward the message itself. In Newmark's terms, utterances oriented toward the sender are termed expressive, i.e., expressive of the speaker's intention, state of mind, etc. Those oriented toward the receiver are called vocative; they may be aimed and informing or persuading the hearer.

Message-oriented utterances are called aesthetic and have to do with creativity in sound, form, metaphor, and the like. Interpreters continually process dialog having all these functions. Unless the interpreter intervenes overtly, all his or her interactions with sender and receiver are covert, accessible only to the other two participants and those who can understand both languages. We see this ongoing, intelligent processing of the dialog as a form of mediation. The interpreter unobtrusively mediates the transaction in order to serve the purposes of the court or s/he serves.

Figure 2 symbolizes the mediator role by means of a logo currently being proposed by the Australian National Accreditation Authority for Translators and Interpreters. If adopted, it will be used to advise the public that interpretation service is available at a given place. The symbol may be understood to represent the interpreter as an impartial mediator between two people desiring to communicate across language barriers.[8]

Figure 2. Interpreter as Transaction Mediator

The two remaining factors are Code and Channel. Newmark's language function applying to the linguistic code is metalingual. Here, the utterance focuses on language itself ... choice and meaning of specific words or utterances, register, etc. The is the communicative framework in which the transaction takes place. For our purposes, this is the special discourse genre of examination of witnesses.[9]

Newmark's corresponding function is the phatic, which may include the overall form of the dialog, the tone and other paralinguistic elements, and how the interpreter assumes the personae of attorney and witness. Here the interpreter faces the task of complying with ethical and legal requirements which either permit or oblige the interpreter to intervene to avoid miscommunication between the parties. In effect, the interpreter is led to assume an overt role as communication facilitator.

Figure 3 below symbolizes an ideal situation whose three participants interact successfully (?) thanks to the interpreter's mediation. The solid black and the solid white figures are the presumably monolingual participants, while the dotted figure is that of the bilingual / bicultural interpreter. This symbol is taken from a Canadian interpreter-training manual where it illustrates a discussion of "triadic" relationships where one member of the triad interprets for the other two.[10]

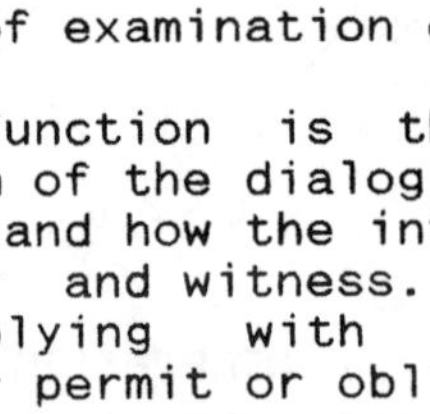

Figure 3. Interpreter as Communication Facilitator

4. Codes of Conduct for Court Interpreters

As court interpretation develops into a recognized profession, numerous codes of ethics or canons of professional conduct have appeared.[11] This topic is a standard feature of textbooks for interpreter training[12] and is often hotly discussed at interpreter meetings. Table 2 below displays four relevant parts of the Code of Professional Responsibility of the Official Interpreters of the United States Courts.

CANON 1	Official court interpreters act <u>strictly in the interests of the court they serve.</u>
CANON 7.	Official court interpreters work <u>unobtrusively with full awareness of the nature of the proceedings.</u>
CANON 8.	Official court interpreters fulfill a special need to <u>interpret accurately and faithfully</u> without indicating any personal bias, avoiding even the appearance of partiality.
CANON 11.	Official court interpreters <u>preserve the level of language used, and the ambiguities and nuances of the speaker,</u> without any editing. Implicit in the knowledge of their limitations is the duty to <u>correct any error of interpretation</u> and to demonstrate their professionalism by <u>requesting clarification of ambiguous statements or unfamiliar vocabulary</u> and to <u>analyze objectively any challenge to their performance.</u> Interpreters have the duty to call to the attention of the court <u>any factors or conditions which adversely affect their ability to perform adequately.</u>

Table 2. Code of Professional Responsibility of the Official Interpreters of the United States Courts

4.1 Canons of Conduct and Interpreter Roles

Canon 1 is ... to use Star Trek terms ... the Prime Directive from which many of the others stem, particularly those enjoining strict impartiality. Canon 7 foreshadows the mediator and facilitator roles, since full awareness of the nature of the proceeding makes it incumbent on the interpreter to assure that his or her own performance does not hinder that proceeding. Canon 8 addresses the language broker role; it should not be misconstrued as to require unthinking, "literal" translation of the linguistic content.

Canon 11 addresses matters that pertain to all three interpreter roles. Preserving the register or "level of language" is done at partly at the language-brokerage level and partly at that of mediation. Stylistics and ambiguity are generally devices having, in Newmark's terms, an expressive or vocative function, perhaps even a metalingual function. Canon 11 finally enjoins the interpreter to clarify linguistic doubts, respond to challenges to the interpretation, correct the record as to any prior errors, and finally to make the court aware of any problems affecting the interpretation. These latter provisions amount to an instruction to the interpreter to step out of the unobtrusive (Canon 7) broker or mediator role and become an overt facilitator of the communication triad.[13]

Communication through an interpreter is subject to many of the same problems as other types of spoken discourse in a single language. Words and expressions may be unheard, misheard, or misunderstood. There may be grammatical ambiguities and uncertainty as to the speech functions described above. In cross-cultural communication, unawareness of cultural facts may impede understanding. This is far from being a simple matter of interpreter error. All parties may be involved and affected. The mere fact of interpretation renders the dialog somewhat unnatural, particularly when the discourse is subject to strict procedural rules. Table 3 below lists typical problems that arise in witness interpretation.

<u>COMMUNICATION</u>
CULTURAL ISSUES HAMPERING COMMUNICATION
INTERRUPTIONS OR DISTRACTIONS
UNFAMILIARITY WITH THE FIELD OR THE SETTING
<u>LANGUAGE</u>
NOT HEARING/NOT UNDERSTANDING SOME PART OF THE MESSAGE
NOT KNOWING HOW TO INTERPRET SOME PART OF THE MESSAGE
FINDING THE MESSAGE INCOHERENT OR UNCLEAR
<u>INTERPRETING</u>
FORGETTING PART OF THE MESSAGE
CRITICISM OF THE INTERPRETATION
AWARENESS OF EARLIER MISINTERPRETATIONS

Table 3. Typical Problems in Consecutive Interpretation

William O'Barr's research has shown that the style of questioning affects jurors' perceptions of the witness and of the questioning attorney.[14] Similarly, as Susan Berk-Seligson demonstrates, the interpreter's performance can have an impact on jurors' assessment of a witness in terms of competence and credibility[15]. In such circumstances, the court interpreter must assume the role of <u>communication facilitator</u> in addition to the demanding roles of <u>language broker</u> and <u>transaction mediator.</u>

4.2 The Communication Pentad

The basic triad of Questioner -Interpreter -Answerer generally follows a strict order of turn-taking. When the communication situation warrants it, the basic triad temporarily develops into a potential five-member unit, a "pentad". The interpreter interacts in an overt facilitator role with one or more of the other four participants and may or may not be the initiator of such interaction. In the López examination, the interpreters at different times interacted in this way with the attorneys, the judge and the witness.

5.0 The López Examination: Interactions

Tables 4a and 4b below display a succinct classification of observed interactions between the interpreter and one or another of the other participants for purposes of facilitation. Most often, it is the interpreter who initiates the interaction. The communication problems that arose in the López examination are common instances of those cited in Table 3. For each general category, the table further classifies the various linguistic, cultural or procedural matters involved. Space does not permit inclusion of specific examples here. Some such examples will be presented in the oral version of this papers.

INTERPRETER-INITIATED INTERACTIONS
Clarifications from witness
Ambiguous sentence structure
Uncertain lexical items
Self-monitoring and correction
Misperceived sentence structure
Gender marking
Control of turn-taking by witness or attorney
Awainting rulings on objections
Avoiding overlapping speech
Witness overrides interpreter
Requesting clarification from attorney
Missed question
Incomplete sentence structure
Clarification of cultural issues
Hispanic naming customs
Spanish spelling conventions

Table 4a. Interpreter-Initiated Interactions

Table 4b below describes interpretation-related interactions initiated by one of the other participants. Challenges were made by both questioning attorney and opposing counsel. Rosa López herself sometimes addressed the interpreter directly, as did Judge Ito.

ATTORNEY-INITIATED INTERACTIONS
Challenges to interpretation
Request to specify Spanish version used
Correction of misheard proper name
Correction of misunderstood word
Admonishment of interpreter or witness
WITNESS-INITIATED INTERACTIONS
Asking interpreter's assistance
Correcting interpreter's version
COURT-INITIATED INTERACTIONS
Comments or questions on interpretation
Introduction/qualification of interpreters
Instructing attorneys on use of interpreters
Responsiveness to interpreter requests

Table 4b. Interactions Initiated by Parties other than Interpreter

6. Summary

CA v. Simpson, the self-styled "Trial of the Century", has doubtless produced materials that language professionals should find valuable for years to come. There have been few instances of televised interpretation of witness testimony arousing the same amount of viewer interest as well as linguistic accessibility to a large sector of the American people. The four days of videotaping provide a wealth of information on the style and content of court interpretation. More generally, they may lead public and professionals alike to recognize that witness interpretation as a special type of communication and to appreciate the communication skills that interpreters must bring to their task.

NOTES

1. Cárdenas, Roxana. "Lost in the Translation: Court Interpreters all too easily are made Scapegoats." *Proteus* 4,2.7, 1995. See also Rainof, Alexander. "Misdeeds of the Press". *Polyglot Extra*, Special Edition 4-5. March 1995.

2. Ruben, Brent. Communication and Human Behavior. New York: Macmillan, 1984 pp. 251-3 and Sluzki, C.C. "The patient-provider-translator triad: a note for providers." *Family Systems Medicine* 2:397-400. 1984

3. Ruben, op. cit. p. 253 and Cairncross, Larissa. Cultural Interpreter Training Manual. Ontario, Canada: Ministry of Citizenship, 1989, p. 120-1

4. Stygall, Gail. Trial Language: Differential Discourse Processing and Discursive Formation. Philadelphia, PA: John Benjamins, 1994, ch. 4

5. Jacobson, Roman. "Closing statement: linguistics and poetics," in Thomas Sebeok, Style in Language. Cambridge, MA: MIT Press, 1960, 350-77

6. Newmark, Peter. A textbook of translation. New York: Prentice Hall, 1988, ch. 4

7. Reproduced from Arjona, Etivia. "Intercultural communication and the training of interpreters at the Monterrey Institute of International Studies". in Gerver, D. and Sinaiko, W. (eds.). Language interpretation and communication. New York: Plenum, 1978

8. The Australian National Accreditation Authority for Translators and Interpreters, in a brochure issued in 1995, authorized use of this symbol by anyone during a one-year trial period.

9. Regarding different phases of a trial as speech genres, see Stygal, op. cit., and Walter, Bettyruth. The Jury Summation as Speech Genre. Philadelphia, PA: John Benjamins Publishing Co., 1988

10. Cairncross, op. cit., p. 121 and 11. Boston City Hospital. Medical Interviewing across Language Barriers. Boston City Hospital Section of Internal Medicine, 1991 pp. 10-2

11. Judicial Council of California. Professional Ethics and the Role of the Court Interpreter. San Francisco: Administrative Office of the Courts, 1995.

12. E.g., Hewitt, William. Court Interpretation: Model Guides for Policy and Practice in the State Courts. National Center for State Courts, 1995; De Jongh, Elena M. An Introduction to Court Interpreting: Theory and Practice. Lanham, MD: University Press of America, 1992, ch. 2-5; Palma, Janis. Introduction to Judiciary Interpreting. New York: National Association of Judiciary Interpreters and Translators, 1987, ch. 3; and Dueñas González, Roseann, et al. Fundamentals of Court Interpretation. Durham, NC: Carolina Academic Press, 1991, ch. 35. The canons of conduct listed here are quoted form the latter text, pp. 585-6

13. Carroll, John. "Lawyer's response to language and disadvantage before the law." in Gibbons, J. (ed.), Language and the Law, London: Longman, 1994 op. cit., 306-15 and Morris, Ruth. "Court interpreter: Record or Participant?" *Proceedings of the 12th FIT World Congress*, Belgrade, 1990, pp. 724-9

14. O'Barr, William. Linguistic Evidence: Language, Power and Strategy in the Courtroom. New York: Academic Press, 1982

15. Berk-Seligson, Susan. The Bilingual Courtroom: Court Interpreters in the Judicial Process. Chicago: University of Chicago Press, 1990. See also Laster, K. "Legal interpreters: conduits to social justice." *Journal of Intercultural Studies*, 11,2,15-32.

THE CHALLENGES OF PROVIDING AN INTERPRETER SERVICE WITHIN A HEALTH CARE DELIVERY SYSTEM: IMPROVING QUALITY AND INCREASING PRODUCTIVITY WHILE DECREASING COST

Maria T. Mesiti
ATA
Laurie Yacuzzo
HIMSS

Keywords: Health Care Delivery System, Quality, Productivity, Cost

Abstract: This session will describe the evolution of the Interpreter Service (the Service) Program at Baystate Medical Center from providing 24 hour Spanish language coverage to supporting nineteen languages for patients in need of medical care. We will discuss processes the Service uses to ensure patients' and provider's needs are met.

Baystate Medical Center in Springfield, Massachusetts is a 725 bed, Level 1 trauma center with multiple site outpatient clinics. In 1978 the Service hired a Spanish interpreter for each shift. Other language requests were often unfilled since employee volunteers were used. Currently, the Service employs full-time Spanish, Russian and Vietnamese interpreters, and has available interpreters for seventeen other languages (e.g., Polish, Portuguese and French).

The Service's time translating for a patient during his/her medical care is called an encounter. From 1990 to 1994 encounter volume has grown in excess of nine times. Fiscal Year 1994 volume (21,703 encounters) was a 79.9% increase from the prior year's level. The actual average cost per encounter for the first four months of Fiscal Year 1995 ($9.93) represented a 12.8% decrease from Fiscal Year 1994 ($11.39). Working closely with the Service's managers, the Management Systems department developed productivity measurements and process improvement mechanisms which would increase interpreter utilization and would foster cooperation between the Service and the clinical areas.

1. BAYSTATE MEDICAL CENTER

Baystate Medical Center is a member of Baystate Health Systems (BHS), which is "one of New England's largest multi-institutional health care systems."[1] Baystate Medical Center is composed of two major campuses that are geographically separate, they are: the North Campus building which includes the inpatient units, emergency department, and ancillary services (e.g., rehabilitation, pharmacy, surgery), and the South Campus which houses our adult and pediatric medicine clinics, urgent care center, psychiatric services and long-term care unit. The Medical

Center also has two outreach primary care clinics. Due to the intensifying development of managed care plans, the anticipated increase of capitated contracts, "and the national health care debate, BHS adopted a vision of becoming an integrated delivery system which will manage a coordinated continuum of hospital and physician services necessary to meet the needs of Western New England residents."[2] Managed care balances price and utilization controls with access to high quality care through coordinating the design, financing and delivery of health care. Capitated contracts provide a payment method to participating providers at a fixed amount per member per month for each member enrolled with the provider. In return, the provider must render or arrange for select health services to these members. To achieve this vision in an era of limited financial resources, BHS leadership determined that the system must "Achieve superior health status, patient satisfaction, and care outcomes for the system's enrolled population at a cost which is competitive with that of other integrated delivery systems in this region and state."[3]

2. INTERPRETER SERVICE PROGRAM

The Interpreter Service Program (the Service) began in 1979 with one full-time Spanish interpreter working each shift (i.e., day, evening, night). The interpreters were managed under the department of Education and Training. Within a matter of months, the night shift was eliminated and the program operated with one medical interpreter working 7 a.m. to 3 p.m. and 3 p.m. to 11 p.m., 7 days per week. The individual covered both the North Campus and South Campus, via beeper (paging) system, traveling between campuses on the Medical Center's shuttle van. Non-Spanish translations were performed by employee volunteers who spoke another language besides English. Requests for non-Spanish translations were often unfilled since there were no committed resources.

During 1987, the Service moved under the direction of Patient and Guest Relations. The interpreters' office was eliminated and their job title was changed to Patient Representative. A Patient Representative was responsible for greeting and assisting patients with his/her visit to the Medical Center. At this point, there was no recognition for their translation skills. An interpreter was added 5 days a week to the South Campus to reduce travel time and to meet the growing needs of ambulatory care in the clinics. A part-time Russian interpreter was hired shortly thereafter.

At the end of 1992, the Medical Center hired a Coordinator for the Service and a part-time Vietnamese interpreter. The Service was transferred to the department of Social Work and Discharge Planning. Because there had been no measurement of an interpreter's competency level, the Coordinator developed a proficiency exam in thirteen languages to validate written competency levels. Eight months later, the Coordinator secured office space for the Service. Spanish, Vietnamese and Russian are the primary languages translated and the Service employs full-time interpreters in each language; however, requests for French interpreters are increasing due to an influx of Haitian immigrants. To meet the community's needs, the Service also employs on an as-needed basis, interpreters in seventeen other languages (e.g., French, Polish). In addition to providing verbal translations, the Service translates written documents from English to foreign languages, and educates Medical Center personnel about cultural awareness.

The biggest obstacle in managing the Service was the lack of external departments scheduling interpreters in advance of a patient's arrival. There was no way to anticipate needs in order for the Service to prioritize requests. The physical and personnel structure of the ambulatory clinics caused problems with overbooked patients; often there was a need for more than one interpreter at the same time. This situation created a great deal of confusion and provider complaints.

In January 1994, more ambulatory areas were requesting that they have an interpreter on-site, continuously, which caused budgeting problems for the Coordinator. In the spring, Senior Administration formally requested Management Systems (MS) perform a time study analysis. In addition, an operations review of interpreter utilization was performed concurrently. The following documents the study's background, methodology, results and recommendations.

3. MANAGEMENT SYSTEMS STUDY

3.1 Background

The Service was a fixed cost center on the productivity reporting system. For a cost center that is considered "fixed," productivity is not based on a specific workload volume; rather the productivity level will be 100% unless the actual worked FTEs exceeds budgeted FTEs. For the period covering the twelve months of Fiscal Year '94, Interpreter Service's productivity report revealed YTD (year-to-date) Budget FTEs - 7.00, Actual Paid FTEs - 11.23, and Actual Worked FTEs - 10.96. This situation resulted in an overall productivity level of 63.8% with an FTE variance of 4.0 worked FTEs more than required. During the month of July 1994 the department reported the highest level of worked FTEs (13.88) which was 26.6% above the YTD average. Because the department did not have workload standards, it was difficult to determine the appropriateness of the staffing levels achieved in FY'94.

Through September 30, 1994 the Service was staffed seven days a week from 7 a.m. to 12 midnight. The Coordinator was notified when an interpreter was needed between midnight to 7 a.m.. MS performed an analysis of encounter volume during the evening hours and found that the volume was not significant. A determination was made that the evening staff could be eliminated by implementing an on-call policy. (Included in the Recommendations section.)

The Service's primary workload unit is an encounter which is the time spent translating for a patient, either face to face or over the phone, concerning his/her medical care. For departmental statistics an interpreter may document more than one encounter per patient visit. (e.g. An interpreter translates for Patient A, leaves and translates for Patient B, then translates discharge instructions for Patient A - this would equate to three encounters over two patients.) Each interpreter documented patient encounters in their log book/calendar, and then transcribed the entries to the departmental "Master Log."

3.2 Methodology

To develop variable standards and to assess interpreter utilization the following steps were taken during the study:

- To determine the study's scope an introductory meeting was held with representatives from Senior Management, Social Services, Interpreter Service, South Campus Adult Ambulatory and Pediatric Clinics, and MS.

• To gain an understanding of current operations (e.g. scheduling of interpreters, work flow) observations were conducted by MS personnel shadowing interpreters (Spanish, Russian and Vietnamese) during the day and evening shifts at the two campuses and Neighborhood Health Center (NHC).

• To derive standards per encounter, by location, a data collection was performed. Over a two week period, interpreters documented the length of each face-to-face or telephone encounter. Each person also documented travel time between sites (e.g. North Campus and NHC) and each patient no-show or cancellation occurrence, or when the interpreter was booked for an appointment but was not needed.

• To improve operations through policy development and procedure documentation, on-going meetings with the Coordinator, Manager and MS personnel were held.

3.3 Results

The results section is broken down into two categories: Productivity Variable Standards and Scheduling.

Productivity Variable Standards

The variable standards that are used to determine the Service's monthly productivity were based on the results of a two week data collection performed during May 1994. Each interpreter documented, for each encounter, the date and the location where the encounter took place, the language translated, the length of the encounter and any related comments.

Results of the encounter data collection are shown in the table below.

Location	Average Hours per Encounter
North Campus (excluding the Emergency Room)	.74
North Campus Emergency Room	.45
North Campus Overall	.64
South Campus (excluding Urgent Care & West 3 Rounds)	.34
South Campus Urgent Care	.23
South Campus West 3 Rounds*	.50

Location	Average Hours per Encounter
South Campus Overall	.33
Neighborhood Health Center	.69
Telephone Call encounters	.09
Overall	.43

*-West 3 Rounds will be included in the Constant per Calendar Day calculation.

To account for personal fatigue and delay each standard documented above was increased by 17.0% (e.g. NC - excluding Emergency Room was increased from .74 hours to .866 hours.)

Personnel's indirect activities and their associated hours per year are accounted for in the Constant per Calendar Day. The activities identified for inclusion were education, translations, West 3 rounds, travel time, patient no-shows, and meetings. The education component allocates the amount of time spent preparing for, and facilitating events that educate health care personnel about different cultures. Translation hours account for the process of translating and/or proofreading written material from English to a foreign language. A Spanish interpreter is responsible for performing daily rounds at the South Campus' West 3 unit. These rounds are primarily a patient relation function: The interpreter asks the patient if he/she is completely satisfied. Sometimes, the interpreter will translate for a medical need. Through the data collection it was determined that each daily round took approximately .50 hour to complete. Travel time hours were computed from the results of the data collection. On average, interpreters spent 1.21 hours per week day traveling between sites. The 1.21 hours per week day was projected over a year by multiplying the number by 261 work days per year, yielding 315.81 travel hours per year. MS calculated the indirect hours per year the Service incurs for patients who do not show, or bring someone to do the translating for their appointments. On average, the Service encounters 5.8 no-shows per week day, each requiring .25 hours of an interpreter's time. Multiplying the 5.8 by .25 and then by 261 work days per year yields 378.45 no-show hours per year. There were a total of 58 no-shows over the two week period, broken down into each category as follows: 28 (48.3%) no-shows, 26 (44.8%) occurrences where the interpreter was not needed and 4 (6.9%) cancellations.

In addition, 1.0 FTE Coordinator (includes supervisory and program development responsibilities) and 1.0 FTE Secretary (administrative) positions were included in the constant. Each 1.0 Paid FTE was multiplied by a Paid Time Off (PTO) factor to convert the Paid FTE to a Worked FTE. The PTO factor (12.0%) accounts for holiday/sickness/vacation; therefore each 1.0 FTE (2,080 hours) was multiplied by .88 to reduce the allocation from paid to worked hours per year (1,830.4 hours).

The Constant per Calendar Day is 20.968 hours and is weighted by the number of days in the month.

Scheduling/Staffing

Based on the observations performed by MS personnel, the process for scheduling interpreters was inconsistent and inefficient. Interpreters could be booked for patient appointments in several ways: Diagnostic Scheduling could call the Service and afterward send an appointment notice; the outreach clinics could call as needed; the South Campus Clinics (Adult Ambulatory and Pediatrics) primarily requested and/or paged the on-site interpreters when patients were in exam rooms; and interpreters, translating for patients during the discharge instruction process, could identify follow-up/diagnostic procedures where the patient would require an interpreter. Because of the inconsistencies concerning advance requests for interpreters, the Service was unable to adequately plan their staffing needs and was reactive versus proactive to customer requests.

When the Service was given advance notice that an interpreter was needed for a patient appointment, they would record the appointment in a calendar book under the language, time and location: At a future date an interpreter was assigned. The Service did not designate individuals who would be the primary interpreters for booked appointments with others filling in for urgent/emergent situations. Because the Service did not book by resource (interpreter) at the time of notification, it was difficult to have an overview of each interpreter's daily assignment load.

The South Campus had several issues unique to their environment and they were discussed at the introductory meeting. MS observed the two (FTEs) Spanish interpreters who were responsible for covering the inpatient and ambulatory services on the Campus. Primarily, each interpreter worked independently of the other, with one stationed in the Internal Medicine Clinic (IMC, C-level) and the other in the Pediatric Clinic (fourth floor). The interpreter assigned to IMC worked 8 a.m. - 4:30 p.m. and the Pediatric interpreter worked 9 a.m. - 5:30 p.m.. When feasible, each interpreter would assist the other if one had a back-log of patients; however, because they were not working from a central location, this situation made it difficult for them to work as a team and effectively cover the entire Campus (including Urgent Care).

The process for requesting an interpreter in the Internal Medicine Clinic created a chaotic environment. There was a tendency for confusion and the process did not provide an efficient mechanism to serve the patient and provider/clinician needs. The determination that an interpreter would be necessary was made at the time a patient was placed in an exam room. Once determined, the provider/clinician would document the request on a log, identifying the time requested, patient's name, room number, and provider/clinician. The log was located on a board across from the main nurse's station. The interpreter would review the log and on a first come first serve basis, would proceed to the first listed appointment. The interpreter would document on the log their name and the time he/she went to translate. Additionally, if the interpreter was needed in an area outside of IMC (e.g., Urgent Care) he/she would document his/her time out/in and where he/she went. While the interpreter was in with the first patient, other providers/clinicians and their patients were waiting for an interpreter. During the interim the providers/clinicians would page the interpreter to determine where he/she was and when he/she would be available. Because of the physical set-up of the IMC, various IMC personnel could be paging the interpreter concurrently since there was no easy way to decipher where the interpreter was located. The interpreter could receive multiple pages (via beeper and overhead) during the time he/she typically was with a patient. The interpreter would respond to the page(s) and inform the pager(s) he/she would be there as soon as he/she completed the translation with the first patient. This process caused inherent delays, disruptions and frustrations for all individuals (interpreter/provider/clinician/patient).

South Campus Pediatric Clinic personnel would page the Spanish interpreter when a patient was placed in an exam room. In addition, if a patient needed to complete forms prior to being seen, personnel would walk over to the interpreter's desk and request their assistance.

Generally there were inherent inefficiencies due to environmental factors. The interpreter who accompanied a patient through his/her medical care from arrival through discharge experienced the same wait times as the patient (e.g. waiting for an exam room or provider/clinician). During the wait the interpreter tried to be flexible where allowable and would translate for another patient or would work on written translations.

In addition, the Service had one Spanish interpreter (1.0 FTE) assigned to work 3:30 p.m. to midnight (Monday - Friday) whose primary responsibilities were to be available for translations, coordinate requests for interpreters in other languages (e.g. Russian) and work on written translations. During the evening shift, the primary requests were from the Emergency Department and inpatient units. Because the Service is geographically separate from the main building, the interpreter spent time traveling back and forth (via the elevator) from the Service to the requesting area.

3.3 RECOMMENDATIONS

Demand for the Interpreter Service has been increasing due to the large and diverse non-English speaking population living within the Medical Center's service area. Improvements in operation control were made in order to enhance the department's customer service role, to improve staffing, to allow for flexibility and to foster cooperation. Recommendations and Action Plans to achieve these goals are outlined below.

1. The Service should obtain via the Patient Care Information Systems (PCIS) advance notice of ambulatory patients requiring interpreters.

- MS worked with the Manager of Ambulatory Systems to create the "*Interpreter Notice*." At the time a resource is booked and the need for an interpreter has been identified, the scheduler should input "Need (Language) Interpreter" in the comments field, and select the "Print Interpreter Notice" option, at which time a "Notice" prints out in the Service. Additionally, the comment prints out on the respective PCIS Clinic Roster.

- The Service needs to stress the importance of advance notification to all scheduling areas and communicate back to each area the number of booked interpreter appointments via the "Notice" versus the number of unbooked interpreter - but scheduled - appointments. (i.e., A clinic visit was recorded in PCIS but the interpreter was not.)

2. Resources (interpreters) instead of encounter site (e.g., IMC) should be scheduled at time of notification .

• The Service should designate individuals by language and site who are the primary resources. When the Service is notified of an impending appointment, it should be recorded in a master schedule under the appropriate interpreter. This scheduling process would allow the Coordinator to have an overview of each interpreter's daily assignment load. (i.e., Who could handle an add-on or would there be a conflict?)

3. Review each PCIS Clinic Roster on a daily basis to ensure all scheduled appointments have an assigned interpreter.

• The Service receives Rosters documenting appointments four days out and they also receive Rosters that print out with the next day's appointments. These Rosters should be compared to the master schedule. If an appointment does not have an interpreter assigned, and needs one, the master schedule should be updated and conflicts resolved at that time.

• Volume statistics should be maintained to monitor compliance levels and to provide a feedback loop to scheduling areas that have the ability to print out "Notices."

4. Review the necessity and financial costs of staffing evening and week end shifts.

• As of October 1, 1994 an on-call policy was developed and procedures implemented to ensure adequate coverage during evening, night and weekend hours and allow for more efficient scheduling of employees.

5. Utilize a standard form to document patient encounters to eliminate transcription and to create a permanent record of each medical translation.

• The "*Patient Service Record*" (an internal document designed by the Coordinator) will be completed for each patient translation encounter. In addition, all patient no-shows and cancellations should be documented on this form.

6. Using the "Patient Service Record" as the source document, an Access database should be created for the purpose of maintaining, reporting and monitoring statistics.

• Data elements input to the database include (but are not limited to): Patient demographics, language translated, type of encounter (visit/telephone), location, method of booking (including urgent, walk-in), reason for translation, and time encounter began and ended. Analyses and reports were developed which document volume by location and language, average length per encounter, primary reasons for translations, and other types of management reports.

7. South Campus interpreters should be stationed in a central location during IMC hours (8 a.m. - 4:30 p.m.) and have one person designated as the interpreter on page to reduce the duplicate and successive pages. When Clinic personnel deem a patient will need an interpreter (but one was not scheduled in advance), the Interpreter on-page should be notified at the time the patient presents.

• The Service has assigned two Spanish interpreters to C-Level, one assigned to booked appointments and one to cover urgent or unscheduled appointments. Additionally, to accommodate the Pediatric Clinic's increase in afternoon volume, one of the interpreters will be assigned to the area from 3:30 p.m. to 5 p.m.. When a patient arrives at the reception desk, the interpreters will be notified and will ensure that there is someone assigned to that patient. The IMC Charge Nurse/Nurse Manger is the designated person to page the interpreter when circumstances arise. Additionally, each Clinic's Charge Nurse is responsible for negotiating appointment conflicts amongst themselves.

4. Current Operations - Increasing Quality and Productivity while Decreasing Costs

The Coordinator and the Manager of Medical Staff Services have made vast improvements in the Service's operation. By redesigning and implementing new processes and instituting a feedback mechanism with external departments, the Service provides a more efficient and effective delivery of translation activities to meet patient and provider needs. For example, the Coordinator spends several hours per week working as an interpreter at the South Campus Adult Ambulatory Clinic providing a management presence to address issues or concerns the Clinic personnel have regarding the Service.

By controlling human resources, the Service has increased productivity while decreasing the expense associated with each encounter. For the tenth month of Fiscal Year 1995 (July 1995), the Services' productivity level was 103.9%, which represented an 0.5 FTE (full-time equivalent) understaffing variance. The department had 2,969 encounters that required 13.40 FTEs; however, the department utilized only 12.90 FTEs resulting in the greater than 100.0% productivity level. July's encounter volume was 17.5% above the year-to-date average encounters (2,527). In January 1995, Interpreter Service achieved its highest productivity level of 127.5%, representing 3.3 actual worked FTEs less than required FTEs. On a year-to-date basis, the Service is operating with 12.15 worked FTEs versus 12.57 required FTEs (based on volume). This situation yielded a 0.4 FTE understaffing variance and productivity of 103.5%.

Concurrently, the actual cost per encounter has decreased since Fiscal Year 1994. As of January 1995, the average cost per encounter was $9.93, compared to $11.39 in Fiscal Year 1994: A 12.8% decrease. Broken down between salary and wage expense, and operating expense, salary and wage declined 10.8% (from $10.33 to $9.21) while operating expenses decreased 54.4% (from $1.58 to $0.72).

Over the past five years the Service's encounter volume has grown from 2,097 in 1990, to over 30,000 projected for Fiscal Year 1995. Even with this tremendous volume increase, the Service must continue to control costs in this competitive environment. While Baystate Health Systems' primary care market expands, there is also an increase in interpreter demand for assistance with home health, postpartum and physician practice visits. The organizational changes made within Interpreter Service should position them well: Budget control forces the Service's management team to continuously look for creative solutions in providing more translations with no staffing increase.

---

[1] Baystate Health Systems, "An Integrated Health Care Delivery System Serving Western New England," p. 3.
[2] Ibid.
[3] Ibid., p. 5.

AN ASSESSMENT TOOL FOR COMMUNITY INTERPRETING

Roda P. Roberts
University of Ottawa

Keywords: Community Interpreting, Standards, Assessment, Testing, Screening, Accreditation, Sight Translation, Consecutive Interpreting

Abstract: While standards for conference interpreting and court interpreting have been established for some time in North America, this has not been the case for community interpreting (also called cultural interpreting). Community interpreters have traditionally been hired on the basis of a short interview, which, at best, gives some idea of language skills. This situation is about to change, thanks to the Cultural Interpreter Language and Interpretation Skills Assessment Tool Project (the CILISAT project), spearheaded by the Cultural Interpretation Services of Ottawa-Carleton. This paper, which will discuss the purpose of the tests, their nature, the process of establishing them, and the marking system, will highlight a number of innovative features of these tests.

1. INTRODUCTION

The establishment of standards for any profession leads to the development of educational programs designed to train potential professionals to meet the standards and/or to the development of certification/accreditation examinations[1] to ensure that would-be professionals have achieved the standards required. Translation, conference interpreting and court interpreting in North America have followed this trend - albeit slowly and timidly. There are now well-known translation and conference interpreting programs in Canada and the United States, as well as translator and conference interpreter accreditation examinations in both countries. And there are court interpreter accreditation examinations in North America, as well as a certain number of court interpreting programs, which are admittedly limited in duration and level. However, community interpreting (also called cultural interpreting, among other names[2]) lags well behind the others in terms of the establishment of standards and of training programs and accreditation exams.

There have been some dispersed efforts to establish standards. Thus, for instance, in British Columbia, a group of hospitals, interpreter services and interpreter trainers that grouped together to form the "Health Care Interpreter Partnership Project" in 1993 established as its first objective "to formulate standards for quality and delivery of interpretation, and the criteria for the evaluation of such standards" since review of the literature had confirmed the non-existence of established and generally recognized standards for health care interpreters. (Ref. 1) And in Ontario, the Ministry of Citizenship that oversees the "cultural interpretation services" in the province, has been working to establish standards for all community interpreters.

However, as indicated above, standards are only as good as their implementation through education and/or accreditation examinations. And education in community interpreting is limited in scope and duration, not readily available everywhere, and often offered in an ad hoc fashion in North America: it ranges from a few hours of training offered by community interpreting services and user agencies to a part-time one-year program (at the Alberta Vocational Centre and at the University of Minnesota) offered to candidates who have been through little, if any, screening. (Ref.2) It
cannot therefore be counted on to produce interpreters who are automatically capable of meeting rigorous standards. In most cases, such training helps potential interpreters to develop basic skills or enables those who already have these skills to improve on them.

Given these circumstances, accreditation tests for community interpreters are essential if the market is not to be flooded with untrained or insufficiently trained professionals. However, no such tests existed, at least to my knowledge, until the development of the Cultural Interpreter Language and Interpretations Skills Assessment Tool (CILISAT) in 1994-1995.

This paper will discuss the purpose of this assessment tool, its nature, the process of establishing it, and the marking system, and will highlight a number of its innovative features.

2. BACKGROUND OF CILISAT

The CILISAT project was the brain-child of Dong Liu, founder and Coordinator of the Cultural Interpretation Services of Ottawa-Carleton (CISOC). When she was first hired in 1993, she called up various community interpreting services to find out how they selected their interpreters. She found that, for the most part, they based their decisions on a resume and an interview. To quote her own words, "What I found most insecure was that I could not tell whether the person had enough skills in actual interpreting." (Ref. 3) After trying a number of screening methods, including selecting people who had passed interpreting tests with other agencies, she realized that none was satisfactory given the lack of uniformity and objectivity.

Convinced that a professional service needs to ensure standards, she envisioned a standardized test that would not only assist in the selection of interpreters, but also set competency goals for interpreters in terms of professional development. With the support of the Ministry of Citizenship and other cultural interpretation services in the province, CISOC thus launched the Cultural Interpreter Language and Interpreting Skills Assessment Tool project, to produce a tool that could be used by all cultural interpretation services in the province.

A core team consisting of Dong Liu and two interpreter trainers (Professor Brian Harris of the University of Ottawa and myself) was set up to devise the test. An advisory board consisting of coordinators of other cultural interpretation services was established. Other interpreter trainers and interpreters as well as individuals from user agencies were brought in at specific stages of the project.

Phase I of the project, which took nine months to complete, consisted of developing model tests for English-Spanish and English-Arabic community interpreters. Phase II of the project, which is still awaiting funding, will involve implementation and testing of the model tests. The discussion below will be limited to Phase I of the CILISAT project.

3. GOALS OF THE TEST

The first question that needed to be addressed was the purpose for which the assessment tool was intended. Was it to screen potential interpreters for training or was it to accredit interpreters who had achieved the standard required to work in the community setting? These two purposes would normally call for two separate assessment tools. Before responding to this question, we analyzed the training available.

We soon realized that the nature and the length of training provided were inadequate to make a great difference in language and interpreting skills between the screening and the accreditation stages. The limited number of hours of training, the many aspects of community work to be covered, the inability in a generic training course to improve foreign language skills, and the fact that many of the trainers are not themselves interpreters means that the training, while good for developing interpersonal and intercultural skills, could not be counted on to take candidates with poor language skills and few or no interpreting skills and turn them into professional interpreters. In other words, the difference that the training would make in language and interpreting skills would be minimal.

In light of this situation, it was unrealistic to conceive of two different assessment tools - one for screening the aptitude of candidates for community interpreter training, the other for accrediting qualified interpreters. So the somewhat unorthodox decision was made to devise one tool that could be used, with some variation in marking, for both purposes.

The second question that needed to be answered was exactly what we wanted to evaluate: language skills? interpreting skills? knowledge of user agencies? terminology particular to agencies? cross-cultural transfer? interpreter ethics? interpersonal skills? It was decided to focus primarily on language and interpreting skills, since the training does not adequately develop them, while it does cover the other skills and topics.

The third question dealt with the type of language and interpreting skills the community interpreter requires. This required taking into consideration the settings in which he works as well as the type of work he is called upon to do. Community interpreters are expected to work in a variety of social service, medical and legal settings (although not necessarily in the courtroom per se). They therefore need to be able to cope with a variety of levels of language. They are called upon to do "short consecutive" (interpreting short spans of speech) for two or more parties and sight translation of documents that may be produced during the communication event in either of the languages involved in the event.

The final question to be answered was how the test was to be administered. This was a critical point since the test is intended to be used throughout the province of Ontario, which is very large. Would all the candidates be summoned to a central test site or would they be able to take the test in their own locality? Given the time and expense involved with obliging candidates to travel to some central site, it was decided that the test had to be simple enough to be administered by a non-interpreter anywhere in the province.

In light of these decisions and findings, the goals of the test were set as follows: (a) the test had to serve a dual purpose (screening and accreditation); (b) it had to verify general language skills (comprehension and production) and interpreting ability; (c) it had to cover short consecutive and sight translation in both language directions; and (d) it had to be "portable" and simple to administer.

4. LANGUAGES USED IN THE MODEL TESTS

Of course, it was clearly impossible to produce tests in all the languages used in community interpreting in Ontario at the same time. Nor was this recommendable, since the goal of Phase I of the project was to produce a model assessment tool that could be tried out and modified if necessary. However, two language combinations were used for the model tests: English-Spanish and English-Arabic.

The decision to use two language combinations was dictated partly by the urgency of the testing needs, but mainly by the fact that we wanted to see if parts of the test could be used for more than one language combination. The two language combinations chosen were selected for three reasons: these combinations are in fairly high demand; one member of the core team (Brian Harris) knows both Spanish and Arabic; and, most importantly, reasonably qualified markers are available for both language combinations.

The choice of Arabic posed particular problems because of the existence of several Arabic dialects. While Standard Modern Arabic is the standard written form, it is rarely used in everyday speech (although it is spoken on formal occasions and in the media and it can serve as a lingua franca among Arabic speakers). Indeed the spoken language differs considerably from one Arabic-speaking country to the other. We chose to use both Standard Modern Arabic and the Lebanese dialect, the latter because of the large number of Lebanese immigrants in the Ottawa-Carleton region.

5. NATURE OF THE TEST AND ITS PREPARATION

Since the test is intended for professional purposes, we decided to make it reflect as closely as possible the linguistic tasks that the community interpreter is called upon to do. Thus, we eliminated the idea of a written section, since writing is not one of his major tasks, and decided on an entirely oral test.

5.1 Test Format

The test consists of two parts, which reflect the two main tasks of the community interpreter: sight translation and consecutive interpretation of dialogue.

The sight translation part consists of two texts, each of about 250 words, one in English, the other in the foreign language. Candidates are required to translate the texts orally into the other language, following a 15-minute preparation period.

The consecutive interpretation section consists of one dialogue in both English and the foreign language, between 650 and 700 words in length. Candidates are required to interpret into the other language what each speaker says.

5.2 Test Content

The texts for the test were determined by the settings in which community interpreters are called upon to work. As indicated above, these comprise social service settings (welfare office, school administration office, employment centre, etc.), medical settings (doctor's office, hospital emergency room, etc.), and legal settings (immigration office, lawyer's office, administrative hearings, and even the courtroom).

Sight translation texts therefore had to include written documents used in such settings - for example, birth and marriage certificates, school transcripts, apartment leases, hospital consent forms, general brochures for immigrants, etc. The dialogues covered the same three settings (with two dialogues for each setting).

5.3 Text Preparation

Once the decisions explained above had been made the actual preparation of the test was relatively simple, although lengthy. It consisted in the selection or creation of texts for sight translation and consecutive interpretation and their revision or modification.

For sight translation, the same English text was used for both language combinations, i.e. for translation into Spanish or Arabic. However, although we considered the possibility of using another English text pretranslated into Spanish and Arabic for sight translation into English, since this would have simplified the marking process, we decided to use authentic documents as far as possible. Two English texts, four Arabic texts, and four Spanish texts were selected.

The dialogues for consecutive interpreting were, for the most part, composed by personnel of the appropriate user agencies. The one notable exception was the dialogue in the legal setting, which was adapted from actual transcripts. Six dialogues were initially prepared: 3 for the social service setting, 2 for the hospital setting, and 1 for the legal setting.

All the texts were reviewed by an interpreter trainer for their level of difficulty, and modifications were introduced to ensure their relative homogeneity from this point of view.

6. PACKAGING THE TEST

As indicated above in section 3, one of the practical decisions that was made at the preliminary stage of the project was that the test should be portable and simple to administer, for in many parts of the province there is only one interpreter working on his own. How could the test be packaged so that any untrained official with no language laboratory in sight could administer it?

The first answer to this question was to require only simple equipment (cassette recorders), which are readily available everywhere. The texts for sight translation could be sent to the untrained test administrator, who would give them to the candidate and have him record the translation on a blank tape.

Administration of the consecutive interpreting part posed greater problems, since, in principle, at least two "readers" would be required for the dialogue. Given the fact that we were unsure of the quality of potential readers and that we wanted all candidates to have the same advantages and disadvantages as far as the source dialogue was concerned, we decided to pretape the dialogue (using good readers and good recording facilities) leaving a blank space sufficient for interpreting between each segment to be interpreted. While the pre-taped dialogue has the disadvantage of the interpreter being unable to ask the speaker to stop or to repeat a segment, the advantages of portability, uniformity, and simplicity seemed to us to outweigh the disadvantages.

To ensure that there would be no misunderstanding concerning either what the candidate was expected to do or what the administrator had to do, a sample test with instructions was prepared for candidates and detailed instructions were prepared for test administrators.

7. PREPARATION OF THE TEXTS FOR MARKING

The next step was to plan the marking of the tests. To ensure maximum homogeneity among markers, the sight translation texts and the dialogue texts were carefully prepared for marking.

Both the sight translations and the dialogues have been divided into units of information. In the example below, the units are separated by two oblique lines.
Well, it was at night.// I can't say the exact time// but around 10 o'clock.// I always look out of the window// to check on my car.//

In the **sight translation texts**, the (major) units of information have been further subdivided when necessary into minor units to help the marker see if anything has been left out of a major unit. In the example below the end of minor units is marked by a single oblique line.
Canada's experience/during and after World War II/ raised awareness/ of the refugees/ and of the need of families to be together.// All of this has been reflected in Canada's immigration policies.//

All the major units in the sight translation are numbered consecutively as they appear in the text, as follows:
15 Canada's experience/during and after World War II/ raised awareness/ of the refugees/ and of the need of families to be together.//
16 All of this has been reflected in Canada's immigration policies.//

At the end of the text, the total number of major units is indicated.

In the **dialogues**, the units of information have been classified in terms of their significance in the text as vital (**), important (*), and less important (@), as follows.
* Well, it was at night.
@ I can't say the exact time
** but around 10 o'clock.
* I always look out of the window
* to check on my car.

Moreover, the units of information in the dialogue have often been reworded in the marker's text to emphasize that what they have to consider is not the wording but the information conveyed by it. Thus, the span of speech above has been reworded in the following way:
* It was at night

@ I don't remember the exact time
** It was about 10 o'clock
* I always look through the window at that time
* to check that my car is O.K.

All the reworded units of information in the dialogue are also numbered consecutively. But those units that go logically together to comprise a set of questions or a complete answer are grouped together numerically, as follows:
* 14.1 It was at night
@ 14.2 I don't remember the exact time
**14.3 It was at about 10 o'clock.
* 14.4 I always look through the window at that time
* 14.5 to check that my car is O.K.

At the end of the reworded and numbered dialogue, the number of vital points, important points, and less important points are totalled.

8. TEST MARKING SYSTEM

The detailed but simple test marking scheme, described below, was developed and applied to the texts prepared for marking.

Both the sight translations and the consecutive are marked for information and for language, since the appraisal of both exercises is based on two criteria: (a) full and accurate reproduction of the original information; and (b) correct language in the target text.

The rendition of the **information** contained in each major unit of the sight translation texts and each unit of the dialogue is assessed in terms of the following appraisal categories:
(1) Conveyed clearly
(2) Conveyed, but not clearly
(3) Incomprehensible
(4) Omitted completely
(5) Added to
(6) Partially omitted
The first four categories are mutually exclusive, as are also categories 4 and 5, and 4 and 6. However, categories 5 and 6 can apply in conjunction with 1 or 2.

Each of the appraisal categories is worth a certain score:
Conveyed clearly = 3
Conveyed, but not clearly = 2
Incomprehensible = 0
Omitted completely = 0
Added to = deduct 1
Partially omitted = deduct 1

Using this scoring scale, the marker assigns a score for each major unit of the sight translation and for each unit in the dialogue.
Up to this point, therefore, the grading for the sight translations and the consecutive is similar.

The difference comes at the point of totalling the scores for each unit. For the sight translation, the marker merely adds up the individual unit scores to obtain a total information score, which is then converted to a percentage of the total information score possible (which is 3 X the number of major units in the text). However, for the dialogue, the scores for each type of information unit (vital, important, less important) have to be added separately and the total score for each type then has to be multiplied by a coefficient attributed according to its importance (X 3 for a vital unit, X 2 for an important unit, and X 1 for a less important unit). Finally, the weighted total score for all categories has to be calculated and then converted into a percentage of the total information score possible.

But, as indicated above, evaluation of the reproduction of the original information is only one criterion of the appraisal. **Language** proficiency also needs to be assessed. This is done globally (i.e. for the entire text or dialogue) in terms of the following linguistic aspects, each of which is worth a maximum of 2 points for a total of 10:
General Vocabulary ..../2
Technical Terms../2
Grammar../2
Appropriate register, level of language and tone.........../2[3]
Pronunciation and audibility....................... ..../2

Total Language Score......................................./10

Each of these linguistic aspects has to be evaluated using the following appraisal scale (which contains categories which are mutually exclusive):

Very good, i.e. native fluency and correctness	= 2
Good for a non-native, but with a few errors	= 1.5[4]
Understandable, but numerous errors	= 1
Definitely too poor to be an interpreter	= 0

The consecutive interpretation of the dialogue is marked twice for language (one language score for English and a second for the foreign language).

At this point, we had subjectively decided on the information and language scores which would be required in each part of the test (a) for screening for training, and (b) for accreditation. However, these and other issues were to be revisited after field testing.

9. FIELD TESTING

All the texts were pretested using a combination of working community interpreters and potential community interpreters: a group of 6 for each language combination, with each participant doing at least one dialogue and two sight translations. They were also asked to express opinions on the test. The pretests were administered using the instructions previously set up for test administrators.

Markers for the pretest were selected using guidelines established for the selection of test markers: 2 with Spanish as their first language, 2 with Arabic as their first language, and 2 with English as their first language. The 2 Spanish or Arabic markers and one English marker graded each of the dialogues. All the English-Spanish and English-Arabic sight translations were graded by the two Spanish or Arabic speakers, while the Spanish-English and Arabic-English sight translations were evaluated by one English speaker.

The raw results were then reviewed and analyzed at meetings of the markers with the core team and of the core team on its own. Some of the conclusions were the following:

(1) Exercises:
- Many participants said they found sight translation difficult as a task. This was only partially borne out by the scores.
- However, Spanish candidates seemed to have particular difficulties with Spanish-English sight translation.

(2) Texts:
- The best average scores were obtained in the three social service dialogues. Then came the two medical setting dialogues. And last in line, well-behind the others, was the legal dialogue. These were the results despite the fact that relatively few technical terms were used and those that were were explained later in the dialogue.
- A high number of units of information in a given dialogue did not adversely affect the results. In fact, candidates obtained a better average score for a dialogue that had a fairly high number of units of information (531) than for a dialogue with a much smaller number of units of information (377).
- A disproportionately large number of units of information in a given language in a dialogue had an adverse effect on the results. Thus, the legal dialogue, which contained far more units of information in the foreign language than information units in English, resulted in the worst average scores. This again could imply problems with expression in English.

(3) Marking
- The markers judged the marking system satisfactory overall and easy to apply. However, this did not prevent a few rather noticeable discrepancies between markers.

(4) Performance of Working Interpreters vs Non-Interpreters
- Overall, the working interpreters performed better in the dialogues than did the non-interpreters. The difference was most marked in the most difficult dialogue (the legal one), with the former obtaining an average score of 80 and the latter 58.

10. DECISIONS MADE AFTER FIELD TESTING

A certain number of adjustments and additions were made to the existing texts and scales after pre-testing.[5] More importantly, however, some significant decisions were made:

(1) Two Levels of Testing:
Given the fact that candidates had the lowest scores for the medical and legal dialogues (i.e. the more specialized texts), it was decided to have two levels of testing: one general, the other specialized. The format will be the same for both levels (two sight translations - one in each language direction, and one dialogue to be interpreted consecutively). However, the nature of the texts and hence their level of difficulty will be different. The General Test will serve both as a screening and accreditation tool. The Specialized Test will be given only to those who are already accredited at the general level.

(2) Pass Marks
On the basis of the field testing, the following pass marks were tentatively adopted:
For screening applicants for training:
- 65% minimum on the information score in each direction
(4 information scores: English-Foreign language in sight translation, Foreign language-English in sight translation, English-Foreign language in consecutive, interpretation, Foreign language-English in consecutive interpretation)
- 6 out of 10 for language proficiency

For accreditation (at both the general and specialized levels)
- 70% minimum on information score in each direction (Total of 4 scores: English-Foreign language in sight translation, Foreign language-English in sight translation, English-Foreign language in consecutive, translation, Foreign language-English in consecutive)
- 75% overall average for information reproduction
- 7 out of 10 for language proficiency

11. CONCLUSION

It is hoped that Phase II of the project, scheduled for this Fall, will allow us to administer the series of tests prepared to a wider range of applicants. This in turn may lead to further adjustments and modifications, especially in the pass marks. Although the idea of having tests for all language combinations is still a dream, the model tests set up, the marking scheme established, and the basic test administration and marking guidelines prepared should make other tests easier to plan.

However, even if the dream of a community interpreting assessment tools for every language combination is not realized, there now exists in North America a general test for screening and accrediting community interpreters and a specialized test for accrediting interpreters to work in medical and legal settings for two very common language combinations.

12. END NOTES

1 Although I use the term "accreditation" in the rest of the text, I make no distinction between "certification" and "accreditation".

2 Some of the other names used to designate community interpreting are public service interpreting, dialogue interpreting, ad hoc interpreting, liaison interpreting, and escort interpreting, medical or legal interpreting.

3. During the pretesting, the last two categories were broken down into 3 distinct ones for consecutive (but not for sight translation): Accent and Intonation, Tone and Attitude, and Voice quality and Volume. This made the total language score twelve for this exercise, as opposed to ten for sight translation. The reason for the difference is that tone and attitude were considered important for consecutive but not for sight translation. However, since pretesting, we are leaning towards a 10-point scale.

4. This category was added to the scale after pre-testing.

5. Texts: The longest dialogue, which was also the most difficult, was shortened. The blanks in the forms used for sight translation were filled with information. A second legal dialogue and two legal sight translation texts were added. Marking scale: The category of "Good for a non-native, but with a few errors" with a score of 1.5 was added to the language scale.

13. REFERENCES

1. Carr, S. "A Three-Tiered Health Care Interpreter System." Paper presented at the First International Conference on Interpreting in Legal, Health and Social Services Settings," Geneva Park, Ontario, June 1-4, 1995.

2. Roberts, R.P. "Community Interpreting Today and Tomorrow," in P.W. Krawutschke, ed. *Proceedings of the 35th Annual Conference of the American Translators Association.* Medford, NJ: Learned Information, Inc., 1994, p. 127-138.

3. Liu, D. "Test Cultural Interpreters: Issues and Challenges," Paper presented at the First International Conference on Interpreting in Legal, Health and Social Services Settings," Geneva Park, Ontario, June 1-4, 1995.

ADDRESSING BIAS IN THE INTERPRETATION SETTING

P. Diane Schneider
U.S. Department of Justice

Keywords: Discrimination, Biased Treatment, Interpretation,

Abstract: Interpreters have long struggled to counter misperceptions about their role, their ethics and the value of their services. Interpreters frequently find themselves expending considerable effort educating non-interpreters and monolingual persons about basic fundamental concepts of language interpretation. Often, monolingual persons lack a sufficient knowledge base to understand the mysteries of interpretation and they may perceive it to be effortless, thus undervalue it.

A disturbing side of such perceptions, however, involves encounters with persons who feel they are "forced" to utilize interpreters by an "overly broad" interpretation of the U.S. Constitution guaranteeing equal treatment under the law. They may consciously or unconsciously hold personal beliefs that immigrants do not "belong" here or have not "earned" equal services or that persons of a given ethnicity tend to share stereotyped behavioral or values characteristics.
These perceptions and stereotypes can impair the equal access interpretation services are intended to ensure. Interpreters of non-European background seem more likely to experience apparently biased treatment against them than interpreters of European background, however, any interpreter may experience it. Interpreters need to be well informed about issues involved and prepared to address such problems when they arise, whether in the court room, the attorney client conference or even the medical interview. Some types of incidents commonly occur in many settings, while some sites present unique challenges due to local or regional demographic movements and geographic location. In some areas, xenophobia is perceived to be increasing in direct proportion to the numbers of "visible" persons of foreign origin moving into the area.

1. PROBLEMS ENCOUNTERED

A significant factor supporting the ability of this society to function rests on the trust its residents have in the fairness of the system. Whenever members of the public perceive that the system is biased or rumors of bias persist, then the ensuing lack of trust may impair the functioning of the system to such a degree as to impair the public good. Some persons may not come forward as witnesses in court or may refuse to testify. Others may be reluctant to cooperate with public agencies such as public health agencies seeking to control communicable diseases. Victims or witnesses may refuse to report crimes or assist police attempting to investigate a crime due to their impaired trust.

Interpreters report experiencing gratuitous comments or innuendos directed to them or in their presence which tend to stereotype, belittle or demean a specific group of foreign nationals. Interpreters also report experiencing different treatment by agencies or offices using their services if the interpreters are perceived to be non-white or non-native born. Occasionally interpreters themselves, due to political or class differences, contribute to biased behavior when their personal feelings interfere with their professional duties. Some innate characteristics of interpreters cannot be changed and if these appear to elicit bias or to present obstacles to carrying out a professional assignment, the interpreter may benefit from preparing those most effective responses to confront the problems. Interpreters must also prepare to effectively confront what they identify as biased behavior affecting their work and to seek redress appropriately when they are unable to prevent or defuse such incidents. Interpreters can have a significant impact on issues of bias by taking steps to recognize it, prevent it, confront it, and to seek redress for it, whether individually in cases involving them or jointly with their association or other community organizations dedicated to fairness and nondiscrimination.

1.1 Types of biased behavior

When we speak of "biased" behavior, we refer to various degrees of behavior resulting from preconceived notions about persons which are not related to the situation at hand. We may assume therefore, that "biased" behavior is behavior which would be different if an individual were involved who did *not* belong to the group about which the person has stereotypes or preconceived notions.

1.2 Identifying biased behavior

It is necessary to identify what kinds of behavior may be based on bias in order to address such behavior. The behavior may be divided into overt and subtle biased behavior. If we draw a horizontal line and at the left end list that behavior which demonstrates the least bias we can conceive of, then at the far right hand list that behavior which represents the most biased behavior that we can conceive of, we can then proceed to list many kinds of behavior and place them appropriately on a continuum of least to most bias.

Least **Most**

< -- >

Bias **Bias**

We can then draw a line between that point where the bias seems to us to be "subtle" and that behavior which seems to us to be "overt". Interestingly enough, we will not all agree that certain behaviors are subtle nor that some behaviors are overt. We may also not agree that some behaviors represent any degree of bias at all. Nevertheless, we will be able to agree that some of the behaviors listed in the middle are all biased behaviors. Of those behaviors we agree are biased or prejudicial, we may then begin to review their rank and assess the degree of damage such behaviors cause and the possibility of addressing such behaviors in a preventive, real time or redress mode. It is illuminating to identify which behaviors reflect bias against the interpreter, against the client or against an entire group or country:

Examples of biased behavior against interpreter:

An interpreter observes she is assumed to be advocating for the accused person if he/she is of the same ethnic background. The interpreter is assumed to condone illegal behavior when simply interpreting the comments expressed in the non English language.

Examples of biased behavior against a client:

The person communicating with the client refuses to look at the client when communicating with him; The judge, taking advantage of the fact the defendant speaks no English, advises him in open court and on the record that a subsequent DWI conviction would result in the death penalty. Then he advises the interpreter not to interpret it and orders the comment stricken from the record. A judge regularly requires higher bail amounts of non-English speaking defendants explaining they are likely undocumented and that they may be deported or escape to their own country, and therefore not appear for trial.)

Examples of situations where something other than bias may cause prejudicial behavior:

(eg. May involve ignorance, lack of resources) We like to believe few judges would consider the juvenile defendant to be a qualified interpreter for his non-English speaking parents especially in a case where his own interests may be adversarial to those of his parents. The same would be true in the case of co-defendants, one of whom speaks no English, the other offering to interpret even should they be tried together and share the same attorney.

The following are examples of incidents interpreters have recently described. Do they appear to represent biased or discriminatory treatment? ignorance? insensitivity? the potential for biased treatment? A hostile climate? What are the indicators of bias? What are the possible underlying assumptions or misunderstandings? What are steps which might be taken in each instance to prevent, identify, confront and/or seek redress and educate to prevent future such incidents?

Situation I

The fire chief of a small town in Eastern Washington was quoted in the local newspaper remarking on the fact the town now has more Hispano/latino residents, many of them new immigrants, than Anglo/Europeans. Asked what accommodations his department had to make to provide services when victims needing his agency's help are increasingly limited-English-speaking. His response was "How can you expect me to save someone's life when they are not speaking English?" The remark was published in the local newspaper and caused an uproar in the community. Some readers expressed outrage, saying one should not be required to be able to speak English simply to receive emergency services. What underlying assumptions may lead to this kind of remark and how does this impact provision of professional interpretation services? How do such assumptions impact confidence in the government?

It is likely such a statement is rooted in the assumption that persons should not come to this country before learning sufficient English to communicate in emergency situations and which will not fail them when under extreme stress. Another underlying assumption may be that persons who do not speak English well do not belong in this country and specialized attention should be denied them if they fall victim to crime, illness or other misfortune. The opinion that foreign nationals of color are overburdening the "welfare" system is frequently expressed in the media by politicians even though statistics do not support this assumption. A fourth underlying assumption may well be that when monolingual "Americans" go abroad, no such services will be offered them because they are foreigners. The national movement to deny public services to persons suspected of being undocumented or even to those immigrants who are not yet citizens appears to reflect these assumptions. By understanding that such beliefs assumptions and attitudes do indeed exist, one can then prepare oneself to confront and/or counter them effectively.

Situation II

A state police officer who has published a phrase booklet to assist officers in the field who speak no Spanish is monolingual himself. He confided to me that people who really do not speak English need help from officers sometimes and while his fellow officers will find such a tool useful, he has a surefire way of determining whether the suspect really needs an interpreter in interrogation settings or is faking such a need. He said he offers the interrogee coffee and asks if he wants cream and sugar. If he is able to understand the question and answer then the officer considers him to be competent to participate unaided in a police interrogation. The underlying assumption seems to be that the foreign born are likely to falsely pretend not to speak English well for their own convenience and not because they find themselves under stress in unfamiliar circumstances. Another underlying assumption may be that people who say they do not speak English must have some nefarious intent.

One cannot help wondering whether that officer would be offended if one assumed he or she were "faking it" when pretending not to understand French since he or she obviously knows many words in French such as: apres vous, merci, croissant, quiche, cafe au lait, demitasse, hors d'oeuvres, and cordon bleu.

Situation III

An interpreter reported she was asked to transport family members of the defendant to the airport after she had interpreted at trial and the defendant had been convicted of homicide. No court officer was sent to accompany her. The underlying assumptions may be construed to be that the interpreter, because of her race, would not mind accompanying the family but the attorney as a "professional" need not waste his time for that purpose. Imagine the response if a bailiff were asked to perform this task for English speaking relatives of defendants. What is the likelihood the medical examiner would be asked to prescribe medication for a defendant's relative? Another assumption may be that the interpreter, because of her ethnicity, would be in no danger of resentment by the family attributing any responsibility to her for the negative outcome of the trial. Still another common misperception is that interpreters, especially those sharing the ethnicity of the person for whom interpreting, are somehow on the "side" of that person when of course an interpreter is an officer of the court responsible for assisting the judge in ensuring that the trial is fair by ensuring the defendant's constitutional right to be "present" at his own trial.

Situation IV

A police officer commented before investigating a latino individual just involved in a minor traffic accident that "They don't have drivers licenses or insurance. None of them do. Most of them are illegal." What impact might such a statement have on the investigation?

Situation V

An agency reported that a specific dentist's office treats "caucasian" interpreters with utmost respect, allows them to use the telephone, gives them access to file material, but when non-white interpreters were assigned, treated them rudely, charged them to make telephone calls or photocopies of medical coupons needed for billing purposes. On one occasion, an interpreter was interpreting for a distraught mother who was concerned that the dentist would not allow the mother to enter the dentists area with her young child and the interpreter was told the police would be called if the interpreter did not leave immediately. The dentist's office obviously received public funding to offer medical services to persons with medical coupons.

Situation VI

A juvenile court probation officer arranged a hearing for a juvenile who spoke English well, having been in the U.S. since age 5. No interpreter was scheduled, though the parents, residing in the country for eight years, had not attended any English classes due to their long working hours and child care responsibilities. The parents arrived in court and the judge ordered the juvenile to interpret for his parents. The parents were determined financially responsible in juvenile court proceeding involving shoplifting. The juvenile told his parents only that nothing of importance had occurred and that the matter had been resolved. What was the responsibility of the judge in this case?

2. ASSESSING THE PROBLEM

2.1 Recognizing the impact of the incident:

Incidents of bias and prejudice may be so subtle they are not always perceived immediately or by all parties present. A facial expression of displeasure may or may not be directed at an individual. They may be so flagrant that they elicit a violent reaction on the part of some party. Requiring all the witnesses of a given ethnic group to be searched for weapons while others are not may elicit such a reaction. Biased behavior may be the result of several factors:

Cultural misperceptions:

Communication styles are culturally linked and interpreted, and then reacted to by the listener in light of his/her own experiences and expectations. This may involve body language or an assessment of the forthrightness of the individual based upon a perception of how directly he comes to the point in describing or answering a question.

Ignorance:

May consist of gratuitous comments, linguistic habits or humor which are not necessary to repeat here.

Bias:

Results from unconscious or conscious assumptions about the parties, positive or negative. Some folks tend to hold their liquor well, some have a tendency to act out violently where some will always be respectful and obedient.

Hostility:

Manifestation of negative bias. It may contain elements of fear or a perception some action is necessary to prevent something bad from happening (Lock up all your daughters because the ____ are in town.)

Systematic:

May occur when any of the above factors has contributed to the development of process, policy or practices which tend to act to the detriment or advantage of specific groups.

2.2 Educating/Informing self

Laws, policies and ethical standards exist which may apply to specific situations. There may be specific and unique laws, policies or even constitutional issues which apply to your field, locality or issue.

3. RESPONDING AFFIRMATIVELY AND SEEKING REDRESS

3.1 Prevention is Crucial

Preventative steps can indeed prevent and educate potential perpetrators of biased and prejudicial actions. They may also serve to protect the victim from counter accusations in the event a complaint must be filed. While such steps may sharply reduce incidents of overtly biased and prejudicial treatment, they will not discourage the hard core from underrating the value of interpreters as professionals and as human beings, however. There is a hierarchy of indicators with which the interpreter must become familiar in order to derail biased and disrespectful behavior before it escalates.

3.2 Factors in Prevention

- Preparation
- Dress code
- Code of Ethics, (of professional interpreters)
- Body language (watch gestures even when not interpreting, subtle, not histrionic or flamboyant)
- Voice (clear and even)
- Demeanor (calm and professional)

Pay particular attention to dress, walk and body movements and a well modulated voice which denote professionalism, adopting the "Miss Manners" persona: pleasant, formal, yet clear about the limitations of the relationship one performs tactical planning before the fact.

Acknowledging the necessity for preventive tactics does not equate with blaming the victim. There is no excuse for persevering with biased behavior even if the individual seems to act in a stereotypical way. Circumstances exist in society which many want to change as they do not coincide with their ideals. Unfortunately or fortunately, not all people agree on the way the world should operate, but while we say persons should not be subjected to biased treatment due to their mannerisms, dress or accent or physical appearance, nevertheless, it occurs. What behavior of interpreters may escalate underlying biased feelings against them? What behaviors and actions may minimize the bias incidents against interpreters or provide a groundwork for effective redress should it occur in this setting? Some preventive tactics may help to minimize the impact of biased behavior against interpreters. Not surprisingly, many are the same tactics recommended by consultants in the business world for clients seeking promotions or employment. Many are tactics already used by interpreters seeking to promote the image of the profession.

Physical appearance

There are more female interpreters than male. This is probably due to several factors. Many interpreters are small in stature and not of Anglo-European background. A strong cultural value commonly held in the U.S. is that one is perceived to be more authoritative if male, white, tall, conservatively dressed, eloquent in the English language, self assured and assertive, and uses upper middle class body language. This is a stereotype which must be acknowledged in order to counter the stereotype. Unfortunately, in spite of efforts in the profession to raise the perceived level of professional standards, interpreters continue to appear in court and other settings who give the appearance they may be defendants rather than professionals. While this is also a common complaint about defense attorneys especially public defenders, interpreters must take advantage of every opportunity to create and maintain the aura of professionalism.

Courts are very conservative places, even when a particular judge may seem to hold very informal hearings.

Common mistakes

Female interpreters must seek to avoid the "cute" or the "tart" look, bright, garish lipstick, bangle earrings, noisy jewelry, fluffy, frizzy hair, sleeveless blouses, low cut garments, undergarments worn as outer garments, or leather garments. Interpreters dress conservatively and professionally even if by so doing they appear more professional than the attorneys, physicians, business persons or social workers with whom they perform services. Male interpreters must avoid the eccentric look, the gang member appearance, or the cowboy look. They should avoid tee shirts with designs or messages, denim pants,(blue jeans, dungarees, Levis) and unkempt Albert Schweitzer hair and rumpled suits. Review books published by John T. Malloy (Dress for Success, women's and mens' versions, and Live for Success. These books are available through local public libraries or used book stores.) Though somewhat dated, they address very precisely the image interpreters need to maintain in order to improve their chances of being treated as professionals.

3.3 Develop an authoritative presence

Interpreters already take steps to promote a professional appearance by dressing as professionals and there are a few more "tricks" which can assist the interpreter in gaining the respect they need. Interpreters who are of non-European origin, small in stature, female, speak in high pitched and/or very soft voices and who have pronounced non-native accents need to develop specific tactics to overcome perceptions of clients that they are anything less than consummate professionals. This may involve identifying a model or mentor and/or attending classes to improve vocal quality, and body language if they are serious about improving the acceptance of their interpretation and avoiding further overt demonstration of biases while they attempt to carry out their official duties.

3.4 Develop affirmative responses to questionable treatment

Although interpreters must think very fast on their feet to perform their job, they may well be caught flat footed if they do not anticipate and prepare some model responses to some of the most common situations involving biased behavior they may encounter. List incidents you have observed and/or experienced which seemed to be improper treatment possibly related to bias. Ask your colleagues about incidents they may have experienced and do not rule out those which you cannot determine for sure if they are associated with bias. One will still be fighting ingrained confusions about the role of an interpreter.

Is the interpreter being asked to do things which would never be requested of the court reporter or an expert witness? For example is the interpreter asked to "guard" the defendant while correctional officers are being summoned to the court room? Is the interpreter subjected to racial and/or ethnic comments not intended for the person for whom the interpreter is interpreting or, in the absence of the client? Draw a continuum of possible on-site responses and note most subtle to most obvious.

Among the possible responses to a questionable comment may be to seek clarification: "I'm not sure I understood your comment, request, suggestion or joke... Could you go over it again? Excuse me, I thought I heard you say "...?" Brainstorm a number of possible responses, both immediate and subsequent to the incident as well as preventive measures. Place them in order of degree from the most subtle to the most outrageous: for example from the significant pause before responding, the raising of an eye brow, the oral responses "I beg your pardon?" "Are you expressing a personal opinion?" to throwing paperwork down and stomping out of the room or to slapping the face of the perpetrator. Draw a line between those responses which one might get away with and still be considered a professional and those which might seem boorish, reinforce racial/ethnic stereotypes and/or involve one in legal troubles.

4. ADDRESSING BIASED TREATMENT IN REAL TIME

It is extremely unproductive for the interpreter to be sensing biased incidents around every corner even if the bias exists. This is why effective responses to such incidents must include a range from very subtle, ingenuousness wherein one does not accuse the individual of anything but simply expresses puzzlement about what the person really intended. With less subtle incidents which carry the potential for formal complaints, documentation is crucial. Complaints will not be taken seriously if specifics are not documented. Document or identify all details and specifics before reacting on impulse or generalities. Be succinct. Immediately bring problems to the attention of the person in charge if they interfere with providing professional service. Use the Miss Manners approach: Genteel outrage: "Your Honor, the interpreter is not able to offer: a.) taxicab services for witnesses and friends of the defendant. b.) custody services while the court is in recess. c.) personal hygiene services". Otherwise, document all suspected incidents you observe, even if not directly involved in the case, but do not disrupt the proceedings.

5. SEEKING CORRECTIVE ACTION AFTER THE FACT

Seeking appropriate redress/correction after the fact is a process which also involves a continuum from mild to very serious. Be informed about options for redress. Research and identify options for seeking corrective action in this setting. Follow the steps beginning at the lowest level where the incident occurred by calling the the issue to the attention of supervisory personnel or oversight bodies.

- Document precisely before taking any action
- Call the issue privately to the attention of the perpetrator.

- Specify what should have been done, what you want done now and for the future and if necessary, why.
- Submit a complaint to the individual responsible for or supervising the perpetrator
- Submit a complaint to an outside agency having jurisdiction over the issue
- Network with other professionals and with service providers who may have similar concerns
- Notify ATA and/or other professional association-write letters and/or speak up at meetings and/or meet informally with colleagues
- Submit a press release outlining the incident and what should be done in the future to avoid discriminatory treatment
- Disseminate information about such types of bias and why it is discriminatory and the impact such incidents have on the community, educate but avoid publishing the name of the perpetrator (Leave such things to attorneys if you are filing a suit or an advocacy agency is doing so)
- Write an article on the issue for publication in a place where service providers and the public will see it such as law and medical journals, local newspapers, agency newsletters for employees.
- Identify any state Human Rights task force, Licensing authority, or commission on bias in the courts or on equal access. Is there a state minority commission? Is there a professional

interpreters association?

- Inform yourself about the code of ethics or code of conduct required for professionals in your state with whom you interact professionally. Attorneys, judges, medical practitioners and others generally must adhere to standards defined by law or their professional association or the agency for which they work. There may be a state or federal entity which provides funding for services offered by an agency, if biased or discriminatory treatment is documented, the entity may be pressured to correct the problem or forfeit funding.

If the problem is seen to be a community-wide problem ie. the community and not one individual is experiencing this problem, the Community Relations Service of the U.S. Department of Justice may be able to offer conciliation services to help resolve the problem short of litigation.

5.1 What ethical questions may be involved?

Interpreters, like attorneys, sometimes must withdraw from assignments for ethical reasons such as when the party hiring them participates in unethical behavior or asks the interpreter to do so. If we consider that biased and prejudicial behavior constitutes a violation of our professional ethics, then under what circumstances and how should an interpreter withdraw from an assignment? Interpreters do not wish to lose assignments any more than they want to compromise their professional and personal ethical standards. Those taking the longer view will opt for withdrawing from assignments where they may be perceived as abetting prejudicial behavior. This may occur, for example, if the interpreter is asked to conceal apparently racially charged comments or if, in a private interview the interviewer makes statements to the interviewee which are demeaning, threatening, intimidating or otherwise unethical. In each case it will be a judgement call and the interpreter may wish to call it to the attention of the interviewer to offer them the opportunity to correct the error before withdrawing.

5.2 Educating

Public education may play a role both in preventing and in correcting wrongs which are occurring. The dilemma surfaces when one attempts to educate the public about the importance of the issue without appearing to be vindictive. How can one call to the court's attention an interpreter should have been used without appearing to seek to unethically drum up business? Many actions, if taken by an individual interpreter or agency on behalf of the profession may be perceived to involve self-interest. If the interpreter intends to attempt to change agency or court practices which abet biased behavior, it is more effective to work within the structure of a professional organization, not limited to interpreter/translator organizations, by calling attention to general policies or practices having a detrimental impact on our concept of fairness mentioning only tangentially vignettes which illustrate the point. Otherwise, the resolution of an individual complaint may have a very limited impact on overall biased treatment within the system and may label the individual interpreter as a troublemaker.

6. SUMMARY

Incidents of bias occur in the interpretation setting and may not immediately be identified as such. Nevertheless, when they interfere with an interpreter's ability to provide a non-English-speaking individual equal access to public services, the judicial system or business transactions, the problem should be identified, addressed and a means found to correct it. There are several steps which can be taken, including preventative activities, on site corrective actions and pursuing redress after the fact. The interpreter's greatest defense against biased treatment is his/her professionalism and knowledge of the system. If bias issues are not addressed effectively, they erode the very foundation upon which the structure of our government was built - a government of the people, by the people, and for the people must be trusted by the people it serves. It should not be feared or mistrusted. Interpreters must not only learn to defend themselves from biased and prejudicial treatment, but they have an ethical and moral responsibility to protect those they interpret for in official proceedings. They also have a responsibility to promote the welfare of the community and to educate the public about dangerous stereotypical beliefs about persons of foreign origin which interfere with the American Way. Interpreters must avoid falling into an inappropriate "norm" of going along with the mistreatment of a minority group but must find a way in their field to correct that very dangerous norm. The Community Relations Service can offer technical assistance to communities and organizations seeking to develop solutions to bias of this type and in other areas related to race, color and national origin.

PROCESSING STRATEGIES IN SIMULTANEOUS INTERPRETING: EXPERTS VS. NOVICES

Marianna Sunnari
University of Turku

Keywords: Simultaneous Interpreting, Processing Strategies, Experts, Novices

Abstract: The paper discusses and compares processing strategies of experienced and novice interpreters. The material analysed consists of a conference presentation and its simultaneous interpretations from English into Finnish taperecorded on two tracks for simultaneous listening. The original and the expert performances were recorded on line during the actual conference. The novice performances were recorded later in a language laboratory. The analysis of the material showed that experienced interpreters apply a condensation strategy which does not result in actual loss of information, i.e. macroprocessing. It seems that macroprocessing is necessary when the source text information is so dense that there is not enough time to convey everything into the target message. When working with structurally different languages such as Finnish and Indo-European languages, macroprocessing is needed even in an "ideal" situation, where the speaker is speaking without a script and the interpreters are familiar with the speaker and the topic. Novices, who resorted to a listing strategy, failed to produce a coherent output message.

1. INTRODUCTION

When we discuss perfomance criteria in interpretation, it is usually taken for granted that the original speaker's and the interpreter's message, i.e. the input and the output, correspond to each other. It is maintained that the interpreter's most important task is to convey the SL information contents as accurately as possible (e.g. Ref. 1). What this correspondence means in practice is something that is not so explicitly defined. There is, nevertheless, as Gile (Ref. 2, p. 41-42) points out, "a large consensus on the general directions, if not on the precise quantitative assessment of the I[nterpretation]/T[translation] quality parameters, such as fidelity, linguistic correctness, stylistic and terminological acceptability".

When we discuss interpreter training, it is likewise taken for granted that the aim is to turn novices into experts who master their task in the multitude of situations that an interpreter may face. There is a more or less general agreement on the elements comprising aptitude for interpreting, e.g. knowledge of languages, good general knowledge, ability to analyse rapidly what has been heard, ability to listen and speak at the same time, and certain personal characteristics (Refs. 3-5), but what the trainees learn and how the skills should be taught and developed remains rather unclear or based on intuitive knowledge or experience of what is needed in everyday practice.

The purpose of this paper is to demonstrate what interpreters do in a real work situation. It focuses on the process leading from input to output, and the *product of the process*. This approach is linked with the idea of capacity management introduced by Gile (Ref. 6). An obvious hypothesis is that more experienced interpreters have learnt to manage their capacity more efficiently and an therefore convey more of the source language message than the novices. In addition, the procedures used by the two groups have also been analysed for possible *qualitative differences*, which might provide us with new insight into the techniques by which simultaneous interpreters handle and process the two languages involved.

2. MATERIAL AND METHODS

One problem of emprical research which particularly hampers research on interpretation is access to data. Much of the research carried out so far has been criticised for methodological weaknesses, i.e. wrong kinds of subjects, materials, experimental conditions and tasks. As a result, the conclusions drawn and the generalizations made have been criticized as being too hasty (Refs. 2, 7). An attempt was therefore made to simulate field conditions as closely as possible. The material of this study consists of an authentic conference presentation (Raising Multiligual Children, a keynote speech of a Child Multilingualism Seminar) interpreted from English into Finnish by two professional interpreters. Both the original and the interpretation were recorded on-line with the permission of the organizers, the speaker and the interpreters. The original and the interpretation were later transcribed for a closer analysis. The recordings of the original and the interpretation could also be listened to simultaneously from separate tracks of the tape.

For the purpose of comparing expert and novice performance, eleven interpreter trainees of the Centre for Translation and Interpreting of the University of Turku were asked to interpret a fifteen-minute extract from the same presentation in the language laboratory. To simulate the environmental conditions, the students were given the same background material as the experts had received before the real situation. They were also given information about the program, purpose and audience of the seminar. In the final briefing before the practice session they were instructed to imagine themselves in a real situation where the audience was looking forward to receiving new information about the topic and being able to discuss it with the speakers and other participants, and where the interpreter had a key role in providing them with this information.

The novices formed two groups. The first group included four MA students of Translation studies with English-Finnish as the main language pair, with conference interpreting as their minor subject. At the time of the experiment, they had completed the courses on offer and were supposed to have gained a basic knowledge of what interpreting is about. The second group consisted of seven trainees participating in a post-MA course of conference interpretation, who had mastered consecutive interpreting and had also been taught the basic routines of SI. All novices were considered to have received the same amount of training before the experiment. The novice performances were also transcribed for a closer analysis; the SL text and the novice interpretations were also available on tape for simultaneous listening.

3. RESULTS

3.1 General observations on the interpreting situation

One of the complaints made against the validity of interpretation experiments is that the input materials are very different from the source texts interpreted in real situations. For example, subjects have either been made to interpret sentences read out of context or texts constructed especially for the experiment. It has been argued that both processes and results may deviate considerably from those that occur in genuine conference situations.

On the other hand, source texts interpreted in conferences also vary, both in terms of their syntactic and terminological complexity and in terms of access to background information. The two extremes could perhaps be described as an *ideal* and a *counter-ideal situation.* In the ideal situation, the speaker is accustomed to speaking to the audience via interpreters freely, without a written script and where the interpreters, who are working into the A-language, are also familiar with the speaker's topic and style of presentation. In the counter-ideal situation, the speaker is reading at a fast speed from a written script which the interpreters have not seen beforehand; the text may be highly technical or culture-bound with complicated syntactic structure, specialized terminology, dates, names, and figures.

In advance, the situation in which the expert performances were recorded was considered ideal. The speaker had written extensively about her topic and was accustomed to speaking in public; the interpreters had been able to prepare for the assignment in advance by reading a book written by the speaker on her topic. However, the real situation turned out to be different. The speaker was adressing her audience freely, without a script, but, unfortunately, without a proper processing plan of her own. This gave rise to a rather incoherent presentation given at a relatively fast speed with frequent false starts, hesitations, and self-corrections, which was much more difficult to interpret than was expected Consider, for example the following extract from what could perhaps be called a *pseudo-ideal situation:*

(1) There're often many trilingual situations in which the mother and the father speak...uhm.. two different languages and the child learns a third language... the... in other words the country language in the pre-school if..if neither of the parents speak the language of the country... There's also examples.. even.. a long time ago there was a pre-school in Spain, and this was an article from 1920, where children learnt three or four languages just by interacting with native speakers. In other words, these were Spanish-speaking children whose parents were interested in foreign language learning and the children...uhm.. (?) at a pre-school where there were English-speaking teachers and French-speaking teachers, German-speaking teachers and the children through play and through natural interaction acquired these languages, so that.. uhm..it's..it's very common and possible to grow, up..uhm bilingually...Uhm.. bilingualism is also...uhm.. important for several reasons at national level...uhm.. there's a... and American congressman called Paul Simon...who..you might...may have heard of him... he ran for the presidency in the...during last election..I mean...he didn't get that far..but he's, he's written a book called "The tongue-tied American" where he..hmm.. is very concerned about the lack of..of foreign.. interest in foreign langauges in America and he cites three reasons on a societal level for why it's very important to take advantage of the bilingual resources that we have...

It became evident that the interpreter could not possibly "say it all"; it was obviously necessary to filter the gist of the speaker's message into a better-formulated target message. How, then, did the interpreters cope with this situation and were there differences between expert and novice performance?

3.2 Experts

It was assumed that the experts would be able to cope with the assignment, i.e. that they would "have built up a certain baseline below which they will not drop" (Ref. 3, p. 61). It was assumed that in order to convey the *ST information content* proper, they would have to aim at concision and synthesis similar to that present in consecutive interpreting, entailing "the deletion of all superfluous words, the choice of shorter syntactic solutions or shorter words whenever they are available, the simplification of SLT structure." (Ref. 8, p. 394-5). In other words, in accordance with the text-linguistic approach to speech processing (e.g. Ref. 9), an expert interpreter would be constructing the *macrostructure* of what she hears by applying certain mental operations (macrorules) to the SL message (microstructure) during comprehension. The macrorules available are *selection*, where important or relevant micropropositions are included in the macrostructure, *deletion*, where unimportant, irrelevant or redundant micropropositions are omitted from the macrostructure, *generalization*, where a set of propositions are merged into one generalizing proposition, and *construction*, where the macroproposition formed entails several micropropositions. The macrostructure resulting from such processing contains fewer individual propositions than the microstructure.

Mackintosh (Ref. 10) has established that macroprocessing is applied in consecutive interpretation; my analysis of expert performance showed that macroprocessing was also applied in the simultaneous mode. The most common strategy was deletion of superfluous or redundant material, as illustrated by the following extract of the original where the segments deleted in the expert interpreter's output are given in square brackets:

(2) SL

And I think this shows, [you know, a lot of] attitudes, about [the...] how provincial people can be and how, [what I call] linguacentric, many Americans are, [for example, and probably this is so in other countries.] And it also has to do with that people don't [really] understand bilingualism, [people think...] in America there's [a lot of] talk about...if people learn Spanish, there will be.. [you know], a separate political movement, people won't want to learn English [and this isn't really true, very often the reason for why] the Spanish language is...[is so] important to people [is] because they don't feel that they..uhm..can] move into the ...uhm...] they're not able to enter the [sort of] English... sphere of the country.

The deleted segments contained false starts, empty fillers and repetitions. While this kind of synthesis is likely to increase the processing load during comprehension because the interpreter must be able to hold larger chunks of information in her short-term memory, the effort needed in the target message production and articulation is reduced, which means that there will be more capacity for e.g. monitoring. That deletions of this kind are *intentional* and result from a *conscious* process is shown by the fact that where the speaker repeated or reformulated something that had been said before, the omitted segment was not always the one that came later in the linear order. Generalizations and constructions are more obvious examples of a conscious synthesis.

There were instances where the interpreter's capacity was exceeded -- probably because of the time and effort needed to go beyond the recognition of words in the speaker's extremely superfluous delivery -- which resulted in a distinct 'failure sequence' (see Ref. 6, 174-6), characterized not only by a loss of words but also by hesitations, pauses and self-corrections. In example (2) below, the interpreter misses completely the segment in the middle (in italics), probably because it is something quite unexpected. Unlike most of the novices (see 3.3), she manages to save what can be saved and complete the message. The expert interpreter's Finnish output is provided below as a backtranslation into English.

(3) SL

I think that for a start in a seminar like this we couldn't have a more qualified person than L. to speak for us or to us. There's one further point; I know that L. doesn't want me to say this, *but one of the nice things with this sort of subject, especially when you're dealing with academics is that the individual on bilingualism has a personal interest or a personal involvement in this. This personal involvement is demonstrated, you can't see it but I know, Leonore hurt her leg...so there's a question of how does she limp to Finland...* she's still brave enough to say that she's going to stand, because she has better contact with the audience. But if you see her sort of sag a little bit...it's because she has hurt her leg. So, I'll come rushing to grab you if you fall. So, without any more comments, I give the floor to L. Thank you very much for coming.

E1 (Backtranslation from Finnish)

I believe that in a seminar like this we couldn't... we couldn't have found a better person to speak to us than L.A. *One.. that..one thing that is very nice in this type of... type... in this type of seminar is that... that... that.. here is... the question is... individ... that the question is that...vvvvv.... multilingual... I'd like, I'd say now, even if* L. might not want me to say this, that she had a small accident before she came to Finland, so that she limps slightly, yet she'd like to stand when she talks, but if something happens, we can then help her to stand up or sit down if she cannot remain standing. All right, I'll now give the floor to L. The floor is yours.

It is interesting to note that at the point where the interpreter finally succeeds in returning to the right track, she starts the new sequence (after the segment in italics) with a proposition which the speaker uttered before the sequence where she lost her grip (*L. doesn't want me to say this).* This shows that the expert interpreter is processing and holding in her short-term memory rather large entities of information and monitoring her performance while listening, comprehending and speaking, in order to be able to understand and convey the speaker's logic and meaning.

3.3 Novices

The most obvious differences between the two experts and the eleven novices were that the novices failed to cope with the speaker's rate of delivery and had more comprehension and production problems than the experts. Not only were there more misunderstandings and uncompleted sentences but the novices also seemed to have fewer satisfactory coping techniques. It became evident that the novices had internalized the importance of trying to convey as much as possible of the source information and of formulating complete sentences. They, too, were able to delete a considerable number of the speaker's self-corrections and redundancies but there were also more omissions of relevant information and more overgeneralizations than in the expert performance. This suggests that the novices' total capacity was exhausted more quickly than that of the professionals when they were interpreting long, interconnected segments of speech. This finding is in line with that of Kalina (Ref. 11, p. 230).

There were also differences between individual novice performances. Those with the greatest production problems fell into two groups: One group obviously understood the input quite well but, for one reason or another, failed to reformulate or articulate their own output in a satisfactory manner. This type of failure sequence could be detected easily, because it included numerous hesitations and self-corrections or incomplete sentences. The other group sounded rather fluent and convincing, but a closer analysis of their performance revealed that they were picking up the segments and linking them together into a new message without an overall understanding of what the speaker was aiming at. This kind of approach could be described as a *listing strategy*, where the propositions understood (or even misunderstood) are rendered in a target language in the linear order they appear after what has not been understood is deleted, as in example (4) below. The original can be found in example (2) above.

(4) N4 (Backtranslation from Finnish)
... this shows something of the attitudes, how people can be and how, for example, Americans are. And this... and here, too, the point is that people don't understand bilingualism. When people study Spanish, they are for example a different.. another political group, but this is not the case here.. people don't necessarily... I think.. when...

In another version of the listing strategy (example 5) propositions were connected with *and*. This trainee also had serious comprehension problems, partly because of his limited vocabulary.

(5) N5 (backtranslation from Finnish)
And this only tells about attitudes, how rejecting people can be and how linguacentric and this is certainly a problem also in other countries and people who understand bilingualism and there has been a lot of talk about they... about Spanish speakers...uhm... also have started learning English and this is also a reason for because Spanish is so important because they don't feel that they would have a possibility... uhm... to adapt to the country in the Spanish-language English-speaking country...

The attempt to produce an unbroken sequence of propositions seems to take up all of the novice's processing capacity leaving no room for self-monitoring, and the rendering hardly makes any sense. This trainee had shown a similar tendency during the practice in consecutive. When instructed to concentrate on its elimination, his performance had improved. Now, under the heavy processing load, he was again resorting to this unsatisfactory strategy.

Another feature common to novice performance was the difficulty to recover from failure sequences. This finding is also consistent with Kalina's conclusion (Ref. 11 p. 229) that problems not successfully solved have more far-reaching effects on other text segments in the student than in professional performances. The novices frequently either left their sentences incomplete or added connecting material that could not be found in the original, as in extract (6) below. The source text can be found in example (3) in 3.2.

(6) N6 (Backtranslation from Finnish)
I think that a seminar like this... we'd hardly have a better person than Mr. N. (sic!) who will talk to us. L., I wouldn't like to say this but she is when she talks about academic subjects, she is.... she has a personal interest in child bilingualism and that I can see this.... but.......... because she wants to stand because it has a better contact with the audience. She has also heard... she has problems with her leg.... Without any further comments I'd like to thank. Thank you for coming here.

All novices had great problems throughout this particular segment, and none of them managed to convey the speaker's train of thoughts properly.

4. CONCLUDING REMARKS

The comparison between expert and novice performance in this material showed that the experts were able to follow the speaker with fewer failure sequences and to produce more of the original propositions than the novices. Examples (4) and (5) show that it is, however, not only the number of propositions produced that matters when the quality of output is assessed. Linkages between the propositions are equally important, and the reproduction of the correct type of link presupposes understanding of the general idea, i.e. the macrostructure of a given sequence. This was also more difficult for the novices than for the experts.

When the experts had problems, they managed to cope with them better than the novices. Even in instances where the solution was to more or less "throw out the information bits in the hope that listeners will perform the further task of properly reconstructing the bits of information conveyed" as Anderson (Ref. 12, p. 116) describes this situation, the experts rarely added information that could not be found in the source text. On the other hand, where the speaker's delivery was not ideal, the experts managed to select into their output only the essential source language information, thus facilitating the audience's comprehension task.

Even though, so far, we have no information about possible changes in the processing capacity and strategies of interpreting students during a period of training, this comparison of experts and novices might suggest that practice and experience result in more independent production. Novices tend to follow the surface structure of the original, while the structures and formal features of the target text produced by experts tend to reflect deeper semantic processing. In other words, novices spend most of their processing capacity on the effort of comprehending and reproducing a list of as many micropropositions as possible, while the experts have learnt to apply macroprocessing strategies.

5. REFERENCES

1 Buehler, Hildegund. "Linguistic (semantic and extralinguistic criteria for the evaluation of conference interpretation and interpreters". *Multilingua* 5:4, 231-235, 1986.

2 Gile, Daniel. "Methodological aspects of interpretation and translation reserach," in Sylvie Lambert and Barbara Moser-Mercer, eds. *Bridging the gap. Empirical research in simultaneous intepretation.* Amsterdam/Philadelphia: John Benjamins Publishing Company. 1994, p. 39-56.

3 Moser-Mercer, Barbara. "Aptitude testing for conference interpreting: Why, when and how," In Sylvie Lambert & Barbara Moser-Mercer, eds. *Bridging the gap. Empirical research in simultaneous intepretation.* Amsterdam/Philadelphia: John Benjamins Publishing Company. 1994, p. 57-68.

4 Longley, Patricia. "The use of aptitude testing in the selection of students for conference interpretation training," in Laura Gran & John Dodds, eds. *The theoretical and practical aspects of teaching conference interpretation.* Udine: Campanotto. 1989, p. 105-108.

5 Bowen, David & Bowen, Margareta. "Aptitude for Interpreting," in Laura Gran & John Dodds, eds. *The theoretical and practical aspects of teaching conference interpretation.* Udine: Campanotto. 1989, p. 109-125.

6 Gile, Daniel. *Basic Concepts and Models in Interpreter and Translator Training.* Amsterdam/Philadeplhia: John Benjamins Publishing Company. 1995.

7 Pöchhacker, Franz. "On the science of interpretation." *The Interpreters' Newsletter*, No. 5, 52-59, 1993.

8 Viezzi, Maurizio. "Considerations on interpretation quality assessment," in *Translation - the vital link.* Proceedings of XIII FIT World Congress. Vol. 1. London: Institute of Translation and Interpreting. 1993, p. 381-397.

9 Singer, Murray. *Psychology of language: An introduction to sentence and discourse processes.* New Jersey: Lawrence Erlbaum Associates. 1990.

10 Mackintosh, Jennifer. "The Kintsch and van Dijk model of dicourse comprehension and production applied to the interpretation process." *Meta* 30:1, p. 37-43, 1985.

11 Kalina, Sylvia. "Analyzing interpreters' performance: methods and problems," in Cay Dollerup & Annette Lindegaard, eds. *Teaching translation and interpreting 2: Insights, aims, visions.* Amsterdam/Philadelphia: John Benjamins Publishing Company. 1994, p. 225-232.

12 Anderson, Linda. "Simultaneous interpretation: Contextual and translation aspects, " in in Sylvie Lambert and Barbara Moser-Mercer, eds. *Bridging the gap. Empirical research in simultaneous intepretation.* Amsterdam/Philadelphia: John Benjamins Publishing Company. 1994, p. 101-120.

SIMULTANEOUS INTERPRETING AS EXPERIENCED BY THE AUDIENCE

Anna-Riitta Vuorikoski
University of Tampere

Keywords: Simultaneous Interpreting, Use of SI, Questionnaire, Evaluation, Expectations.

Abstract: The users of simultaneous interpreting services and the expectations which they have about SI are one aspect of current interest in interpreting. The paper aims at providing new information for the on-going discussion by reporting on how Finnish users respond to simultaneous interpreting. The discussion is based on a survey carried out in five seminars in Finland in autumn 1992.

1. RESEARCH QUESTIONS

The users of simultaneous interpreting services and the expectations which they have about SI are one aspect of interpreting that has been examined recently by scholars in the field (Bühler 1986, Kurz 1989, 1994, Gile 1990, Moser 1995). These studies provided the stimulus and framework for the study I carried out in Finland in the autumn of 1992. The primary task of the study was to discover to what extent mediated communication, such as is carried out in the simultaneous mode, is satisfactory to listeners. It was hypothesised that communication in the SI mode is satisfactory if the addressees use this service, and if they give positive feedback on the content and quality of the oral translation that they have received. Secondly, attention was directed to the equally important issue, whether certain addressee characteristics determine or influence their attitudes toward SI and their use and evaluation of it. These questions were linked with earlier empirical studies (Kurz 1989, 1994) which investigated the expectations of users of SI in different types of meetings. These studies emphasised the importance of the communicative situation, and its relationship to the expectations of SI listeners. The communicative situation was not discussed in much detail in the previous studies, however, and therefore an important aim was to analyse one type of meeting more thoroughly, i.e. the seminar and its participants .

The structure of five Finnish seminars was first described. In a typical national seminar there is an audience of not more than 200 participants, who come from different parts of Finland to acquire new information and inspiration for their work. Such new information is often provided by guest lecturers from abroad. The communicative roles of the speakers and listeners tend to be formal; the speakers deliver their presentations, and members of the audience listen to the speakers in silence. After the presentations, there may be a few short questions to the speakers, but because of the time constraints and the purpose of the seminar, the time allotted for questions tends to be no more than ten to fifteen minutes.

THE QUESTIONNAIRE

1. I have previously taken part in meetings with foreign experts, where the lecture(s) have been interpreted into Finnish.
/_/ 1 - 5 times /_/ 6 - 10 times /_/ more than 10 times
2. I think that interpreting helps me to concentrate on the actual message, because one easily gets tired of listening to foreign languages while trying to get the essential information of the presentation.
/_/ I agree /_/ I disagree /_/ I do not know
3. It does not make any difference whether I listen to the original speaker or the interpreter, the main thing is that I understand what is being said here.
/_/ I agree /_/ I disagree /_/ I do not know
4. I think it's a good idea to provide interpreting for those who want it.
/_/ I agree /_/ I disagree /_/ I do not know
5. I listened to the English presentations using the interpreting.
/_/ Yes /_/ No /_/ Part of the time

If you were listening to the interpreting, please move on to point 6.

I did not listen to the interpretation, because
/_/ I could understand the foreign language presentations
/_/ I could understand the main points of the foreign language presentations, and that was sufficient for me
/_/ Some other reason ____________
If you did not listen to the interpretation, please move to point 10.

6. I listened to the interpreting for the following reason(s) (please tick the most appropriate points)
/_/ I do not understand spoken Swedish well enough
/_/ I do not understand spoken English well enough
/_/ I am not very familiar with the English/Swedish vocabulary of this special field
/_/ some other reason ____________________
7. The use of the conference equipment (headsets, receivers) did not distract me
/_/ agree /_/ disagree /_/ do not know
8. The interpreting in this seminar was
1. <u>informed</u>
/_/ agree /_/ disagree /_/ do not know
2. coherent, or <u>easy to follow</u>
/_/ agree /_/ disagree /_/ do not know
3. <u>accurate</u>
/_/ agree /_/ disagree /_/ do not know
4. had a pleasant <u>speech rhythm</u>
/_/ agree /_/ disagree /_/ do not know
5. <u>fluent</u>
/_/ agree /_/ disagree /_/ do not know
6. <u>terms</u> were used correctly
/_/ agree /_/ disagree /_/ do not know

9. For the listener, the order of importance of these features (1 - 6) is
(starting from the most important feature and finishing with the least important one)
I __ II __ III __ IV __ V __ VI __

10. The presentations were interesting
/_/ agree /_/ disagree /_/ do not know
11. I think I was able to understand what the speakers wanted to convey us and what their message was
/_/ agree /_/ disagree /_/ do not know

Please turn over ->

Page 2

I am /_/ male
/_/ female

Date of birth 19__

Education:
____

Occupation:
____

How would you evaluate your knowledge of English:
- how many years of English did you have at school (short, long course) ____________
- additional skills in English acquired out of school:
____
- is it possible to follow presentations in a foreign language with the skills acquired at school?
Yes / / No / /
- if skills acquired at school are not enough, what else is needed?
____

I give permission for a short interview by phone, preferably
between ______________ hours

____
Name
____
Phone number(s)

Thank you for your contribution!

Please drop the questionnaire in the box next to the door. Thank you.

One of the reasons for conducting this study was to obtain answers to the questions that had occupied my mind for some time: why is it that conference participants do not always listen to the interpreting, particularly if it is from English into Finnish? The underlying hypothesis of the study was that Finns may feel uncomfortable about their skills in foreign languages; seminar participants may feel that they should be able to cope without SI because of their language studies and any other contact they have had with English, and this is reflected in their non-use of SI. Moreover, the members of the audience are experts in the field that is discussed in the seminar, and consequently they have sufficient common knowledge to be able to follow the original speaker. Another reason for not making use of SI that has been provided could be the inadequate quality of the interpreting. The non-use of SI when the conference language is English seemed to be a relevant issue for study from a professional point of view. One evident consequence of an increasing non-use of SI would be that conference organisers may eventually decide not to provide SI at all, assuming that everyone knows English. This would have repercussions for the profession and also for the training of interpreters.

In order to learn more about the reasons for either using or not using the SI available to the audience it was judged important to solicit information about the general attitudes of seminar participants toward the use of SI. In addition to opinions and attitudes, information was also gathered about participants' use of SI. The seminar participants' answers about their use of SI were then be compared with their evaluation of the interpreting they had heard.

With reference to translation theories (Reiss and Vermeer 1984) and the general quality concepts that are applied in the industrial world, the question that needed to be answered was whether SI fulfils its purpose from the user's point of view. In order to answer the questions posed was deemed necessary to consult the users themselves. This was done with the help of a questionnaire (see adjacent page) which was supplemented with telephone interviews.

The questionnaire was distributed in five seminars that dealt with different topics: foster care, marketing, economic crime and agricultural policy. Out of the 480 questionnaires that were distributed, 173 were completed and returned.

2. SI AS EXPERIENCED BY THE USERS

(1) It turned out that for the large majority this had been the first time that they had listened to SI. This fact may explain why some respondents demonstrated ignorance about SI, which was revealed, for example, in the frequent comment that the interpreters had lagged behind the speakers.

(2) Conversations with conference delegates had evinced the idea that it may be tiring to listen to presentations in foreign languages all day long. Delegates referred to the fact that interpreting often helps them to focus on the message, because they do not have to concentrate on trying to understand the language and the message simultaneously. Such comments were used as the basis for the formulation of Statement 2. While the majority of respondents agreed with it (70%), the reasons for disagreement with this statement proved to be quite interesting. A number of the respondents added comments after the statement, such as [Interpreting helps] "if the interpreting is fluent and the language [of the speaker] is not very familiar", or "it depends on the language [of the speaker]", or "if the language is not quite familiar, or if there's a strange accent". The reference to 'less familiar language' may mean languages other than English. For example, one interviewee said that if the language had been German or some other language he did not know so well, he might not have been distracted by the SI in the same way he was in this case, when he could keep comparing the original with the SI.

The comments by respondents written on the questionnaires together with those given during the interviews show that when SI had been made available in a meeting, participants may choose to use it or refrain from using it for a multitude of reasons. When the conference language is English, the reasons for disagreement with Statement 2 seemed to relate to how important the topic of the seminar was for them, to the performance of the speaker (native/non-native speaker of English, technical/non-technical English, rate of speech, etc.), or to the listeners' familiarity with listening to lectures in English. Disagreement with Statement 2 did not therefore only reflect dissatisfaction with the quality of the interpreting.

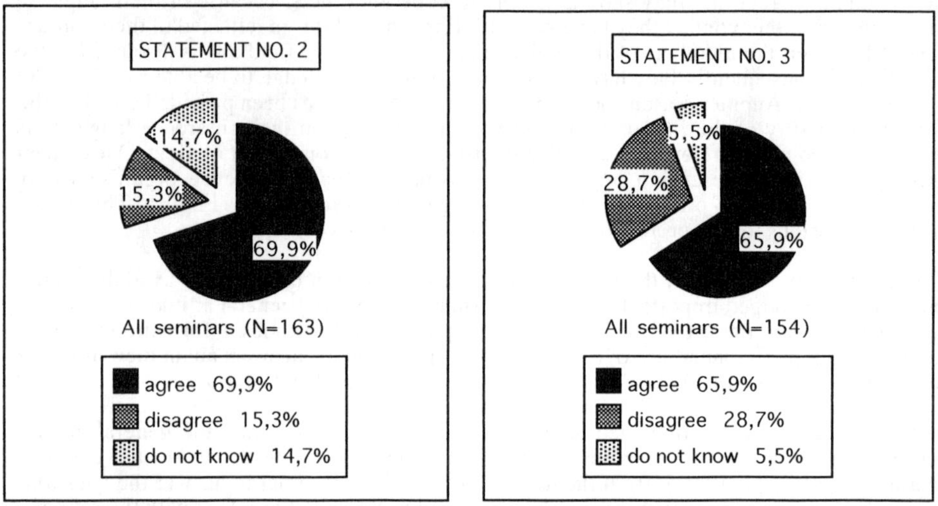

Figure 1. Proportional share of answers to Statement 2 "Interpreting helps me to focus on the message," and Statement 3 "It does not make any difference whether I listen to the speaker or the interpreter, the main thing for me is to understand the message."

(3) Statements 2 and 3 aimed at obtaining information about the attitudes of the Finns toward listening to presentations in English, either directly or with the help of SI. The more specific aim of Statement 3, "It does not make any difference whether I listen to the original speaker or the interpreter, the main thing is that I understand what is being said here," was to see if the participants felt that they could trust the interpreters and their ability to translate and convey the messages of the speakers. It was assumed that if the respondents agreed with this statement, it could be concluded that they had not experienced the situation as a test of their language skills. Moreover, agreement with this statement might indicate that from their point of view communication had been efficient and they had not paid much attention to the fact that instead of listening to the speakers directly they had received the message via the interpreters.

While the majority agreed with Statement 3, a high proportion of the respondents, nearly one-third, disagreed with it. This corresponds to the results of the telephone interviews, which confirmed that if the language of the speaker was English, and the user of SI had studied it for at least 3 years, there was a tendency to listen to both the original speaker and the interpreter at the same time. Seminar participants wanted to listen to the original speaker, and used SI as a support (cf. 3. Reasons for listening to SI). Again, in many cases the reasons for disagreement had much more to do with the lack of knowledge of SI than the interpreters' performance. This point came up in the telephone interviews, which also showed that the formulation of this statement was not very clear. The main reasons are listed below:

- headsets are uncomfortable
- listening to both the original and the SI is distracting
- speaker's voice is audible, which can be distracting
- interpreters are not able to convey all the nuances
- interpreters are not informed about the topic the same way as the speaker and the audience are
- direct participation in the event is not possible unless one knows the language of the speakers
- hearing a female voice when the speaker is male is distracting.

The fact that the majority of the respondents from the five seminars agreed with the statement implies that the main concern of a large portion of a typical seminar audience is to understand the message, regardless of whether it is heard directly from the speaker in the original language or rendered by interpreters into Finnish.

(4) Statement 4, "I think it's a good idea to provide interpretation for those who want it", aimed at getting a clear answer as to whether audiences find interpreting a useful service. Out of the 159 answers to this statement, 156 agreed with it; only one male participant disagreed with it, and another had no opinion. Neither of these persons had listened to the interpreting, because they were able to understand the presentations in the original language. This result could be interpreted as a consensus on the adequacy of SI in the five seminars concerned. However, the statement did not refer to the cost-effectiveness of the service; had there been any indication of the rather high costs involved and of how they are reflected in participation fees, there might have been a wider distribution in the answers to this statement.

(5) Statements 5, 6 and 7 were concerned with the addressees' use of SI. The first part of Statement 5 was designed to find out whether the participants had listened to SI or not, or just part of the time. The second part sought to obtain information about the reasons for not listening to SI, and Statement 6 focused on the reasons for listening to SI. These questions represent the main motivation and stimulus for conducting the study, since the results can be used to explain the diverse behaviour in relation to the use of SI observed in conferences. Moreover, the results could be used to assess the need for SI in meetings of the kind discussed here. The total number of the answers to this statement was 170.

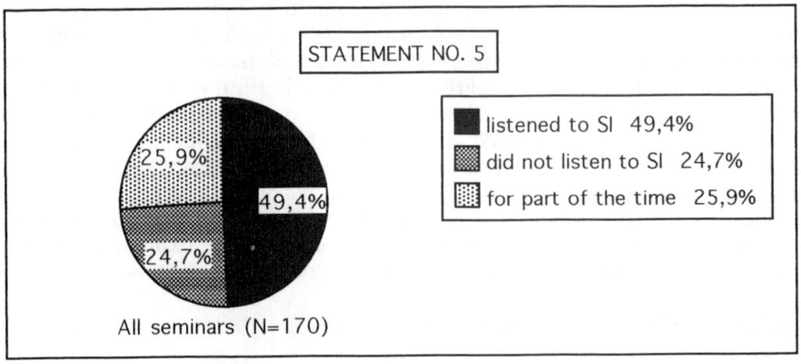

Figure 2. Proportion of respondents who listener to SI continuously, or for part of the time, or who did not listen to SI.

The response shows that one half of the respondents in the five seminars had listened to the SI continuously, and one fourth part of the time. There was some variation between the five seminars, but observations in the seminars confirmed that the majority of the participants used the SI at their disposal.

Respondents' level of education and their use of SI

Comparing the level of education with the use of SI produced results of particular interest. Out of the total number of 170 conference participants who answered the first part of this statement, 157 indicated their level of education by giving their title or profession; about one half of these respondents were university graduates.

The result showed clearly that the respondents who were not university graduates were much more dependent on SI from English into Finnish than the respondents with a university degree.

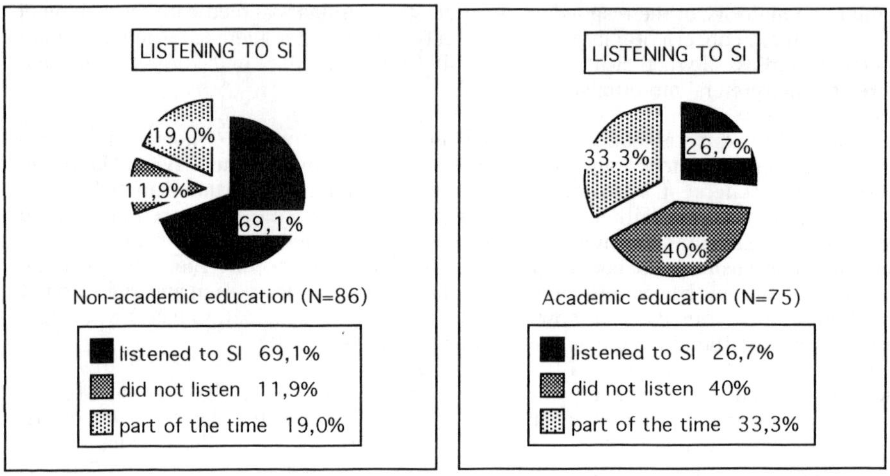

Figure 3. Respondents' use of SI and their level of education.

This result was confirmed by the other parameters surveyed (the number of years that participants had studied English, and other forms of practising English). These results indicate that if SI is provided in a seminar-type meeting, a large number of participants will use it, regardless of their formal educational background. A closer look at respondents with a university degree revealed that slightly less than one third had listened to SI and one third had listened to SI part of the time, while slightly more than one third of the respondents with a university degree had not listened to SI. This would seem to indicate that a university degree alone does not automatically lead to the competence to listen to extended spoken presentations in English — a fact that may not always be easily accepted by conference organisers when they ponder whether or not SI is needed in a given meeting. The addressors and the addressees in the seminars were experts in the field discussed in each of the seminars. Thus the need to use SI continuously, and more particularly, part of the time, had to be explained through other components of the communication situation than the respondents' common knowledge.

3. REASONS FOR NOT LISTENING TO SI

A number of the participants who did not listen to SI filled in the questionnaire anyway. Even if their number was not very large (41),the answers give some indication of the reasons why some conference participants decide not to use SI.

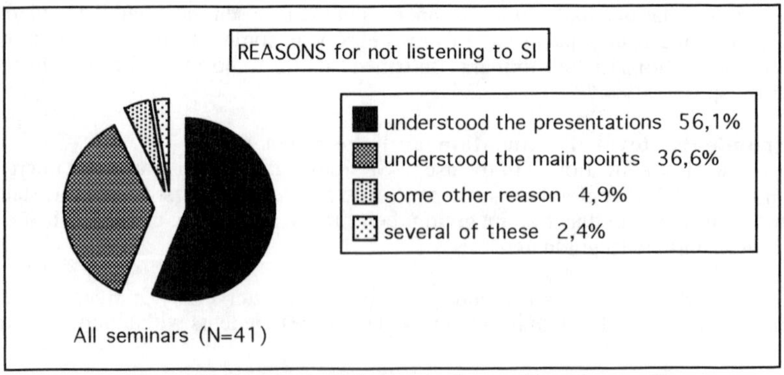

Figure 4. Respondents' reasons for not listening to SI.

Figure 4 indicates that the main reason for not using SI was that the participant was able to understand the addressor's presentation. Secondly, for quite a large proportion of the participants it was sufficient to understand the main points of the English presentation. Thus, even if these addressees were aware that they did not understand every word of the presentation, they still preferred to rely on their ability to receive and comprehend the sense of the message. A number of respondents indicated that they used SI as an aid, listening to it "with their left ear," as one respondent put it, and using SI as an 'on-line dictionary', or 'subtitling', an expression used by another respondent.

In terms of the age distribution of the respondents, it can be seen that younger participants felt that they were able to understand the English presentations, or they felt that they understood the main points, which was sufficient for them. The response corresponds to the development of the teaching of English in Finland and also to the presence of English in the audio-visual media in Finland. Younger Finns have been much more exposed to English, and this may explain to some extent why the respondents in the age group 30 to 39 years were particularly confident about their language comprehension in the seminars concerned. More detailed information about the respondents contacts with English was not included in the questionnaire.

4. REASONS FOR LISTENING

The working hypothesis of the study was that Finns tend to be conscious of their language skills. It was thus intriguing to hear from the seminar participants what their specific reasons for listening to SI were. Statement 6, "I listened to the interpreting for the following reasons: I do not understand spoken English well enough ..." aimed at being tactful about the respondents' knowledge of English (and Swedish in one seminar). The formulation was based on the idea that we all know English, but we may encounter problems with the language of some particular special field. Figure 6 shows that one half of the respondents indicated inability to understand spoken English (Swedish) as their reason for listening to SI.

On the basis of the present survey it would seem that in a seminar-type setting, where people come mainly to learn and to receive new information, the use or non-use of SI will depend on the listeners' language skills, i.e. to what extent they understand extended spoken presentations in a foreign language. Moreover, the use of SI will also depend on how fully members of the audience want to understand the presentations.

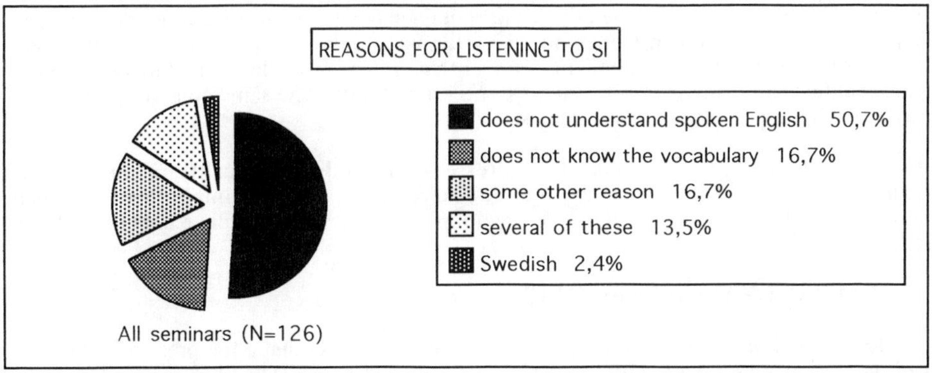

Figure 6. Respondents' reasons for listening to SI: (1) I do not understand spoken English; (2) I do not know the vocabulary; (3) some other reason; (4) two or more of these; (5) I do not understand spoken Swedish.

Seminar participants can be divided into three groups with respect to the second alternative, i.e., emphasis on the full understanding of the message. There is one group of participants who wants to listen to the source language directly in order to hear the nuances as presented by the speaker. Addressees in the second group do not understand the language sufficiently well in order to follow the presentations, or they have not had practice in listening to extended spoken presentations in English, and they listen to SI, or they listen to it part of the time. If the speaker is clear and easy to follow, they may decide to not to use SI. However, if the speaker has a difficult accent, or if the addressee is not familiar with the topic and the vocabulary, then SI is resorted to. The third group of seminar consists of participants who listen to presentations very carefully; they want to make sure that they receive the complete presentation by listening both to the original speaker and the interpreter, using SI as a support.

The interpreter's obvious target group of seminar participants consists of persons who find that their foreign language comprehension skills do not enable them to follow presentations in the language used by the addressor. This group also consists of participants who wish to concentrate on the message rather than try to divide their attention between trying to comprehend the foreign language presentations and processing the message.

The response to Statement 7, "The use of interpretation equipment (headsets, receivers) did not distract me", was uniform, the large majority of the respondents agreeing with the statement that the receivers did not cause them any problems. However, there were some distracting features such as the fact that users were not always accustomed to operating the equipment, or felt that the headphones were uncomfortable (they were "too tight", or they "pressed the ears").

5. ADEQUACY OF COMMUNICATION -- THE USER'S POINT OF VIEW

Statements 10 and 11 were designed to evaluate the overall effectiveness of communication as experienced by the participants.

Statement 10, "The presentations were interesting", was formulated on the assumption that the participants might have found the presentations uninteresting, if communication had not been satisfactory as a result of low quality interpreting.

The last statement, 11, said: "I feel that I understood what the lecturers wanted to tell us." The statement was designed to check whether listeners felt that they had understood the various nuances and subtleties of the message, either on the basis of having listened directly to the speaker or on the basis of what the interpreters had rendered of the message.

The telephone interviews confirmed the results obtained from the questionnaires, according to which SI had only a minor influence on the participants' impressions of how well they had received the message. If the respondents were critical, it was often due to features inherent in the original text or the manner of the addressor's presentation. The same conclusion could be drawn from the response to both statements.

If the language spoken is English, the addressees tend judge the quality of the original presentation rather than that based on the SI. The response seems to imply that in Finnish seminars, even if addressees listen to interpretation, they have an immediate contact with the original addressor.

6. USERS' EVALUATION OF SI

While focusing on how SI was used by Finnish seminar participants, the present study also asked them to evaluate the quality of the SI that they had heard. The choice of criteria was based on those employed in previous studies (Bühler 1986, Kurz 1989), although some were used in a modified form. The participants were also asked to rank the criteria in an order of importance from the user's point of view.
The present study did not attempt to deal with the question of sense consistency with the original message, the criterion used in Bühler, Kurz and Moser, assuming that listeners of SI expect professional interpreters to perform in such a way that sense consistency with the

original message is guaranteed. The criteria provided for evaluation were: [In this seminar, interpreting was] "informed", "accurate", "fluent"; and [interpreters demonstrated] "correct use of terms" and "pleasant speech rhythm". The rating was on an agree-disagree-do not know scale. The purpose of the evaluation was not so much to obtain a precise evaluation of SI, but mainly to see what the listeners' overall impression of the quality of SI was. With the exception of the criteria "fluent" and "accurate", the majority of the respondents indicated agreement with the statements using the given criteria. This was particularly interesting with a view to the fact that the same interpreter(s) worked for three of the seminars, and yet the evaluation of SI varied considerably between them.

After the evaluation the respondents were asked to prioritise the same criteria to see whether there was any correlation between the evaluation and the expectations. The criterion of "informed" was rated highest in order of importance, and the majority of the respondents also agreed with the statement according to which interpreting in the seminar concerned had been informed. A high correlation between the two sets of answers suggests that, like conference participants in other parts of the world, the Finnish seminar participants, too, tend to concentrate on the general content of the message when they listen to interpreters.

7. CONCLUSIONS

The results of the five seminars surveyed for the study showed that not only are there different expectations between different audiences, as Kurz (1989, 1994) suggested, there are also significant individual expectations among the members of one single audience. Other scholars have not as yet referred to the finding that was discovered through the present study. According to the analysis carried out on the basis of the response, the expectations of SI in two audiences varied considerably in two seminars where the topic of the meetings, and thus also the purpose of the meetings and the shared background knowledge of the members of the audiences were all more or less the same. Moreover, besides differing expectations among these two audiences representing two agricultural sectors (grain and horticulture), there were also different evaluations of the same interpreting by the same interpreters in the two seminars. Thus one wonders to what extent it is possible to assume that a certain audience has certain expectations of SI. There will be members in every audience who expect to receive an informed, fluent and accurate interpretation where terms are used correctly. An interpreter should, to my mind, always have the most demanding listener in mind and aim at fulfilling all the criteria of good quality interpreting.

Writers on translation studies, such as Nida, Reiss & Vermeer, and Holz-Mänttäri have underlined the function and purpose of the text to be translated, which requires that the target person or group should be taken into consideration in the decisions concerning the translation assignment. These authors have suggested that the same is true for conference interpreting. It could be assumed that the participants in small (technical) seminars in most cases share a high degree of common background knowledge to which the interpreter may not be party. By and large this may be true, but the present study of five seminars shows that there are also considerable differences in the composition and structure according to age and education among the members of a given audience, and such factors influence their use of SI to a considerable degree. Consequently, the concepts of the function and purpose of simultaneous interpreting, with special emphasis on the needs of the users of SI, have to be discussed within the framework of Interpreting Studies, as has been pointed out by Franz Pöchhacker (1994). Interpreters may need rules different from those used by translators of written texts to guide them in their choice of the right strategies and approaches.

New angles on the function of SI were brought up by the results of the study that has been discussed here. There are nonetheless numerous issues that are beyond the interpreters' control. Some of these are related to the addressee's lack of familiarity with SI, resulting in unrealistic expectations. Some comments referred to the addressors, their manner of speaking, and, in many cases, the fact that they were non-native speakers of English. However, instead of taking steps to educate listeners of SI, as was suggested by Bühler, it might be more appropriate to educate the organisers who are responsible for the arrangements in conferences. Many speakers are not aware of how to get their message across in conferences. Conference organisers should make it clear to the speakers that their messages will be communicated much more effectively if they take the audience into consideration in their planning of how to present their papers or their ideas. This includes providing interpreters with copies or outlines of their presentations together with a list of the key concepts they will be using. Conference interpreters are frequently faced with a situation where a speaker reads a manuscript which the interpreter has not had the opportunity to read beforehand, or has barely had enough time to glance through. The study showed that depending on how well the audience (and the interpreters) have been taken into account there can be a significant difference in how successfully the speaker's message is communicated to the listeners.

8. REFERENCES

Bühler, Hildegund 1986. Linguistic (semantic) and extralinguistic (pragmatic) criteria for the evaluation of conference interpretation and interpreters, *Multilingua* 5, #4, 231-235. Amsterdam: Mouton de Gruyter.

Gile, Daniel 1990. L'évaluation de la qualité de d'interpretation par les délégués: une étude de cas, *The Interpreters' Newsletter* 3, 66-71.

Holz-Mänttäri, Justa 1984. *Translatorisches Handeln. Theorie und Methode* (Annales Academiae Scientiarum Fennicae B 226)Helsinki: Suomalainen Tiedeakatemia.

Kurz, Ingrid 1989. Conference Interpreting -- User expectations, in Deanna Lindberg Hammond (ed.) *Proceedings of the 30th Annual Conference of the American Translators Association 1989*, Medford, NJ: Learned Information, Inc., 143-148.

Kurz, Ingrid 1994. What do different user groups expect from a conference interpreter? *The Jerome Quarterly* 9, #2, 3-6.

Moser, Peter 1995. *Survey on Expectations of users of conference interpretation.* Final report January 1995. Commissioned by AIIC, International Association of Conference Interpreters. Produced by SRZ Stadt + Regionalforschung GmbH, Vienna, Austria.

Nida, Eugene A. 1964. *Toward a science of translating*. Leiden: Brill.

Pöchhacker, Franz 1994. *Simultandolmetschen als komplexes Handeln.* Tübingen: Gunter Narr.

Connections

Literary Translation

NONFICTION TRANSLATION CAN BE LITERARY TRANSLATION TOO

Clifford E. Landers
Jersey City State College

Keywords: Literary translation, translation of nonfiction

Abstract: Translations of scholarly works have a reputation, often justified, for being dry and stodgy. But the techniques of translation of fiction can be profitably applied to nonfiction. This paper discusses the use of such techniques with nonfiction works to enhance their appeal to target language readers without sacrificing the information content of the source language text.

Translation, said Walter Benjamin, is not "the sterile equation of two languages" but is "charged with the special mission of watching over the maturing process of the original language and the birth pangs of its own."[1]

A common, if mistaken, notion posits an unbridgeable gulf separating literary translation from other areas of translation, maintaining that the skills involved in translating fiction–the subset of literary translation with which I am best acquainted–have little or no bearing on translating so-called "nonliterary" nonfiction. One of the purposes of this paper is to argue that the literary/nonliterary distinction can be imprecise and that many of the techniques associated with translation of fiction can be profitably used in other areas of translation. I also hope to show that a mind-set that separates "literary" and "nonliterary" translation as polar opposites is shortsighted at best, can be conducive to poor translation habits, and can compromise the overall quality of the work.

The focus of this paper is on translation of nonfiction in the sense of intellectual, academic, or expository writing–e.g., a historical study, biography or autobiography, perhaps a recounting of the rise of some new superstar on the business or entertainment scene. For this discussion, "nonfiction" does not include commercial, financial, scientific, or technical translation.

In what follows, the terms "fiction" and "nonfiction" usually refer to *translation* of the two genres. To be sure, many of the observations apply equally well to original fiction and nonfiction, but that is another paper.

The illustrations for this paper come from three nonfiction projects that, like the fiction that I more often translate, have as their source language (SL) present-day Brazilian Portuguese. All three are scholarly works by professionals in their field, which happens to be anthropology.

Let's begin with an example that seems at first glance straightforward enough:

> Os números, por falsos que sejam, deixam, no entanto, claro que em todo o Brasil do século XIX, até a Abolição, a população livre excedia – às vezes, à justa – a população escrava, mas a população negra e parda sempre excedeu a população branca.[2]

A quasi-literal translation might read:

> The numbers, however false they may be, leave it, nonetheless, clear that in all of Brazil of the 19th century, until Abolition, the free population exceeded–at times, equaled–the slave population, but the black and brown population always exceeded the white population.

The changes in an improved, more "literary" version are minimal but telling:

> However spurious, the numbers nevertheless leave clear the fact that in all of Brazil in the 19th century, up to Abolition, the free population exceeded–or was sometimes equal to–the slave population, but the black and mulatto population always exceeded the white population.

The reader may object that there is nothing "literary" in this example at all, that it is merely the application of standard techniques of good writing to an area where they are often less than a primary consideration. I would agree in part, pointing out that while the conveying of information is the foremost concern in technical and scientific writing, the same information can be conveyed while simultaneously observing the stylistic niceties that ideally characterize *belles lettres*.

It may already be evident from this short sample that I belong to the school of "target-oriented" translators, with all due respect to my "source-oriented" kindred. That is, my focus is on the eventual receptor of the information rather than its producer. If a choice must be made–and one always hopes that no such dilemma arises–between total fidelity to the author and comprehension by the intended audience, my tendency is to privilege the latter. Needless to say, this in no way justifies censoring, adding to, or bowdlerizing the SL material; it merely means that a tie goes to the reader.

If a rationale for this approach is necessary, it is this: The author desires his or her ideas to come across in a new language, and if the translation process itself–whether because of slavish literalism, syntactical inflexibility, or frustrated (and misguided) efforts to reproduce every last unit of meaning–gets in the way, the author has been badly served. Better to sacrifice some relatively unimportant detail, without compromising the main thrust of ideas, than to make the author appear tongue-tied in the target language.

To what extent is a translator justified in taking small liberties with a nonfiction text? The answer depends on the aspect of the text under consideration. Malcolm Coulthard, of the University of Birmingham, touches on the question:

> [M]y Brazilian mestrado student, Ivan Burity, . . . attempted to translate an American text about World Bank employees, who were said to be earning "$96,000" and were facing redundancy "in the fall." Brazilian salaries are typically expressed monthly, American salaries yearly, so should he translate the salary as "$8,000" or gloss the "$96,000" with "por ano"? Similarly, "in the fall" in the original text meant "October/November," so should it be rendered as "na primavera" or would that cause a double-think on the part of a sophisticated . . . reader, who might then interpret "primavera" as April/May, or should he play safe and paraphrase to "later in the year"?[3]

In terms of factual content, there is no leeway. Even what appears to be an erroneous statement must be translated as is. It is, however, good policy–not to mention the polite and professional thing to do–to contact a living author or a deceased author's publishers and point out the possible error of fact. No one wants to be caught in a misstatement, even posthumously, and your extra care will be appreciated.

One recent example of this, taken from a novel I translated, had a Brazilian singing an American popular song–and getting the words all wrong. Since the use of *sic* would be inappropriate in fiction–it would destroy memesis–there was no way to tell if the author intended these errors (to show, say, the character's lack of command of the foreign language) or if they were inadvertent, coming from the novelist. In such cases there is only one recourse: consultation with the author. (It turned out the mistakes were unintentional, and they were corrected for the English translation. The author was quite grateful for my intervention.)

When dealing with matters of style, on the other hand, the translator has more leeway. This of course assumes that the text is not "authoritative." As is so often the case, Peter Newmark is of help in understanding this translation principle: "[I]mportance may be conferred on a text or a quotation by the status of whoever is responsible for it–I refer to such a text as authoritative; thus the phrase 'to be or not to be'; *sein oder nicht sein*; *être ou ne pas être*; *ser o no ser* (which limits its meaning), or the nouns in the phrase 'Water consists of hydrogen and oxygen,' where 'consists of' is not important, since it may be replaced by 'is' . . . or 'is composed of' . . . with negligible semantic loss."[4]

As Newmark has said elsewhere, "In a non-authoritative text, if the content is more important than the style–the more badly written the text, the more it has to be rewritten and restructured in the appropriate register, . . . with a sensitive choice of language."[5] It is safe to say that much, if not most, nonfiction writing is not authoritative.[6]

What features normally distinguish fiction from nonfiction? In nonfiction the paramount concern is to convey information, and style is secondary. Also, while a nonfiction SL author may have a distinctive style–though this is less likely than with a writer of fiction–the translator must avoid introducing any distinct style not the author's own. Subtext features, elements like alliteration and rhythm, and symbolism are normally the province of poetry and fiction rather than nonfiction. But, as I hope this paper demonstrates, there is no justification for remaining oblivious to such niceties when translating nonfiction.

Another example, this time emphasizing the role of a few key words:

> Se algumas doutrinas **como** o positivismo, em sua penetração sobretudo nos meios militares, mereceram uma atenção mais **pormenorizada** por parte dos cientistas sociais, . . . **esse** não foi o caso das teorias raciais. Tais modelos deterministas acabaram recebendo um tratamento particularmente caricatural no que **se refere à seu conteúdo, pairando** uma espécie de **má consciência** em relação à **larga** adoção dessas doutrinas em **território nacional.**[7]

Again, a close translation yields something like the following:

> If some doctrines like positivism, in its penetration above all in military circles, merited a more detailed attention on the part of social scientists, . . . that was not the case with the racial theories. Such determinist models ended up receiving a particularly caricatural treatment in that which refers to its content, a kind of bad conscience hovering in relation to the wide adoption of those doctrines in the national territory.

Surely we can do better than that (it barely reads like English), so here's an example of an improvement that retains the informational content but introduces stylistic considerations–while at the same time enhancing the intelligibility of the passage:

> If some doctrines such as positivism, especially in its penetration in military circles, warranted more thorough attention from social scientists, this was not the case with racial theories. The content of such determinist models would receive a notedly caricatural treatment, engendering a type of heavy conscience toward the widespread adoption of these doctrines in the nation.

As can be seen, it is sometimes necessary to gloss the text (e.g., the interpolation of *engendering* in this example) in order to make the meaning clearer for the reader. If this is a translational sin, it is surely a venial one.

A final example displays certain translational principles in action:

> É generalizável esta concepção da escatologia? Sim e não. A escatologia é, em todas as culturas, uma utopia no sentido literal, etimológico da palavra: descreve uma sociedade que não existe em lugar nenhum. Mas este sentido de "utopia" é diferente, por exemplo, do de Mannheim, . . . Seja como for, esse mundo sem lugar pode ter diferentes funções. A vida depois da morte pode funcionar como um sistema de recompensas que sustenta normas morais, já que julga condutas; pode, também, de forma mais simples, garantir a continuidade ou a hierarquia das linhagens, ou a relação entre grupos sociais e seu território, etc. Em tais circunstâncias, é duvidoso que a escatologia possa ser caracterizada como uma zona de fantasia desabrida. . . .[8]

> Is this conception of eschatology generalizable? Yes and no. Eschatology is, in all cultures, a utopia in the literal, etymological sense of the word: It describes a society that exists noplace. But this sense of "utopia" is different, for example, from that of Mannheim, . . . In any case, that world-without-place may have different functions. Life after death can function as a reward system that sustains moral norms, because it judges conduct; it can also, in simpler form, guarantee continuity or hierarchy of lineages, or the relation between social groups and their territory, etc. In such circumstances, it is doubtful whether eschatology can be characterized as a zone of unrestrained fantasy. . . .

It is evident from this example that the adage "If it ain't broke, don't fix it" applies to translation as well as other areas of life. This passage is clear, well written, and direct. Few adjustments are necessary to bring it into another language.

Some additional observations, dictated by the principles elucidated earlier:

• *noplace*. This word is decidedly less frequent in writing than *nowhere*, and the option to use it may seem idiosyncratic. However, because Portuguese does not distinguish between the two, the choice comes down to the translator's feel for the TL language, and what tips the scales (ever so slightly, perhaps) in favor of "noplace" is the author's phrase "the literal, etymological sense of the word": *utopia*, Thomas More's coinage, comes from the Greek *ou* 'not' + *top(os)* 'a place'.

• *In any case*. My ideolect privileges this rendering over the more literal "be that as it may." Either is acceptable.

• *world-without-place* is admittedly a somewhat uncommon construction, but consider the alternatives. "Placeless world" ? "World without location"? "World found nowhere"? "Nowhere world"? One possibility would be to gloss the phrase as something like "world to be found nowhere," but that becomes awkward when the phrase is encountered frequently. Note that the use of hyphens in "world-without-place" serves to alert the reader to the special function of the term.

• *reward system, moral norms*. These are standard anthropological/sociological jargon and should be translated as such. "System of rewards" or "standards of morality" would lose the academic flavor of the text.

• *whether*, which enjoys a higher register than "if" or "that," has a slight edge because of the academic tone of the passage.

In most of these choices (and translation, especially literary translation, is in many ways the process of narrowing down choices from a near-infinite range of possibilities) I would have no quarrel with translators who prefer the alternatives that I rejected, or possibly others not mentioned here. But the point of this exercise is to urge translators of "nonliterary" works to apply the same stylistic consideration to the text that one would use in rendering Thomas Mann or Gabriel García Márquez.

NOTES

[1]Quoted in Susan Bassnet, "Writing in No Man's Land: Questions of Gender and Translation," *Ilha do Desterro*, No. 28 (1992), 65.
[2]Manuela Carneiro da Cunha, *Negros, estrangeiros: os escravos liberetos e sua volta à África*, 19.
[3]Malcolm Coulthard, "Linguistic Constraints on Translation," *Ilha do Desterro*, No. 28 (1992), 14.
[4]Peter Newmark, *About Translation*, 2
[5]Peter Newmark, *Paragraphs on Translation*, 92.
[6]Some borderline cases obviously exist–e.g., Derrida.
[7]Lilia Moritz Schwarcz, *O espetáculo das raças*: *cientistas, instituições e questão racial no Brasil, 1870–1930*, 15.
[8]Manuela Carneiro da Cunha, *Antropologia do Brasil*, 2nd edition, 80–81.

CORRESPONDENCES: MUSIC, PAINTING AND LITERATURE

Camilla Bozzoli Rudolph
National Geographic Society and Georgetown University

Marguerite Shore
Freelance Translator

Keywords: Literature, Music, Painting

Abstract: As suggested by Baudelaire's poem, *Correspondances*, the boundary between one art form and another often overlaps, and it is not rare for a writer to "invade" the world of a painter or a composer, with the special liberty to which an artist is entitled. Works in numerous languages, translated into English, will be examined, with audience participation encouraged.

1. INTRODUCTION

La nature est un temple où de vivants piliers
Laissent parfois sortir de confuses paroles;
L'homme y passe à travers des forêt de symboles
Qui l'observent avec des regards familiers.

Comme de longs échos qui de loin de confondent
Dans une ténebreuses et profonde unité
Vaste comme la nuit et comme la clarté
Les parfums, les couleurs et les sons se répondent.

Il est des parfums frais comme des chairs d'enfants,
Doux comme des hautbois, verts comme les prairies
Et d'autres, corrompus, riches et triomphants

Ayant l'expansion des choses infinies,
Comme l'ambre, le musc, le benjoin et l'encens,
Qui chantent le transport de l'esprit et des senses...

This famous poem by Charles Baudelaire (Correspondances) puts into words what one always instinctively knew: that the senses (and the arts) are bound together in a dark and deep unity. In the creative chaos, it makes no difference what is mine and what is yours. The idea of property and distinction belongs to another world, a "bureaucratic" world, where things are neatly filed away and files do not mingle. but in the creative world all senses are involved: touch (some perfumes are as fresh as children's skin), hearing (as soft as oboe music), sight (as green as prairies).

That there is a close link between sensations has long been known. Its most famous example is the episode of Proust's "petite madeleine," the small cake that, dunked in a cup of tea, conjured up for him all the world of his childhood. The old house, the town square, the streets where he used to go shopping, the trails where he used to hike, they all came to the surface from the subconscious. As in that Japanese game in which one dunks in a bowl of water pieces of special paper and they come out in the shape of flowers, houses, people, in the same way all the flowers of the garden of his childhood and all the inhabitants of the village, their houses, the church, etc. took shape and form out of his cup of tea.

2. LITERATURE AND PAINTING

One of the threads that runs through Antonio Tabucchi's short stories (indeed, they are so short I would prefer to call them "ruminations") is a fascination with the creative impulse itself. As a writer he often dwells on the initial spark, the quirky situation or condition that ignites the artistic process. The context or stage where he chooses to present his ideas is often the work of a well-known artist (painter, writer, composer). While the reader's familiarity with the subject is not required, a knowledge of the work of, say, Fra Angelico, Goya, Caravaggio, can deepen the reader's response to Tabucchi's insights and ironies. On the other hand, Tabucchi's readings of the past can also become a tantalizing gateway that brings the uninitiated to new vistas.

Tabucchi takes an unexpected sliver of life, wanders in at an atypical angle and then lets the reader enter. It is as though he has discovered, as if by accident, a fragment from which a world can be reconstructed. In his introductory note to *The Winged Creatures of Fra Angelico*[i], Tabucchi describes his tales as follows: "Extraneous to any orbit, I have the impression that they navigate in spaces that are confidential and yet of unknown geometry; we might call them domestic fractiles: the interstitial zones of our everyday necessary being or certain swellings of existence." He refers to these writings as "quasi-racconti" or "racconti mancati" (almost-tales or unfulfilled tales), "hummings that have accompanied me and accompany me still" [...] "a 'background noise' turned into writing."

In the title story of the eponymous collection, Fra Angelico, the 15th-century Tuscan painter and monk, is tending the monastery garden, going about his habitual chores, when he hears a voice call out, "Guidolino," summoning him by the name he had left behind in the secular world. "It was a rosy little creature, soft looking, with small yellowish arms like a plucked chicken's, bony, and with two forepaws, these too very skinny, with prominent joints and callused fingers like a turkey's." He rescues the creature from the pear tree in which it has become entangled, and with his Father Superior's permission, shelters it in a makeshift refuge covered with netting, to prevent the exhausted creature's giving in to the "force of ascension to which it was prone." Two other creatures arrive the following dawn, one shaped like a roundish ball with a thick, plumed tail, the other resembling a big, loose-jointed dragonfly or cricket, but with the delicate face of a woman. Indeed, this last creature reminds the pious monk of a girl he once knew and loved. "That night the big dragonfly paid him a visit. {...} Fra Giovanni smelled an intense fragrance of basil so strong he felt sort of intoxicated. He sat down on the bed and said: "is it you that smell of basil?". The winged creature brought one of its very long fingers to its mouth, as though to indicate that he shouldn't speak, then it drew close and embraced him. And then Fra Giovanni, confused by the night, by the fragrance of basil and by that white face with its long hair, said: "Nerina, I am dreaming about you." The winged creature smiled, and before leaving him, with a rustling of wings, said: "tomorrow you must paint us, that is why we have come." And paint them he does, inserting them into the frescoes he has created in the small vaulted cells where the monks meditate and pray. The odor of basil, the embrace from the long haired otherworldly creature--this is a potent seduction scene. Fra Angelico was known in his own lifetime, for his piety, for the purity of his vision, the sweetness of his colors. But Tabucchi's intuitive undercurrent or "background noise" tells him that it is an earthly, even sensuous memory, a recollection of a youthful carnal escapade, that puts him in touch with his vision. Art historians typically have described Fra Angelico's work and life in terms of his piety. In the 16th century, Vasari, in his *Lives of the Most Eminent Painters, Sculptors, and Architects*, wrote that Fra Angelico was ""said never to have painted a Crucifix without tears streaming from his eyes," his saintly life reflected in his work. While later historians are more grounded in their analysis of this undeniably great painter, they still emphasize the painting's roots in theology. As a fervent admirer of Fra Angelico, I find Tabucchi's stance particularly interesting and exhilarating. According to Tabucchi's (albeit fictive) version, the sacred is rooted in the profane, piety emerges from real life, and the crystalline, unearthly purity of this painter's figures, on closer examination, has a homely, all-too-human underpinning. Having read Tabucchi, I now look at Fra Angelico with new eyes. Toward the end of the tale, Fra Angelico uses the winged creature that had arrived first as the model for an angel of the Annunciation. This painting has always had a powerful impact on me; within its harmonious setting, there is a perfect and painful tension of approach and withdrawal, subtly conveyed by the tilting postures of the figures. But how does Tabucchi view it? "...he posed the winged creature in a kneeling position, leaning it against a bench to keep it from falling; he had it cross its hands over its chest in a reverential expression and he said: "I will cover you in a pink tunic, because your body is too ugly." Aha, the anticipatory tilt of the angel's posture is, in fact, a result of its awkward propped up position, its inability to stand on the ground. And that gorgeous pink tunic conceals an ugly, decidedly unheavenly body.

"Dream of Michelangelo Merisi, known as Caravaggio, painter and hot-tempered man,"
one of the tales in *Dreams of Dreams,* recounts a night in early January, 1599. Caravaggio, who has a reputation as a brilliant painter and violent hot-head, is sleeping with a prostitute when he dreams that God, in the figure of Christ, pays him a visit. In the dream he is gambling in a tavern, surrounded by scoundrels. "And he, he wasn't Michelangelo Merisi, the famous painter, but an ordinary adventurer, a brigand. When God called upon him he was taking the name of Christ in vain, and laughing. You, the finger of Christ said wordlessly. I?, asked Michelangelo Merisi with amazement, I'm not a saint by vocation, I'm only a sinner, I cannot be chosen." Christ insists, and Caravaggio finally gets up and follows him out of the tavern, stopping to urinate in a grimy alley before continuing on to a deserted piazza. "I'm sad, said Michelangelo Merisi. Christ looked at him and didn't answer. He sat down on a stone bench and took off his sandals. He massaged his feet and said: I'm tired, I've come by foot from Palestine to look for you." After vomiting, Caravaggio screams at Christ that he must be mistaken. "Christ approached him and touched an arm. I've made you a painter, he said, and I want a painting from you, and then you can follow your destiny's path. [...] The visit I paid you tonight in the tavern, only that you will be Matthew." Caravaggio agrees, the dream and the tale are over, he turns over in bed and is embraced by the prostitute. Caravaggio has long been appreciated as a realist, whose down to earth portrayals make Biblical episodes immediately accessible, yet are characterized by dramatic lighting that conveys epiphanic moments of faith. Tabucchi delves into this mysterious sacred/profane dichotomy. His Caravaggio lies abed with a prostitute, in his dream he, almost unthinkably, urinates, vomits in the presence of Christ. And the Christ figure is touchingly afflicted with human frailty; he's tired, his feet are sore. Tabucchi's quick, off-center peek into the painting/painter's world shows us the conflict that is never quite resolved and that gives this Baroque master's work

much of its power. Perhaps Tabucchi, better than any art historian, has reconciled the life of this violent, bawdy man to the transcendent body of work. The spark of inspiration slices through the dark, through a dream, commanding a response, after which the artist is destined to continue on his dissolute way.

In another tale the dreamer is Francisco Goya. The painter dreams that he is pushing a lover on a swing at midday. She falls, and he rolls after her, down a hillside, stopping against a yellow wall. Looking out over the wall they see soldiers, illuminated by a lantern, firing on some men. "The lantern was incongruous, in that sunny landscape, but it lividly illuminated the scene. The soldiers fired and the men fell, covering pools of their blood. Then Francisco Goya y Lucientes removed the painter's brush sashed at his waist and moved forward, brandishing it in a threatening manner. As if by magic, the soldiers vanished, frightened by that apparition. And in their place there appeared a hideous giant..." Goya asks the giant who he is and the giant answers, "I am the monster that dominates humanity, History is my mother." Goya once again brandishes his brush, the giant disappears and is replaced by an old woman, a toothless hag. "Who are you?, Francisco Goya y Lucientes asked her. I am disillusion, the old woman said, and I dominate the world, for every human dream is a brief one." Again Goya wields his brush, the old woman vanishes and a small dog appears, buried in the sand with only its head sticking out. "Who are you?" [...] "the beast of desperation and I make fun of your afflictions." Vanishing beneath the brush, the dog is replaced by a man, "a fat old man, with a weak and unhappy face." "Who are you?" [...] "The man gave a tired smile and said: I am Francisco Goya y Lucientes, you will be helpless against me." "And at that moment Francisco Goya y Lucientes awoke and found himself alone in his bed." Here Tabucchi's dreamer is caught in his own hallucinatory nightmare. The first image is Goya's *The Third of May, 1808*, a gripping depiction of the execution of Spanish peasants by Napoleon's troops. The second image, the giant, could be Goya's *Colossus*, a terrifying figure looming above the frightened populace. The third and fourth images, the toothless hag and the half-buried dog, are typical of Goya's *Los Caprichos* etchings, which offer a devastatingly cynical, black view of man's folly. Tabucchi has chosen some of Goya's most troubling images. Yet his lead in, the dream-stage from which action will unfold, is a springtime idyll of two lovers on a hillside. This brief tale, less than 500 words in length, goes to the heart of the conflict that informs Goya's art: the blackness and cruelty of human nature that lay behind the sunny facade of 18th-century Spain during the so-called age of reason.

In his fictive exploration of the world of painters he loves, Tabucchi reveals much about their work and about his own attitude toward creation. Art and life, dream and consciousness, faith and doubt are inextricably intertwined. For Fra Angelico it is a memory of his flesh and blood past that inspires him to sublime heights. Caravaggio, the brawling prototype bad-boy painter, is momentarily shaken out of his torpor to use his God-given gift. And Goya, the quintessential artistic split personality--society painter and voyeur into the darkness of the human soul--unveils the shadows and terrors that lurk behind the sunshine.

3. LITERATURE AND MUSIC

Literature, visual arts and music are indeed connected in a dark and deep unity, as big as the night or the light of day. Music in particular is difficult to detect, as it is the most abstract of all arts. But to the sensitive ear, it is ever present: It could be the rhythm of a sentence, the sound of a word or a landscape that has become a musical vision, as exemplified best in Thomas Mann's novels.

Sometimes a piece of literature is organized like a musical composition, in a crescendo that culminates in a climax of almost unbearable intensity. I am thinking of works like *Crónica de una muerte Anunciada* by Gabriel Garcia Marquez. Tomasi di Lampedusa, in his novel *Il Gattopardo* (The Leopard), compares the love story between Angelica and Tancredi -- which takes place in the baroque palace of Donnafugata and is contained in a single, exquisite chapter -- to a movement of a musical composition, which survives as a movement, while the major work of which it is a part fails miserably and is soon forgotten.

Other musical techniques can be consciously applied to literature. For instance the technique of the Leitmotif, used by Wagner and successfully picked up by Thomas Mann, first, very openly, in *Buddenbrooks*, and then, in a more subtle and concealed way, in all his work.

Music can play yet another role, the one of transforming the normal person, the bourgeois, into an artist. Again in Thomas Mann's work, when a regular person turns to art, and especially music, his will to live is weakened and the end is near. The experience of music is similar to the experience of death, in the sense that it creates that state of oblivion that Schopenhauer, borrowing from Indian philosophy, called Nirwana. This destiny befalls, among others, a young woman who, once initiated by a fellow patient to the music of Wagner (more precisely *Tristan and Isolde*), loses all interest in life. We are not informed of her death, but we know that the doctors start treating her as a desperate case and the rest is left to our imagination.

Music holds the power of sublimating the most common experiences and lifting them to a higher, universal level. I myself experienced something similar a long time ago, when I overcame and completely forgot the pain of being stood up by a date by listening to a movement of Beethoven's Piano Concerto Number 5, the *Emperor Concerto*, which kept playing through my head. On a less trivial level, young Nikolai Rostov in *War and Peace* experienced the same, when, anguished over the loss of a huge sum of money he had gambled away, temporarily forgot all about it while listening to Natasha sing an Italian aria...The whole world centered for him on anticipation of the next note, the next phrase, and everything in the world was divided into three beats: O mio crudele affetto...One, two three... What were gambling losses ... --he was thinking -- it is all trash..., but this is real!

Music can detach one from one's particular situation and deepen joy (or sorrow). In Proust's immortal novel, it is sorrow and regret that overwhelms Swann while, out of the blue, he happens to hear at a concert the "petite phrase musicale" that reminds him of a woman he loved. It is worth going into that passage that describes so accurately how Swann -- but it could be me and you, in fact anyone -- can be caught in a tangle of mixed emotions, evoked by the sound of music he had heard when he was happy. Faced with the sound of all that happiness that he was again experiencing, he was overwhelmed by the vision of his present despair and felt so much pity for that man (himself), that he had to cast his eyes down, so that people might not see that they were full of tears. The feeling of unreality is such that the emotion is projected outward and he feels as if this were happening to somebody else. But the best part is the description of how the musical motif is first detected and recognized and the process is followed in all its stages of surprise, anticipation and final reconciliation. I mean reconciliation with destiny, which ceases to be Swann's own particular destiny but becomes deep understanding of all human suffering. Starting from this moment, Swann understands that the feeling Odette had for him will never come back, and his hope for happiness will never come true... Love comes and goes, but music stays. What was the most flimsy, most abstract of things becomes the real thing, or even the only thing that is real.

NOTES

[1] All translations from Tabucchi are my own. A translation of *I volatili del Beato Angelico* has been published by Chatto & Windus, but as of this writing I have been unable to obtain that translation. [M. Shore]

TRANSLATING A KILLER

Laura Esther Wolfson

Keywords: Russian underworld, literary translation

Abstract: Killer for Export is a picaresque, post-perestroika tale about an endearing and emotionally bruised KGB agent who, in the spirit of the times, leaves the state sector and becomes a private hit-man. He dreams of domestic tranquility and a house with a white picket fence but like many of his countrymen, he cannot shake off his past. The story moves with gumshoe grace through a series of clearly delineated and evocative spaces: a prison camp orphanage, the emigre community of Brighton Beach, teeming with shady fortttune-seekers and bright commodities, and the blank streets of suburban America. Killer contains enough bilingualisms and underworld slang to make a purist groan and a translator weep. The author, Edward Topol, claims the distinction of being one of a miniscule number of Russian writers who earns his living by writing alone. His books are bestsellers in Russia and are frequently issued in pirated editions by fly-by-night Moscow publishers.

1. INTRODUCTION

Early in my literary translation career, that is to say about eleven months ago, I found a message on my answering machine from a Russian-speaking man with a raspy voice. He said, "Hello, this is Edward Topol. I have a novella which I would like translated into English. You have probably read my books." For the word "books" he used a Russian diminutive--"knizhki"--which reflected an off-handed, disparaging attitude toward his own oeuvre, contradicting his assumption that I had read his works.

In fact, not only had I not read his books, but I had never heard of him, a fact of which I did not hasten to inform him. I started putting out feelers to find out who he was. My emigre acquaintances in Brighton Beach all knew of him. The translator who had given him my name said simply, "He writes potboilers." An issue of The Economist which came out around that time said that he had two books on the Russian best-seller list. Browsing over the booksellers' tables some months later while in Ukraine on business, I had the opportunity to see that indeed there seemed to be no other fiction writers whose works were as widely available as his.

The proposed project was called Killer for Export. It had been published serially in a Russian newspaper and he told me that he was hoping to place the English version, with the help of his agent, in a magazine such as Playboy or Penthouse. After some haggling we agreed on the price and then discussed the deadline. Estimating how much time I needed to finish my current project, I suggested a date about six weeks away. There was a long silence at the other end of the phone line. Then he said, "Look, I sat down and dashed this story off in two weeks; I think the right translator can translate it in the same period of time."

2. PLOT SUMMARY

Killer for Export is a picaresque, pornographic, post-perestroika tale about a KGB agent who, in the spirit of the times, leaves the state sector and becomes a private hit man. He dreams of domestic tranquility and a house with a white picket fence but, like many of his countrymen, he cannot shake off his past.

The story opens with the hero, Nikolai Umansky, in a pick-up truck with a shady emigre named Nathan, speeding northward from Manhattan to Scarsdale, toward an appointment with murder. Through a series of flashbacks we learn how he got this job which brought him to America. A casualty of post-Communist downsizing, he received a free-lance murder assignment from his former bosses at the KGB who were now running a private security firm. (During one of our early telephone consultations Topol told me firmly that as underworld professionals they would never call what they do "murder" but prefer such terms as "knock off," "rub out," and so on. My Russian vocabulary grew to encompass such terms as "ubrat'," "ustraneniye," "shlyopnut'," and "mokroe delo.") He hatches a plan to use the job as a ticket to America, and decides to come here, not commit the murder, go against a lifetime of habit by not violating any laws here, dump the client who has brought him over, and start a new life in America, breaking with his criminal and KGB past.

Next, we learn about Nikolai's past life and see it set in the context of Soviet history: he grew up in a prison camp orphanage after WWII. It is not stated whether his parents were political prisoners and victims of Stalinism or part of the criminal underworld, known in the Soviet Union as "thieves" [vory, in Russian], but his childhood setting provides him with the education he needs to become a ruthless criminal, recruited as a teenager by the KGB because he exhibits cruel aptitudes that his peers do not. His c.v. with the KGB parallels most of the major events of the Brezhnev period and perestroika: he combatted various types of dissidence, including Zionism, samizdat, Seventh-Day Adventism, and Crimean nationalism, served in Afghanistan, in Transdniester and finally was retrained to combat racketeers. This is the life he wants to leave behind.

Unlike the emigres and refugees who stand in line at customs, their luggage carts piled high with bedding, laundry detergent and other domestic items, Nikolai brings nothing except a small sports bag and the clothes he has on. But even that is too much baggage. The Soviet Union, symbolic of his past, greets him as he steps off the plane in America. Already the reader suspects that it is too late for him to realize the American dream of changing his life:

"He stepped out of customs and toward the door to America with a trembling in his stomach and his chest like the feeling you have as you go to your first assignation or to your first murder. And those doors opened before him right away, moving automatically apart, and maybe that was nothing in itself, but, oh, how good it felt! However, once through the door, he discovered not America but just a crowd of Jews holding cardboard signs over their heads that said, "Schwartz, here we are!" "Welcome, Rosa!" and so on."

The juxtaposition of assignation and murder is no accident, as it turns out, for Nikolai's intended murder victim throws herself into his arms and begs ardently for sex, saying that her husband won't touch her. Nathan had told Nikolai he could engage in a little recreational rape before the murder, but instead he gets propositioned and seduced. He resists the habits ingrained by years of KGB training and does not kill her. When he leaves her, she is so distraught that she shoots herself. He has satisfied both his client and the law to the letter, for when the police find her, they conclude correctly that the case is a suicide committed after sex, while the clients are thrilled because they think that Nikolai has managed to do a clean job with no traces of violence.

The plot thickens as Nikolai knocks out Nathan with a blow to the neck, takes his money, leaves him lying face-down by the roadside, seats himself at the wheel of the pickup truck, and goes whizzing down the highway with no plan except to discover America.

After a brief stop to visit the thoroughly American nanny service, called Nice, Clean & Perfect, Inc., which serves as the front for the U.S. half of the hired killing partnership, we find ourselves on a beach in Massachusetts where Nikolai marvels at the fabulous wealth of the American middle class and feasts his eyes on a well-endowed and scantily-clad sunbather. When he rescues the sunbather's drowning son, he becomes the boy's spiritual father. He plucks him from a car sinking in a vile polluted bay, grabbing him by his hair and pulling him out like "a carrot from a vegetable patch." In fact, he is giving birth to himself, tearing himself out of the sordid world--which is now going under too, like the car--in which he had lived all his life. He moves in with the boy's mother, embarking on a short-lived suburban idyll, learning from her of the joys of love, innocent eroticism and domesticity.

But almost as soon as he reaches his goal, he must leave it; the chief executive officer of Nice, Clean & Perfect and his Russian partner, a figure known as the King of Brighton, are onto Nikolai. Nikolai is catapulted out on a journey of escape which takes him all over New York State, finally looping back to Brighton Beach to confront his pursuers, settle scores with Nathan and preserve his honor. There he meets his end on the dark shores of Brooklyn, shot by a former comrade of his flown in from Russia on a similar job. "You've lost your form here in America," says his colleague to Nikolai's corpse as the waves lap it gently, and then he leans down to remove Nikolai's wristwatch, a memento of service in a special squadron in Afghanistan.

"Trash fiction!" you cry, having read or heard this summary. That is what I thought after reading the tale through once. But in fact, it is rich with linguistic and cultural challenges, and it traces an interesting and little-known vein of Soviet history, having to do with the criminal underworld and its relation to the prison camps and the government security forces. As I went through the story again and again, my respect for it and its creator increased. What follows are some literary, cultural and linguistic notes based on my thoughts as I grew more intimate with this work.

3. STATE SECURITY AND THE CRIMINAL UNDERGROUND

In recounting the story of Nikolai Umansky, Topol is using a seemingly frivolous fictional genre to take up a serious post Cold-War problem little-anticipated in the euphoria surrounding the collapse of the Soviet Empire: what becomes of the hundreds of thousands of men who were trained in force and ruthlessness to keep the Soviet Empire in place, men who served in the military, KGB, police, Special Forces or Internal Affairs Ministry troops? The state has shrunk and no longer needs them. Unemployed and well-schooled in the art of handling weapons, they are potentially as harmful to developing democracy as anyone in the new order. The skills they have learned to preserve law and order have very few constructive uses outside of the government. Most of them melt away into vague commercial structures where their skills serve them as security guards or racketeers. Trained to provide government security, they perform a flipflop and become criminals.

In the person of Nikolai, we see the incestuous relationship between the government and the criminal underworld so typical of the Soviet and post-Soviet periods. Nikolai is one of "the ancient and noble companionship of thieves"[1] to whom Solzhenitsyn devoted so much space in The Gulag Archipelago. In the camps, they were thrown in the same cells with political prisoners, where they terrorized them and stole their money, clothing and food to divide up with the camp management, with whom they were in league.

"The tattooed chests," as Solzhenitsyn called the criminals, "were merely the rear ends of the blue caps,"[2] or camp trusties. Sometimes they were the camp trusties themselves, "cleaned up and dressed up as representatives of the state." [3] This description depicts aptly Nikolai's role when he worked for the KGB. He was put to work combatting anti-Sovietism. While we do not know who his parents were, whether they were political criminals or thieves, we do know that his spiritual lineage is that of the common criminal, or blatnoy, as they have been known in Russia for generations.

Topol uses another touch to let us know how closely intertwined the criminal classes and the forces of law and order are in Russia. In the final part of the story, we are introduced to Boris, who turns out to be Nikolai's foil, or double. He is the doorman at the restaurant in Brighton Beach where Nikolai goes to settle scores with Nathan, unaware that the latter has been drowned for letting him get away. Boris is captured in a brief and pungent description: "Nikolai went into the restaurant and saw right away that he had come to the right place; by the entrance to the coatroom a hulk with false teeth, crooked bangs, a tattoo and a large ring on his left hand sat smoking a Marlboro and reading Soviet Sport."

As Nikolai waits for Nathan [in fact, the tattoed and hulking Boris has orders to detain Nikolai until a hitman arrives] the two of them get acquainted over vodka:

"After they downed two shot glasses, they discovered mutual acquaintances in Moscow and after the third, they discovered that their destinies had been similar. In 1952, during the last wave of Stalinist arrests, Boris' parents had been arrested and at the age of three he was sent to an orphanage. After that, he became a street hustler. True, that was where the similarity ended, because then Boris, in contrast to Nikolai, went into organized crime."

There it is: similar destinies, the only difference being that one stayed in the criminal world while the other joined the KGB. Hardly any difference at all, Topol seems to be saying, between the two careers.

Nikolai does change his life and his criminal habits. But that is what does him in, for he has lived all his life in a setting where it is a crime not to commit a crime, and he is killed for trying to leave the criminal milieu.

4. Bilingualisms

Bilingualisms are the scourge of the language purist. But the translator (whose task it is to transfer meaning from one language into another language) who suddenly finds that she must render meaning from two languages into something understandable to the reader (in one language) and yet that retains the double flavor, does not have the luxury of purism.

When emigres begin sprinkling their native language with words like "inshurance," "velfair," and "garbidj", they gradually begin to communicate in a dialect consisting of Russian generously sprinkled with English or with Russian versions of English, a dialect incomprehensible to non-immigrant Russians and Americans alike. In doing so, they lose Russian without gaining English. They are thereby paradoxically locked into their emigre community and in practice have less usable language than before emigration.

Nikolai finds himself at this bewildering linguistic crossroads almost as soon as he exits the plane at JFK. Topol conveys the alphabet soup sensation of bilingualism quite effectively, demonstrating it with numerous and varied types of language-grafting throughout the story. In addition to various combinations of Russian and English, Topol invents two dialects which I call as Cyrillo-English and Latino-Slavonic. The different types of language mixing (some of which take place in Nikolai's head) are indicative not only of various types of cultural overlap in the world but also of Nikolai's level of comprehension of the world and culture and of his relationships with various people.

The Russian text is sprinkled with English brand names and remarks made by Americans given in the Latin alphabet. There are also some English words which are transcribed into Cyrillic. Russian signs in Brighton Beach are presented in the Latin alphabet. Topol indicates that certain conversations take place in English but they are presented in Russian for the comprehension of the Russian reader. Other conversations are given directly in English, some with and some without translations into Russian.

Some of the most interesting examples of bilingualism occur in conversations between Russians and Americans. Take this one, for instance, between Bill Longwood, the American representative of the company which provides the hitmen, and his Russian counterpart, Zinovy Bloom, the King of Brighton. Longwood comes to Brighton Beach to tell Bloom that he has information about Nikolai's whereabouts. We are told his route from Manhattan to Brooklyn. Battery Tunnel and Belt Parkway are spelled in Cyrillic, followed by Ocean Park Way in Latin letters, for some reason, with a mistake: "parkway" is given as two words, rather than as one. When he enters Brighton Beach, he feels he is entering a different domain, and sees Gastronom Stolichny, Cafe Arbat, and other signs, but now the Russian names are given in Latin letters, perhaps to remind the Russian reader that he is still in America, if in a Russian-speaking part of America. Bill asks for directions to the restaurant where he is to meet Bloom; he is a foreigner here; he does not know his way around. He asks in English [of course], "Where is the Sadko Restaurant?" In the text, however, the word 'where' has the 'h' missing, making it "were is Sadko?" The pedestrian whom he asks admonishes him to speak Russian, answers in Russian (the English is given in brackets) and points him in the right direction.

He goes to the restaurant, where the doorman does not want to admit him. The doorman speaks in Russian, considerately providing an English translation immediately for each sentence he says. At least, I assume this is what is happening, as the English is also given inside the quotation marks. It also provides a justification for a mistake in the English; the doorman says 'close!' instead of 'closed!' However, other English words and phrases which appear in the story are simply strewn with mistakes with no such justification. The version of the story I worked from was a copy of it as it had been published in a newspaper in Russia. Whether the mistakes came from Topol himself or from the publisher, I do not know. I corrected all of them.

Now he finds Bloom playing poker in the restaurant. Bloom gestures for him to wait a moment, then tosses his cards down on the table, saying, "Shit!" an expletive used by several Russians in the story, and never translated into Russian, presumably because Russian readers will understand it. (By the way, the f-word seems to need no translation either in this story.) His first words to Bill are given in English without a translation for the Russian readers. Bill's answer is given in Russian, but his words are followed by the phrase, "Bill said in English," although we already know that

Bill speaks no Russian. When they met in Manhattan, their conversation was also printed in Russian, but it was simply assumed that the readers understood that they were in fact speaking English. The author even noted that Bloom's English was accented and imperfect. Now the whole conversation is also presented in Russian, but perhaps because they are in Brighton Beach, the author feels it is necessary to insert a reminder that they are in fact speaking English.

Now let's look at Nikolai's conversations with Leslie. He seems to have no trouble understanding her, and mention is made of her schoolteacher diction. When she speaks (English, of course) her words are given in English but spelled phonetically in Cyrillic. For example, "Ai vona maik lav to yu. Ken ai?" and, when he saves her little boy, "Yu a e gud men, yu noy?" (Here, I have back-transliterated her words into the Latin alphabet, because I do not have Cyrillic software.)

However, while they are on the beach, before they get acquainted, she is crying and speaking ordinary English, spelled with Latin letters, only it sounds a little off, not quite like the English of a native speaker: "God! Save him! Save him! Please! I'll do everything!" (I changed this to "I'll do anything!") and "Johnny, sorry! Johnny, excuse me!" which I changed to "Johnny, forgive me!" None of Leslie's utterances are translated into Russian for the Russian-speaking reader, implying that Nikolai understands her without a translation. Topol manages to write the dialogue and action so that each of her statements is followed by an action description in Russian which explains what she has just said without the necessity of providing translation of her words into Russian.

Apparently, Americans who are closer to Nikolai tend to speak Cyrillo-English (English spelled phonetically with Cyrillic letters) because it is easier for him to understand them than other Americans. Leslie's son also has a fluent grasp of this dialect.

When Nikolai and Leslie are in bed together, he curses at her in Russian according to his old habit, calling her 'pascuda,' 'suka,' and 'blyad' ['scum,' 'bitch,' and 'slut,' respectively]. Leslie learns these Russian words, but when she speaks them they sound tender and loving, not hateful. When Nikolai leaves her to flee the goons who have come for him, their last conversation goes like this:

Nikolai: "Ai hev to goy. Sorri." [Cyrillic.]
Leslie: "You are pascuda." [All in Latin letters.]
Nikolai: "Es. Ai em." [Cyrillic.]
Leslie: "You are blyad!" [Latin]
Nikolai: "Es." [Cyrillic.]
Leslie: "Suka!" [Latin.]
Nikolai: "Es." [Cyrillic.]
Leslie: "I love you!"
Nikolai: "Sori. Ai hev to goy." [Cyrillic. Note the difference in spelling of 'sorry' from previous use.]

Here, no translations are given in the original for the Russian reader. Later on, we see a variation of this type of conversation between speakers of the mutually-comprehensible dialects Cyrillo-English and Latino-English (English spelled with Latin letters).

This time, however, Topol sees fit to translate the Latino-English into Russian, while he does not translate Nikolai's Cyrillo-English into Russian. The translation of the former comes across as somewhat ponderous. The conversation takes place between Nikolai and a policeman who is telling him that it is illegal for him to hitchhike or walk along the highway. At the end of the conversation, when Nikolai is thoroughly frustrated and all avenues seem blocked, he shifts from Cyrillo-English into a mixture of English and Russian, saying: "No mani. Ai vok. Peshkom. Nogami, ponimaesh, suka? Ai vok to Boston peshkom. O'keu?" This means: "No money. I walk. On foot. Understand, bitch? I walk to Boston on foot. Okay?" I translated it as follows:

"No money," said Nikolai. "I walk." In Russian, he added, "On foot. With my feet, understand, pig?" And in a final bilingual flourish, "I walk to Boston on foot. Okay?"

When he meets up with foreigners the language soup becomes even more piquant. He gets a dish-washing job in a Korean restaurant because the owner, who studied in Moscow, overhears him cursing in Russian. He hires him on the spot and delightedly curses at him half-coherently in Korean-accented Russian (written in Cyrillic) which he then translates into Korean for his employees. (The Korean is not given in the text.)

As Nikolai rides his stolen motorcycle back toward Brighton Beach, he meets up with a Spanish woman who has a flat rear tire. He helps her with it, pumping it up till is it as hard as steel. He follows her back to her apartment, shining his headlights on the rear tire to make sure it is all right (a clue that they are soon to have anal sex). "Would you like a <u>cap</u> [sic] of coffee?" she asks in English, followed by a translation into Russian. What? thinks the baffled reader. Is this how a Russian writer portrays a Spanish woman speaking in American English? Or is it merely another typo?

When Nikolai leaves her apartment the following morning after showering and drying himself off with a towel printed with a portrait of Fidel Castro, she says, "By [sic] darling! <u>Gracias</u>! Thank you for everything!" Nikolai thinks to himself: there <u>is</u> a difference between America and Russia. Here they thank you for sex, and in two languages, at that!

How to deal with all of this language mixing in an English translation? Of course, it is impossible to maintain the texture of two languages in the original when you are translating from one of those languages to the other. The contrasts in the original get smoothed out. The English words spelled in Cyrillic can be spelled phonetically in the Latin alphabet so that they sound as if they are being pronounced with a Russian accent. Since Topol himself at many points resorts to interspersing bracketed translations in the text and stating what language is being spoken, I felt no qualms about interpolating similar information into my translation when necessary. What can I say? Bilinguality can be as messy and as awkward in the original as it is in the translation.

As I worked through these issues, I felt fleeting envy toward early translators of War and Peace, who simply left the long French passages as they were, assuming that their readers were educated and could understand the French without translation. Of course, Tolstoy was writing about and for a higher class of people than Topol is. The situation reminded me of a remark by literary critic Edmund Wilson[4] on the French passages in War and Peace. He recommended W&P to beginning and intermediate students of Russian language as a work they should plow through in the original as early as they were able to. One of the reasons he gave for this recommendation was that since there was so much French in the novel, that would give the novice an opportunity to rest from all the Russian. Alas, I cannot say the same about the English in Topol's work; it often requires more mental effort to untangle than the Russian.

5. SLANG

Surprisingly, the slang in this tale was easier and more fun to deal with than the bilinguality, perhaps because I recognized early on that the best way to handle it was to render it with the utmost freedom and naturalness, as the context allowed. Slang is the jazz of language, and the translator must improvise on the theme as a musician does a riff in order to render it with the proper flippancy and lightness. And so, with apologies to the non-Russian-speakers listeners and readers of this paper, here follow some English renderings of Russian slang which appeared in the original:

1. "What are you farting around for?" for "Fufli ty vremya tyanesh?"
2. "The less you see of America, the less you'll be able to blab about it!" for "Tem menshe ty uvidish Ameriki, tem menshe smozhesh fizdet o poezdke!"
3. "And you're a piece of shit! Cut the crap!" for "Sama ty gadina! Ne fisdi!"
4. "What a wimp!" for "Nu, slabak!"
5. "You think I have time to waste on chicken shit like that?" for "Neuzheli ya budu zanimatsya takimi melochami?"
6. "I'm up shit creek! I'm screwed!" for "Ya vlip! Ya vlip!"
7. "If the police find Nathan, he'll spill the beans to them right away!" for "Esli politsia vozmet tvoego Nathana, on im zapoet u nikh vperviye zhe den!"
8. "Dead men don't talk." for "Pokoinniki ne poyut."

6. DEATHS AND ENTRANCES

The most important events happen to Nikolai while he is at the ocean's edge. That is where he saves little Johnny and experiences spiritual rebirth, where he falls in love with Leslie after she gives him the kiss which launches his blissful domestic life with her, and it is where he is killed. Just as he is about to be killed, he looks out to sea (toward Russia, incidentally) and thinks of Leslie, a wave of happiness breaks over him, and we see a classic literary twinning of love and death. As he moves further away from the ocean into what Topol calls the 'meteorological garbage pit' of upstate New York, his strength seems to ebb out of him; he no longer drives a car, but travels by Greyhound Bus, he is no longer a killer but a mere dishwasher.

But when he heads back toward the ocean and Brighton the portents of death pile up. He buys himself an expensive silk shirt the color of sand; it will match very well when he is shot and his corpse is bobbing in the shallow waves at Brighton Beach. He hurries to meet with Nathan, unaware that Nathan has been killed, and eats Nathan's favorite dish, solyanka. In my repeated readings of the story and of Nikolai's ingestion of the dead man's favorite dish, I came to think of this Russian cabbage soup or at least this particular serving of it, as the solyanka of death. He enters Brighton Beach and over the din and babble of commerce and street life, he hears the raspy voice of the beloved Russian balladeer Vysotsky, "They're hunting for wolves, the hunt is on!" Topol pauses to note that these words are threatening, then passes on.

And as Nikolai stands out on the beach waiting for Boris to return, he glimpses a familiar figure which he cannot distinguish: is it Nathan? Nathan, we know, is dead, and therefore Nikolai seems already suspended somewhere between the living and the dead. When the hitman comes, it is Nikolai's own past rising up to kill him in the form of an old comrade, not to be swayed from his assignment by sentimental reminders of good times together. He does his job and goes back into the restaurant to finish the vodka started by the dead man and to sup on the solyanka of death.

Nikolai's fate can be read as a particularly pessimistic take on what has happened in the former Soviet Union; attempts to change, to be reborn and pull oneself up out of the foul murk, whether it is the foul murk of corruption in the former USSR or the foul murk of the bay water where Johnny almost drowns, attempts to forget the brutal past and turn Westward, are foiled because the past will not rest but drags down the present and the future with it.

No doubt Edward Topol, dashing off this work in two weeks, would scoff at such weighty readings of his work. A well-known figure in Russian emigre cultural life in Philadelphia says quite accurately that Topol has a 'lyekhkoye pero,' i.e. he wields his quill with a light hand. That may be all that he aims for, but in doing so, he achieves much more.

NOTES

1. Solzhenitsyn, Alexander, The Gulag Archipelago, Harper & Row, 1973, p. 438.

2. Ibid., p. 503.

3. Ibid., p. 543.

4. Wilson, Edmund, A Window On Russia, Farrar, Strauss & Giroux, NYC, 192, p. 168.

Connections

Portuguese

MEDICAL TRANSLATION INTO PORTUGUESE/ ENGLISH: REFLECTIONS

Regina Alfarano
SINTRA

Key words: Medical sciences, biology, parmaceutical, health, translation, Portuguese, English.

Abstract: Medical translation seems to immediately refer to lengthy, heavy texts, tiresome reading, detailed descriptions of surgeries and drug directions. Of course all those hold a strong share in the so-called medical translation - which of course bears a diffuse interface with technical translation - but they are joined by a number of other kinds of texts, equally important and sometimes even more important. For a country like Brazil, expanding its scientific and technological stratum at a surprisingly quick pace, and yet lacking the infra-structure to keep up with such progress, there is at least a twofold angel when considering medical translation: to bring experiences into the country - therefore translate into Portuguese, and to send the results of such progress overseas - and therefore translate into other languages, mainly English.

Medical translation seems to immediately refer to lengthy, heavy texts, tiresome reading, detailed descriptions of surgeries and directions for drug treatments. This is the scientific aspect *stricto sensu,* which undoubtedly holds not only a strong share of the so-called *medical translation* - bearing a diffuse interface with *technical translation* - but its intrinsic core value. Equally important, though, and at times much more urging, are a number of other categories of texts, related not only to pharmaceutical industries, government health agencies and research agreements, but also fact sheets, leaflets, flyers, scientific committees, etc. etc. For a country like Brazil, expanding scientific and technological stratum at a surprisingly quick pace, and yet lacking the infra-structure to keep up with such progress, there is at least a twofold angle when considering medical translation: to bring experiences into the country - therefore translate into Portuguese, and to send the results of such progress overseas - and therefore translate into other languages, mainly English.

A very important reminder here is the fact that the scientific community reads English; so, how wide is the need for the translation we called *stricto sensu?* Statistics are rather sketchy, and the amazing 75% translation represents in whatever is read in Brazil - all fields included - will not get us anywhere close to an estimate. Evidences show translation has been of assistance to medical students at early stages; students doing medical sciences, biology, and related areas; and above all, to give support to the original texts.

On the other hand, while comfortable readers, the same members of this community cannot write English, and so feel hindered when wishing to send research to international periodicals, papers to conferences, and even crucial mail for protocols. Again, they will comfortably read once the translation has been done, and therefore comment on it; but not write themselves.

So, the translator is faced with at least a threefold task when dealing with medical texts in Brazil: 1- the concern with the domestic reader; 2 - the concern with the foreign reader; 3- the huge variety of fields under the umbrella "scientific" or even "medical".

The domestic reader is very heterogeneous, ranging from medical doctors, researchers and students to housewives, general population, blue-collar workers, native Brazilians in the Amazon area who may hardly understand Portuguese, and rural communities. The translator must have in mind, then, who the text is addressed to. Government offices, educational campaigns, AIDS prevention programs, and Malaria epidemiological trials run side by side with the pharmaceutical industry, the food and beverage industry, the cosmetic industry. One names it, one has the "umbrella" covering a very wide circle surrounding each and every one. Brazil is widely known for its hospitality - languages being the obvious core element - and importing technology is part of the whole structure. And importing technology does not only mean manuals, books, instructions. It also means training, learning, adapting, adjusting. When one cosmetic product is launched in Brazil, a long path has been trodden, dozens and dozens of clinical trials have been presented. When a new beverage or a new food product has to be registered, a similar process takes place. Each product will have its own implications and will have to give clear evidence of its safety - in its original language. Medical translations will, then, include as wide a variety of fields as one may think of - and at this very peculiar point in time, this is not measurable. Brazil has been experiencing a more open economy, quite a receptive approach to imported products (part of the hospitality element), and therefore, Brazilians have included imported deodorants and toothpaste, moisturizers and shampoos, etc. etc. in their basic personal products These have a straight correlation with medical texts, once adaptations are necessary due to climate, skin characteristics, regular diet, local regulations, and other specificities related to human responses.

So, as it concerns our domestic reader, the wide varieties of needs will determine the wide demands upon translators. One basic guideline will be that of quality control, which as far as medical texts are concerned, take up a different dimension: the quality control of texts being read practically in two languages by health professionals, the quality control of four- or six-handed texts once professionals are crucial to monitor the final version, the quality control of an objective, clean, surgically-revised style.

The translator's second major concern is his reader from foreign cultures. Translated texts cover papers to be presented at conferences, articles and research work sent to international publications, research reports, bi-national and multinational agreements for project development, work committees for product development and launching, flyers, folders, brochures, directions. Which means, medical translation is in the very core, permeating not only scientific communities but also industrial projects, commerce developments, government procedures -- that is, a citizen's day by day routine. To give a current example for Brazil: Mercosur. The idea that Portuguese and Spanish are the languages involved is quite misleading. As everybody knows, geography has played a different role, and geographical concepts have been redesigned. Consequently, the languages that express this geography have followed the same route. The English language spreads out and branches out faster, reaching different layers in society, and thus creating new niches. The Mercosur reaches out to the world, as does NAFTA, and we should not forget where the fastest-advancing technologies are being developed. Mercosur takes care, for example, of the hygiene requirements of containers for food transportation. The food contained is produced in Portuguese or Spanish as the top layer (often with more than one layer beneath), but the container may have been manufactured in another language. They have to be put together and meet the health requirements of that specific entry-country. The scenario may be routine to European countries, but quite a novelty to Brazil -- a monolingual, continental-sized country, whose protectionism jeopardized foreign trade for decades and decades. The moment the doors opened - at a very timely occasion of easier transportation, faster media means, computers and internet - the in/outflow proves to be enormous.

All this leads us to the wide-reaching umbrella of the so-called "medical translation" in Brazil today, where people from different backgrounds translate, "part-time" or even "parallel-time" translators also translate, as do professionals from each specific area, all of them running along the professional translator's lane. For this very reason, not only readers cannot have a profile, but translators themselves lack identity. The question of visibility is crucial and prevailing - readers' profile, translators' profile, medical and technical areas/interfaces. The much talked about invisibility of the translator is an integral part of the whole scenario, as it were. Our concern here is the translator *stricto sensu*, though. And there are cases when publishers call upon us for a second translation, or a revision (whenever feasible) of texts translated by area professionals. This is specifically the case I am going to use here to illustrate not only the intersections where translation and medical sciences meet, but also the point where each professional follows specific jobs according to specific skills and responsibilities. Such illustration will be carried out through hand-outs from excerpts of Outline of Fractures, by Adams and Hamblen, first translated in Brazil in 1984 by a group of physicians, with a second revised, enlarged edition in 1994, when the publisher decided a translator *stricto sensu* would do the job and then one or two physicians would revise it. My job was then quite unusual in the sense that I had the original text to my left, my translation on the screen, and the first translation into Portuguese on my right - at the same time revising the first translation as a whole and incorporating the updated information. The result was gratifying, although not lacking the usual problems: as the book was published in England, I could not have my final reading, and everybody knows how tricky floppies can be... or new formatting, etc.

So how does the translator face the challenge? What are the techniques to be resorted to? How does this multichallenged professional work?

Translation techniques are directly linked to the applicability of a theory, of many combined or alternating theories, or even of a permeating theory, which may navigate through and along different areas. Gentlzer (1993) quotes Ted Hughes: "if the 60's was a period which experienced a boom in literary translation, the 90's might be characterized as experiencing a boom in translation theory". Whether we navigate from macrostructure to microstructure; whether we agree with Mary Snell-Hornby that translation is the interaction of two cultures or not; whether we consider transferences and not targets; or else Lambert's descriptive research work from the 80's - where cultural complexities would define and describe the act of translating by emphasizing systematic research; whether we look into 'cannibalism' or deconstruction in translation -- we are for ever faced with the need for a more consistent, more outreaching approach to make translation VISIBLE. The translator's invisibility is undoubtedly the core issue. Segmental theories, regional theories or updated theories do not seem to solve the question. As Lawrence Venuti points out, in his book Rethinking Translation (1992), translation is "philosophical, but also political, engaged in questions of language, discourse, and subjectivity, while articulating their relations to cultural difference, ideological contradiction, and social conflict." And he goes on to state: "... it becomes essential to recognize that translation in its many aspects wields enormous power in the construction of national identities and hence can play an important geopolitical role."

If translation is political, then it implies power in its most intrinsic meaning. Is the translator powerful, then? Translators are extremely powerful, but have never invested that power of performing geopolitical activities, intercultural practices, the crucial bridge for the exchange of technologies, sciences, politics, and so on. The translator will have to conquer his space - and here, again, we mean the space *stricto sensu*. To be powerful, the translator must be visible, and that means, to take up his geopolitical role.

And how does that apply to scientific or medical translation? How can the translator be visible when facing a medical text?

Power (power here meaning professional authority, professional grounds, professional status conferred upon) is conveyed by competence and responsibility. The translator will then incorporate the respect he is conferred upon to the other professionals he works with - as peers. By exchanging knowledge, by true reciprocity and responsibility he will respected and therefore, be visible. The translated text does not fall from heaven, does not emerge from a cloud, is not produced automatically -- there is a professional, a name, a history, a background, a profession behind it.

Responsibility will again be a wide umbrella. For medical translation, it will mean from information accuracy and reliability to the language style: the style is surgical. Which means, accurate, objective and economical. Never excise more than necessary, never less than necessary; always save the patient (reader) from further trouble, always fight for his health and well being. After diagnosis, translation therapy is never clinical -- we do not "observe", we do not wait for "responses", we do not wait until the next call -- our diagnosis serves to show the best surgical technique -- we solve the problem as immediately as possible. We are "word surgeons".

The scientific texts, medical ones especially, can paradoxically be the easiest ones and the hardest ones. No better or worse solutions, no half-solutions, no re-writing, no changes in style, no text "melioration". The text is descriptive -- be it academic, didactical or informative. The observation period can be extended up to the moment the professional is consulted. Translators are much more protected here from what Allan Duff (1982) calls "the third language", or from what he shows as the different interpretations of the same painting. The risks the translator is exposed to can be as serious as medical malpractice, though. So, the reciprocity we mentioned previously is crucial , so as to build the bridge over which both the health professional and the translator will overcome difficulties and reach the desired end. Visibility goes for both sides, as does respect. These are two professionals, side by side. As word surgeons, we cannot play around with the knife. We must use it carefully, reliably. As cultural agents, we cannot play around with cultural identities. We must portray them as sensibly and as completely as we can. As responsible professionals, we cannot let translation be trespassed. We must try and fight for its place in the sun.

References:

ADAMS, John Crawford, HAMBLEN, David L. Outline of Fractures. 1992.

DUFF, Allan. The Third Language. 1981.

GENTLZER, Edwin. Contemporary Translation Theories. 1993.

VENUTI, Lawrence (ed). Rethinking Translation. 1992.

Connections

Public Advocacy, Legal and Regulatory Affairs

PROMOTING PROFESSIONALISM IN TRANSLATION

Chris Durban
SFT, MITI

Translators' associations in many countries face increasing pressure from members to raise the profession's profile with client industries and the general public. Only then, goes the argument, will users of translation services begin to realize how critical our input is to a host of cross-border activities — and how much more recognition/respect/ compensation we deserve.

The author agrees. But as the movement gathers momentum and PR efforts are launched, many translators might do well to take a long hard look at their own professional practices. Unless these are of the very highest order, publicity may be the last thing they want.

This paper examines some of the factors behind translators' difficulties in being perceived as — and perceiving themselves as — professionals: lack of contact with the day-to-day business world, a reluctance to address clients' implicit as well as explicit needs, and a failure to specialize. It looks at how these can be overcome by shifting our focus past the word on the page to the end-users of the documents we create. Finally, it suggests a number of concrete initiatives that translators' associations and individual translators alike might find useful.

(1) Public perception
(2) Client (+ end-user) perception
(3) Translators' perception of themselves

Since (3) determines our ability/enthusiasm for tackling (1) and (2), and since it is an issue far easier to address in a room full of translators, this paper will initially focus on **translators' perception of themselves** — how it shapes our professional lives, and how I believe it can be improved.

Hearing a business client dismiss a text with a scornful "oh, that's **just a translation**" irritates me no end — not because the client is unreasonable (which of course they often are), but because more often than not that client's scorn is **not** misplaced: for a demanding end-user, someone who is an expert in his/her field, many, many texts produced by translators simply do not pass muster. They cannot stand on their own as forceful, stylish renderings of the source text.

Or, perhaps more accurately, in view of the uneven quality of the source texts we work from: **regardless** of the weaknesses of the source text, many translations fall woefully short of **achieving the author's intended impact on end readers.** That is one definition of what makes a good quality translation (or source document, for that matter): the extent to which output/product/document satisfies the customer's needs. It is what I think we as translators should strive for, and it is most definitely what makes the type of client I want value my input — and pay a premium for my contribution.

Why do so many translations fall short of this mark?

Aside from purely linguistic factors, e.g., failure to master either source or target language (or both), there are many reasons.

The two I will mention today interest me because they both lie within our grasp. Each of us — individually, and through our professional associations — can, and in my view should, take a stand. Immediately. What better way to demonstrate our status as professionals than through a commitment to defending our clients' best interests? Because in the end, if we view translation as a career and profession, our clients' best interests are also our own.

Why, then, do so many translations fall short of the mark?

1. **Because many translators spread themselves too thin, accepting work that goes far beyond their subject knowledge, on the "muddle through" principle.**

Having got in over their heads, they then do a "professional" job in that they look up all the hard words, spend hours at the keyboard, sweat buckets, and generate a text in the target language in which virtually all the explicit bits of the original can be found in one form or another.

True, readers will be unlikely to mistake the resulting translation for a text written directly in the target language. But the explicit information will be in there.

Unfortunately, many of the implicit messages will not even be identified, much less conveyed. And as any professional will tell you, it is precisely those implicit messages that determine whether or not the text achieves its purpose.

No wonder the demanding client is disappointed.

2. **Because we too often fail to accept full responsibility for our work and its impact (or lack of impact).**

With the possible exception of translations done for information purposes only — for which there is certainly a market, at least for some language combinations, and where a swift rendering of text-on-the-page may be enough — translators are too willing to get out of the line of fire.

Time pressures are one factor. But there is also a lack of confidence in our own abilities, and in some cases sheer ignorance.

For whatever reason, we often fail to see the forest for the trees; worse yet, we neglect to draw our clients' attention to the forest — we, who, with our cross-cultural insights and skills are uniquely qualified to do so. Instead, we limit our input to processing strings of words on a given page.

To take just one example, last year a client asked us to translate an 80-page compilation of post-graduate courses in finance and stockmarket techniques on offer in France into English. The aim was to impress foreign markets with the range of expertise available in France, and make the pool of finance graduates available to French brokerage firms and banks more international. The text was not particularly difficult, and there was lots of repetition (course descriptions, fees, entry requirements, venues, etc.).

Yet for whom was this document intended? Potential students, one would assume. Significantly, almost without exception, the courses were **taught in French**. How could a student who could not understand even the syllabus in French hope to pass the exams? Why was an English language translation being commissioned in the first place? On the other hand, the target audience would probably appreciate a one or two-page introduction in English — and perhaps a few other languages as well — setting out immigration and registration formalities for foreign students coming to France.

Forgoing 80 pages of easy, well-paid translation to make that sort of suggestion is what I mean by "taking responsibility" for the impact of our work. On a smaller scale, it can be something as straightforward as recognizing that a metaphor is too culture-bound, and replacing it with another so that the person whose speech we are translating is understood — or, more importantly, does not offend.

Unfortunately, many translators do not have this reflex. And when, belatedly, they realize that the end document "does not work" — read: cannot hope to achieve the author's aim of explaining, convincing, selling, promoting a set of ideas to a target group — it is simply too late in the day to write up and defend a more effective alternative. So they shrug, send off their proposal (muttering "garbage in, garbage out"), perhaps mention how frustrating it all is to a colleague /spouse/drinking buddy, make out their invoice... and move on to the next text rolling off the fax. (They may, in passing, thank their lucky stars that their name doesn't appear anywhere on the text they've just sent out).

Of course by refusing to take responsibility for the impact their work has (or fails to have), they also forfeit any opportunity to take credit.

Both of these criticisms raise issues related to **professional integrity**. They are nothing new. But dealing with them sets the stage for an approach that can bring translators vastly more intellectual satisfaction, enhance our image immeasurably with those who use our services, and earn us a lot of money.

To restate what we all know: ours is a scattered profession. We often work on our own, with no reliable yardstick to measure our work against. In some cases we have chosen to work this way; in others, we simply find ourselves in what is a very isolated job. Which is neither good nor bad, and translators who accept this state of affairs end up feeling at home with it. But they can also end up losing touch with the rest of the world, and this has serious consequences for the quality of our work. If, through lack of feedback, we ourselves are not really certain of the value of our input, it goes without saying that we will not be able to make a strong case to our clients.

I am convinced, increasingly, that it is the knowledge — or at least suspicion, deep down inside —that the documents we produce may not stand up to scrutiny by "the experts" (subject-matter specialists, end-users) that saps our self-confidence. This, in turn, traps us in a low-status position that, if not of our own making, is at the very least maintained by the attitudes we adopt. How striking, for example, that even translators who are well-established and recognized, swamped with work from apparently satisfied clients, rarely raise their prices to "test the market". As if they feel, somehow, that they would be pushing their luck.

The solution I see starts with investigating ways to ensure that we are, in fact, providing a high-quality product. For example, there is no doubt in my mind that specialization brings a quantum leap in our ability to produce premium work. Equally, both specialists and generalists should actively and continuously seek confirmation of the value of their input from clients and colleagues — even if this means learning to swallow hard and accept some harsh criticism. Or acknowledging that practices we have endorsed for a decade or more are simply not effective.

My aim is not to blast translators for the inadequacies of their work. For one thing, I know that any number of practitioners will respond by citing cases in which clients have made truly unreasonable demands. But that is not what I am talking about.

What I would like to do is to invite translators to take a fresh look at how we work, to draw attention to differences with modern business practices in other fields, and to offer some suggestions as to how each of us can narrow the gap. If we don't, we remain part of the problem of "non-professionalism" among translators. And in any case, we will be unable to prepare for, contribute to and take full advantage of whatever marketing initiatives and public relations drives our associations do get off the ground.

Not all of my suggestions will be applicable to all contexts, but they have been very useful and effective for my colleague and myself. I offer them as possibilities, and would be interested in your reactions.

Things individual translators can do

1. Our aim must be to be secure enough — and knowledgeable enough — to sign the work we do, with pride.

If you are not prepared to sign your work before you send it out into the world, you have failed to address the two issues I mentioned at the beginning of this talk. And complaints about clients' ignorance begin to sound rather hollow.

Professional associations should actively promote this, and explain to clients and translators alike why it is in everyone's interest: the client gets a service provider prepared to state "the buck stops here"; the translator gets her name on high-quality documents that will serve as references for future jobs. If she has any doubts at all about her ability to deliver the goods, she may hesitate to take on a job that goes beyond her skills, or at least arrange for extra proofreading and checking with a colleague — a salutary move by any measure.

Of course, since you do not want to sign something that will reflect poorly on your skills through no fault of your own, it is in your interest to install safety nets. These will help educate clients as to what translation is, which, again, is in our interest.

To take one example, our English-language texts are regularly set by monolingual French printers. As a result, mistakes creep in even when we supply data on disk. The public — and potential clients — attribute these mistakes to the translator. To counter this, our standard estimate includes a clause specifying that we **must** read and sign final proofs before printing — failing which we apply a 25% surcharge for potential damage to our reputation. Figures like that draw clients' attention to our priorities, and help ensure that these are shared by all of us.

2. Devise and implement procedures for quality assurance.

The days of "translation-as-a-life-style-choice" (work the hours you like, at your own pace) are numbered — assuming this option ever existed as a viable alternative for anyone without independent means or a partner in secure employment.

If you aim (and claim) to deliver top-quality text, it is in the interest of you and your client that you **team up with another translator** — in the same room, or across the state, or on the other edge of the continent, using modern telecommunications. On a part-time or full-time basis, i.e., to proofread particularly important texts, or all texts prior to delivery. There need not be any exchange of money, and the correction involved can range from detailed suggestions on vocabulary and style to a scribbled "DQW" ("doesn't quite work") in the margin. ATA conferences would no doubt be an ideal place to find a just such a partner.

Note: the aim is not to boost capacity, but to read the other person's work and offer suggestions for improving it. Perhaps also to learn to step back from our own work and view it less emotionally, with a more critical eye. Good clients want outstanding quality; it is, in fact, their main criterion in selecting a translator. And "I ran it past my wife/husband/partner/neighbor" is **not** a professional approach to quality assurance. If you think someone who is not a professional writer (and we, as translators, **are**) will bring the same critical sense to a text as we would, you rank your own professional skills quite low.

3. Spruce up your documentation.

The aim is to present yourself to the world as a professional.

You should have business cards, letterhead, a one-page sheet of "terms of business", and standard estimate forms. You should also have bilingual fax cover sheets with international calling codes for your phone/fax number.

Documents in your source language — and you **should** have these if you work with direct clients, as a courtesy to those clients — must be impeccably written, for obvious reasons. Team up with your opposite number in that language to do this.

You should have a clean, crisp CV, or better yet, several, highlighting your strengths for different clients. Most important of all, you should have a stock of printed copies of source and target texts signed by yourself (see point 1); be sure to ask clients for these when you are particularly happy with something you've done. It may be hard to get multiple copies later on, and you will want to be able to send them out to new prospects. Except in extreme cases, it is unprofessional to ask clients to return such samples to you.

4. Hone oral skills in your source language.

Oral skills may not be so important to translators who churn out large volumes of "translations for information".

But to succeed in the negotiations on text form and content that I find critical to finalizing an English text for my French clients, I must be absolutely fluent in spoken French (and, of course, track events of interest to them daily).

Why? Because while the product I deliver is text on a page (or disk), my job goes beyond the written word; I explain, discuss, convince the authors of my source texts that one rendering in English is preferable to another.

If I can do this in their language, I gain points: I establish myself as a credible professional in one blow, even as I pay them the courtesy of using their language. Result: I can devote all of my energy to winning them over to my translation.

I would add that I am not aware of any translation programs in France or Britain that address this aspect of the translator's job, and that actually train young translators to be able to do it effectively. It is essential.

5. Cultivate social skills and exposure to the world our clients live and work in.

There are several reasons for this somewhat broad heading. To adapt texts successfully, we need access to information. To assess the options open to us, we need to know what purpose our text is supposed to serve. And to earn a good living, we need access to good clients. For all three, you need to be able to interact with the world out there.

Ideally, we must be as enthusiastic and knowledgeable about the subjects we translate as the clients who hire our services; we must produce work that meets (or surpasses) their professional standards, and present it in a form that demonstrates our awareness of their concerns. To do this, there is only one option: we must spend time with them.

I feel very strongly that **all** translators benefit from being dragged out of their caves and forced to interact in a business-like manner with potential clients. The meek and cowed emerge with a more upbeat view of themselves and the profession, and the aggressive and overbearing are given someone they can bounce their mighty egos off (and/or identify with). Everybody gains a better perspective of where they stand.

I'm talking, of course, about networking— one of the most positive exercises I can think of.

That said, it is important to avoid two traps that have been my experience of unstructured translation events: (1) group therapy in the form of collective moaning and groaning ("clients just don't understand/pay us enough/give us enough time...") or, even worse, (2) successful practitioners getting into a round of one-upsmanship, e.g., feigning amazement that other translators have difficulties.

You might get around these problems by arranging for a fairly prestigious "outsider" (i.e., non-translator) to join you.

This year, for example, several translators in my field in Paris have organized lunches with a high-profile guest speaker. The lunch format simplifies things: there is no paperwork aside from the initial mailing (and compiling a list of participants' names & addresses), since each individual pays for his/her own lunch and gets a receipt directly from the restaurant.

Significantly, our guest speakers talk to us not about translation, but about their jobs. It is up to us to connect their subject up with our concerns, which we do in a question and answer session at the end. There are two advantages: translators are brought out into a situation where they **must** interact with businesspeople — not to win a contract or a job, which gets many people tense, but simply as fellow businesspeople/adult members of the community. Many translators have lost (or never acquired) those skills. And assuming we have managed to ask intelligent questions (and otherwise conceal our eccentricities) the businesspeople we invite leave our lunches with an improved image of translators. In very concrete terms, they think of us when something language-related comes up.

Things professional associations might do

Translators often see their professional association as an interface with the business world that could/should/might promote their skills to business. Yet an interface works both ways: it is equally important that the ATA — or regional associations — act as a channel to bring the business world's skills and knowledge to members. As I have already mentioned, the focus should be not only on convincing business that it needs translators, but on educating translators in the ways of business.

1. Events for translators should regularly feature non-translators as speakers or contributors, with the association using its weight to bring in high-profile participants. This can be the first step in a "virtuous circle": such speakers can be expected to emphasize their concerns, forcing even those translators who are reluctant to re-examine their own business practices to do so. If Bill Gates, or the PR person for General Electric, or a representative of DATAR (the body that promotes French exports) or the head of the local Chamber of Commerce talks to a meeting of translators, participants feel buoyed by the attention. They become more aware of their (potential) role in the national economy.

At the same time, the speaker gains insight into what we do — or could do — for his or her industry. And with luck the media sits up and takes a second look at us, too.

2. Associations can build up their own roster of dynamic, articulate extroverts to represent the profession's many competent introverts at business gatherings.

I mean no criticism of the volunteers who do come forth to represent our profession. But I think some individuals are more effective than others in presenting our role to industry, and we should organize ourselves to make best use of everyone's skills.

Assuming a PR campaign works, the association will be contacted for speakers to represent its point of view. And when a panel discussion on imports from eastern Europe or a TV debate on immigration comes up, an academic or literary translator may not be the most forceful spokesperson.

The time to start thinking about this is now — before the PR campaign gets under way. ITI, ATA's sister organization in the UK, has taken the first steps towards compiling a database of potential spokespeople by issuing a call for volunteers in its bi-monthly newsletter. The response, I'm told, has been very encouraging.

3. Appoint a press officer/team to issue timely press releases and provide a swift, targeted response to each and every mention of language/translation /interpreting-related skills (or non-skills) in the press.

Members should be invited to participate by drawing the PO's attention to articles that need an official response. Aim: within 18 months, an ATA spokesperson should be quoted regularly in such articles at the outset. (I noticed in a recent Chronicle that Alex Gross has launched a "Media Alert" project, and assume it is something along these lines).

4. Prepare regular, articulate submissions to business pages/generalist op ed columns in the local and national press.

If necessary, commission these articles from translators with case studies, or journalists.

The aim must be to shift the focus of current (scant) press coverage of translation matters away from "can you top this?" silly translations in pidgin, to reports on cases in which successful translation has saved the day or gained a client competitive edge.

With these last two points, of course, we are getting into the public relations campaign itself. And here I would like to express a single strong opinion.

We translators often complain that clients use non-professionals to do their translation work, with paltry results. Let us not fall into the same trap: while some of the ideas I've mentioned could probably be implemented by members on an ad hoc or volunteer basis, they would have far more impact as part of a structured, coordinated and professional approach. That can only be achieved by hiring PR professionals with the experience and expertise to draw up and implement an overall public relations strategy.

A successful PR campaign hinges on creating an environment in which users of translation services know what a professional translation is and have a clear idea of how to go about getting one. Surely a prerequisite is that translators themselves be aware of what professional translations are, and be committed to producing just that.

THE OLD AND THE NEW CANADIAN TRANSLATOR CERTIFICATION EXAMINATIONS AND THE FEASIBILITY OF MUTUAL RECOGNITION OF CREDENTIALS BETWEEN THE MEMBERS OF THE REGIONAL CENTRE FOR NORTH AMERICA (RCNA) OF THE INTERNATIONAL FEDERATION OF TRANSLATORS (FIT).

Julien Marquis
Past President of the Canadian Translators and Interpreters Council (CTIC) and Certified Translator by ATIO

Keywords: Accreditation system, certification examination, reserved title, mutual recognition of accreditation, professional recognition, standard certification examination.

First Part

The Canadian Translator certification examination: the old and the new

1. The existing Translation Certification Examination.

The CTIC standard Certification examination is meant for experienced translators. It seeks to recognize competence, not aptitude or potential. In other words it is not an educational exercise resulting in a Diploma but rather a post-graduate recognition by peers of the level of competence and professionalism of a translator. CTIC does not provide copies to failing candidates so that they can improve their performance as it is the case in Translation Schools Programs etc...

In a nutshell, the examination consists of two mandatory texts of about 180 words each minimum, to be translated in two and half hours using all the dictionaries and documents that a candidate wishes to bring with him or her. In order for the candidates not to be tempted to bring their libraries with them, the themes of the texts are communicated to them at least two weeks before the exam (example: Politics and Geography or Economics and Environment).

The day they sit the exam, the candidates are given a number to write on each page of their examination papers when they have finished writing. Their papers are marked in a correction centre by well-seasoned certified translators and the marking procedure has been refined over the years. The marking criteria and scale are the following: no major message transfer error is allowed and a few minor errors related to language style and meaning are tolerated. A marking scale is prepared by a selected team of markers who mark a random selection of papers in order to average out the most common mistakes. Each paper is marked by two markers. The passing mark is 70% and the failing candidates are sent their result together with a report which is written in agreement by the two markers. The report gives some of the typical errors made by the candidate (copied from the candidate's own exam paper); it gives also a corrected version of the passage. Comments related to the main reasons for the failure are stated in the report.

2. CTIC National Correction Centre

The National Correction Centre brings together during a week-end about twenty certified translators with a number of years of experience in the profession and a good knowledge of the standard CTIC examination procedures. Their task is to mark all the exams from candidates across the country.

Before the Centre existed, teams of two markers used to be sent a number of exams to be corrected; then, they would all meet by a certain date, to decide a uniform marking scale and to proceed with the correction. The results were then forwarded to the CTIC member associations which, in turn, sent them to the candidates. Appeals were dealt with by a separate correction process.

Such a system was quite efficient, according to a 1989 report, and produced a less than 8% variation in marking between markers. There were nonetheless some objections about the flexibility of the system, the lack of opportunity to consult with fellow markers, the rigidity of the marking scale once established, and the lack of possibility to adjust for borderline cases or to discuss the impact of the importance of an error on the final meaning of a translation etc. The main reason for change, then, was a wish to accelerate the marking process and ensure greater uniformity. In order to achieve this, CTIC established the National Correction Centre in 1990.

The main difference between the old system and the new is that now the markers correct the exams simultaneously, in the same location, over a weekend. Two weeks prior to the meeting of the Correction Centre, the markers receive the translation texts that appeared on the examination as well as a few sample examination papers to preview. This is only a preparation stage, not a marking stage.

The purpose of the first meeting of the Correction Centre is to establish the marking scale for the exams, based on difficulties found in the texts and on errors detected in the previewed papers. All markers participate in the discussion about a draft marking scale for the first mandatory text prepared by the Centre's co-coordinators, for the English-French and French-English combination of languages. The same principles of correction apply to other language combinations, including native languages, except that the marking takes place in regional correction centres on the same day as the Centre.

Next, teams of two markers set to work on the actual correction of the exams. They have each an average of 16 to 18 papers to mark. The marking starts at 9 a.m. on Saturday morning. The markers have the opportunity to consult with each other and to adjust the marking scale, if required, once all have agreed. Breaks to readjust the marking scale are called, on an as-needed basis.

At the end of the first day, the markers hold a meeting to establish a marking scale for the second mandatory text, still following the same principles.

On Sunday the correction resumes, and at the end, the markers calculate the average mark obtained by the candidates. As the passing mark is 70%, the co-coordinators review any exams with marks between 65 and 70%. For all failing papers, they prepare a single report showing samples of the principal serious errors. The report is intended to give the unsuccessful candidates an idea of the reasons they failed so that they can take corrective action for the next time.

The markers are selected to represent the public, private and academic sectors as well as the member associations of CTIC, whenever possible. In each team of two markers, one marker should have already participated in a correction session.

This system ensures a correction process that is as objective as possible, and that allows, at the same time, the required flexibility for constant exchanges of views between markers. The systematic reviewing of all papers with a mark of between 65% and 70% ensures that borderline cases are duly assessed.

The appeal procedure is still in place for unsuccessful candidates who feel that they have been treated unfairly. In this case, a third marker, who has participated in the Centre's activities but was not on the team marking the unsuccessful candidate's examination, reviews the candidate's paper and produces a final assessment.

All candidates' claims or complaints are to be sent to their own provincial or territorial associations as CTIC does not deal directly with individual members. Suggestions for improving such procedures are welcome and should be sent to the CTIC member associations who will in turn present them to the Board of Examiners.

3. <u>The format of the future Certification examination</u>

The CTIC Translation Certification Exam format, as well as the marking process and standard criteria, has acquired greater credibility with the advent of legal recognition of the profession in several provinces of Canada.

CTIC, for instance, has signed protocols of agreement with the former Secretary of State Translation Bureau for Translators and Terminologists in Official Languages of Canada. These agreements constitute, in effect, de facto recognition by the Federal Government bureaucracy of the validity of CTIC Standard Certification criteria for Translators and Terminologists in both official languages of Canada.

Such an affirmation of the profession was long overdue. But CTIC did not stop there. Three years ago at the ATIO convention in Toronto, the executive officers in charge of the operations of the Board of Examiners at the time were told that, in spite of the reorganization of the CTIC standard exam, "... no professional translator writes his or her translations by hand any more and that there is more than translating skills to assess in order to certify a professional translator".

The subject was brought back to the Board Meeting table along with the whole question of assessing the skills of a professional translator for the purpose of bestowing on him or her a "reserved title". The member associations of CTIC agreed unanimously to review the whole question.

A committee was established, chaired by CTIC vice-President, Christian Després. The members of this committee were: André Sénécal, Mark Villeneuve and Jean Delisle. One year after, the committee produced a report which, basing itself on the past standard exam principles, proposed a brand new approach for testing the skills of professional translators. Such a new approach would take into account far more than professional aptitude for providing a professional translation service of an acceptable quality.

But first a short comment about the review methodology. No stone had been left unturned, and the Committee started with a hard, in-depth look at the way in which the Board of Examiners currently operates.

First they examined the profiles of the candidates applying for certification and they listed the following: the candidate with "academic credentials in translation studies", the "full-time translator in the public sector", the "free-lancer or private practising translator" and "the professional who starts a second career" (e.g. a lawyer, a doctor, a biologist).

Next, the Committee examined the question of the candidates' attitude towards the certification exam such as it now stands. The new perspective adopted for the review led to the Committee's conclusion that "there must be a link between the "daily practice" and the "royal sanction" of the reserved title. Such a conclusion prompted a review of the certification exam, together with its pros and cons for the profession in general, and the review of the daily practices of professional translators across the country. From this emerged a new set of principles to help further the review. The old assumption that ''the certification exam is meant to confirm the candidate's aptitude for translation" has become now " the certification exam is meant to confirm the candidate's competency in translation'', thus recognizing the capabilities of the certified translator to provide his or her customers with a quality product (the translation) that is ready for use.

Furthermore, as the certification exam is the key to the provincial and territorial association's bestowal of professional recognition (reserved title by law) on the translation practitioners, it should not be limited to measuring the candidate's competency in translation but should also assess the candidate's ability to ensure a proper level of quality control for the translations and his or her degree of awareness of ethical principles, etc.

To achieve such a comprehensive goal for the assessment of a professional translator's overall capacity to provide a professional service at an acceptable level of reliability, the Committee recommended a certification exam in three parts:

- Part I is meant to verify the candidate's translation competency (translation part);
- Part II is meant to verify the candidate's capacity to analyse his or her own practices to explain them (analytical part);
- Part III is meant to verify the candidate's awareness of the ethics of the profession.

The CTIC Board approved such a proposal, and CTIC is now conducting a test run of such a certification exam. The test run will result in the validation of the certification of the candidates. If the test run is conclusive all the current operations of the Board as explained above will be reviewed and adapted to the new exam format.

The development of the new certification exam for translators will also entail the use of a more comprehensive and integrated approach for the purpose of certifying professional translators who offer all the required guarantees of knowledge of translation skills as well as sound professional practices.

It is indeed essential for CTIC and other professional associations to enter the next century ready to face all the challenges issued already by new ethics' rules and the use of technology in the daily business of translation. For us in Canada, the review of our Certification exam was long overdue and we are now facing the choice whether or not to implement the new procedures on a full scale.

Second Part

About the feasibility of mutual recognition of credentials between the members of the Regional Centre for North America (RCNA) of FIT

1. Is it feasible to reach an agreement between ATA and CTIC on the mutual recognition of their accreditation processes?

First of all, let's say that I 'm not going to speak of the ATA accreditation process. I will simply assume that you are familiar with the criteria and the requirements to pass such an examination. I will rather touch on the basic criteria of discussion that should perhaps constitute the basis for a round of negotiations between the CTIC and the ATA.

For the last six years I have heard all kind of arguments related to the accreditation exams of ATA and of CTIC. From members of ATA I heard all the praise you can imagine about the ''Canadian system'' and from Canadian translators a mix of opinions on the various merits of the choice of passages used in the ATA system of accreditation.

I also took the extra step to interview Canadians translators who have taken and passed the ATA accreditation exam. These persons were already Certified by CTIC in Canada and they had different opinions about the ATA exam. The first criteria that seemed to matter in their view was related to the degree of difficulty of the texts proposed to the candidates and the second argument concerned the comparison of the texts with regard to the specific requirements of the French language for Canada.

Unfortunately, I did not have time to interview American translators who have taken the CTIC exam, although one of the persons with whom I spoke held dual citizenship.

But I will not go further in that direction; I would prefer to take another route to establish a rather more objective and rational approach to the problem and also one which is more encompassing in terms of global business expansion.

Last spring I was invited to speak to the New York Circle of Translators. The topic was: *The effects of NAFTA on the translation industry in North America.* I obviously did a little research on the matter and prepared my presentation. In spite of all my thinking, I did not find a satisfactory concluding punching line for the topic. I decided to deliver my presentation and then to finish summarizing after the questions and answers period .
The conclusion was, that if within NAFTA's framework we do not put forward professional standards for our services, some bureaucrat in Washington, Ottawa or Mexico City will do it for us and impose them on our profession.

But coming back to our subject, the same kind of objections and difficulties as mentioned above were raised when CTIC started to speak about the negotiations with the Secretary of State Department, Translation Bureau regarding the recognition of the CTIC Certification exam as equivalent to their accreditation test for free-lance translators.

The negotiations took place using a methodology based on a cross-exchange and a cross-marking of examinations papers written by candidates on both sides. As these papers were the property of CHIC and of the Translation Bureau, and furthermore, bore only code numbers, they were perfectly anonymous. The only thing that counted was the cross-marking session and, in the end, the marking criteria used on both sides. The results were so close and so comparable that there were only a few adjustments to make to reach an agreement; the most difficult step before the final signing of the agreement was to convince the provincial associations of the benefits of such an agreement for their membership.

It would make sense to me that, without reinventing the wheel, our respective professional organizations could sit around a table and decide to conduct more or less the same kind of negotiations. It could be as simple as that in this case, as the negotiations would not require any changes to both the current accreditation processes in Canada or the U.S..

That kind of work could be done by a bi-partite Committee ATA-CTIC or by a Special Committee within the activities of the RCNA. As a matter of fact, there is funds in the RCNA which could be used for such a purpose. Obviously at this point, the Mexican associations would probably like to join in the negotiations. That could be possible only provided they establish an accreditation system of their own which would be comparable to the ATA and CTIC processes.

An action plan could be developed along the same lines as the one used in the above-mentioned negotiations for such a project, specifying on what basic criteria the negotiations could be initiated. The resulting agreement would enable individual duly accredited or certified members of ATA and of CTIC member associations to be able to obtain full membership benefits and privileges and to justify the same standards of quality of service to their customers in the U.S. and in Canada. That would also give us a chance to ask our respective governments to be added to the Schedules on cross-border lists of professionals which are attached to the NAFTA agreement.

This proposal would not only make sense under the North America Free Trade Agreement but it also could be a concrete project to be carried out within the mandate of the Regional Centre for North America (RCNA) of FIT and could thus revive the focus of the now dormant RCNA.

I will leave the issue open to consider by ATA and CTIC and the other members of the RCNA.

ACCREDITATION OF MINORITY LANGUAGES: THE CASE OF FINNISH[1]

Anja Miller
Independent Translator

Keywords: accreditation, minority languages, ATA, FATA, FIT, Finland, Finnish, international cooperation

Abstract: The paper describes the step-by-step process started in 1989 and leading up to the 1994 establishment of Finnish as a new language in the American Translators Association (ATA) Accreditation Program. Positive regional, national, and international collaboration in the process involved both formal actions and informal support from the Finnish American Translators Association (FATA), Finnish Translators and Interpreters Association (FTIA) in Finland, the International Federation of Translators (FIT), and Northern California Translators Association (NCTA). Problems were encountered mostly in communication with universities and the ATA Accreditation Steering Committee.

1. THE PROBLEM

Recent discussions in "cyberspace" (CompuServe's FLEFO and Internet's LANTRA-L) have directed particular attention to the need to reform the ATA Accreditation Program. Central to the many issues surrounding accreditation in its present form is ATA's two-tier membership structure. In the interest of maintaining control over their organization's affairs in the hands of professionally qualified translators, voting membership has been based mainly on accreditation in the ten languages adopted over the years into the program. For those ATA members who translate languages excluded from the program, a peer-review process has been offered as an alternative road to active membership and to listing in the *ATA Translation Services Directory*. However, a persistent conception prevails that for translators of "minority" languages, peer review is not comparable to the formal qualification achieved through an accreditation examination under conditions equal in all languages.

Another vexing problem, at least for U.S.-based Finnish translators and many others, is that we are barred from taking the national certification exam in Finland, because by law candidates must be residents in that country. Yet, for instance when translating legal documents, a translator often must certify that s/he is "qualified." ATA accreditation still serves as one of the best vehicles for this purpose. Therefore, several of us Finnish translators and ATA members decided to pursue what we felt was our right to equal access to ATA services: an accreditation program in Finnish.

The following step-by-step account is intended to inform our colleagues and to help others contemplating similar action. It is the history of our work towards the establishment, in 1994, of Finnish as a new language within the ATA Accreditation Program. No doubt there are other ways; this is how we did it to achieve our goal and to advance our professional status.

2. FIRST STEPS WITHIN LANGUAGE GROUP

The process began at the founding meeting of the Finnish American Translators Association (FATA) in 1989. FATA brought together both professional and amateur translators at FinnFest, the annual national gathering of U.S. Finns. Though the main focus of those founders was to make more Finnish literature accessible to the American public, we managed to include the professional goal of national accreditation in FATA's original mission statement.[2] The next step was to form an Accreditation Committee consisting of ATA members Eija Pick Reed, Hilkka Schroderus-Rowe, and Anja Miller (this writer) as chair. The committee then designed and conducted a needs survey of all FATA members. The results indicated support for further efforts by the committee. We felt encouraged to pursue the goal through contacts with academia, our colleagues in Finland, and ATA.

3. UNIVERSITY CONTACTS

The FATA committee thought it natural that our need would be understood by faculty and top administrators at U.S. universities where Finnish is taught. We regarded the prospect of developing a translator test a potential boon to an all-too-often suffering academic program--in the search for grant funds, for instance. We approached both a large state university and a small struggling ethnic college, but the idea did not strike a responsive chord. They were not even interested in discussing the possibility of collaboration. So we had to turn elsewhere.

4. FIRST APPROACH TO ATA

Pursuing our mission at the ATA Annual Conference in Salt Lake City in 1991, I approached ATA officials (one of them a Finnish translator) with preliminary inquiries about adding Finnish to the existing languages in the Accreditation Program. To my considerable chagrin, their emphatic advice was to "forget about accreditation." Instead, they recommended peer review as the road to active, voting ATA membership. In keeping with our original FATA charge, however, our committee had no choice but to reject this discouraging advice and pursue accreditation in Finnish. The next step took us to Finland.

5. COUNTERPART IN FINLAND RESPONDS

Because language is a "living, growing thing," serious translators make an effort to return to the "place where it grows" as often as possible. On the next one of these trips to Finland I spoke with the International Affairs Coordinator of the Ministry of Education and visited the offices of the Finnish Translators and Interpreters Association (FTIA). I asked for the organization's help in developing a FATA exam fully comparable in rigor with both the ATA and the Finnish state exams. After all, FTIA, like ATA, is a member of FIT (International Federation of Translators) and thus committed to international cooperation. Later Eija Pick Reed, during her visit to Finland, further investigated the chances of former Finnish residents being allowed to take the state exam, or at least to use their old test passages for training. This did not succeed; however, the FTIA Board saw our plight and responded with a formal offer to assist. They appointed a highly qualified member of their Board of Directors, Lecturer of Translation Studies at Turku University, Simo Merne, to develop and administer the first examination under either ATA or FATA auspices. The FTIA President also pledged in writing (in English) that ATA Accreditation exam guidelines would be followed. The Ministry of Education awarded a grant for Dr. Merne's travel to the U.S. With these important documents in hand and costs covered, we were now sure that ATA would see the full international scope and seriousness of our effort and agree to collaborate.

6. ACCREDITATION COMMITTEE DROPS THE BALL

Later that same year (1992), at the ATA Annual Conference in San Diego, a delegation of Finnish and Swedish translators met with the then-Chair of the ATA Accreditation Committee to pursue the matter further. The meeting was more productive than previously: We explained our efforts to date and described the collaboration of FTIA, as well as our readiness to have an exam in the summer of 1993. Because the exam would be prepared, administered, and graded by experts from Finland unknown to any of us ATA members, those of us who would pass the test could then take over and continue the program entirely within ATA. We were told that if five ATA members requested an accreditation exam in any new language pair, the Accreditation Committee would cooperate. Even though no written guidelines were offered to back up this rule, we felt encouraged and thought we knew what to do next.

We then contacted all ATA members with Finnish as one of their listed languages. In December 1992, a total of six of them wrote or faxed the Accreditation Committee formally requesting that the exam be held in 1993. For several months, no answer came. We then repeatedly faxed and tried to call various members of the Accreditation Committee. Still no written response. On the phone, we were advised that the Committee was "working on the guidelines" and that these would be published in the Chronicle in due course. At that point, when it became clear that no definitive ATA response was forthcoming in time for our planned first accreditation exam, we decided to go to Plan B: The exam would be conducted under the auspices of FATA at the 1993 FinnFest in Thousand Oaks, California. At the same time we wanted to make sure that one day ATA would accept the exam. Failing that, those passing the exam would still have an accreditation from FATA.

7. FIRST FINNISH ACCREDITATION EXAM

The announcement of the exam was prepared, widely publicized, and sent using the language and procedures of similar ATA announcements. Registration for the exam was similarly controlled, and the exam itself conducted in strict adherence to ATA procedures. In order to prove the fact that ATA procedures were indeed followed to the letter, we invited the President of the nearby Southern California ATA chapter, SCATIA, to attend the exam as an impartial observer. She promised to come or send a representative, but neither showed up.

A total of nine people sat for the exam, four from English into Finnish and five in the opposite pair. As arranged, the examiner flew in from Finland with the support of the Finnish Ministry of Education. The candidates' test papers, consisting of three out of five passages translated in only one language direction by each candidate, were graded by an FTIA committee in Finland, again adhering to ATA grading guidelines. Two candidates in each language pair passed the test. An official FATA Certificate of Accreditation was prepared and sent to them. Douglas Robinson, one of these successful candidates, became FATA Chair and was therefore in a unique position to pursue the cause further with ATA.

8. PETITION IN PHILADELPHIA

Getting ready for the 1993 ATA Annual Conference in Philadelphia--our third in this process--we Finns knew we could not succeed alone without the help of the rest of the membership. Even though we had discovered that the ATA Accreditation Committee had finally come out with some guidelines,[3] they differed from those previously given to us and which we had followed to the letter. We still had not received any direct reply nor any explanation of the changes. Beyond this, we still felt that the full scope of the minority-languages issue had not been properly addressed by the ATA leadership. At this point, we had to reach out.

Even before the conference, my sympathetic fellow members of the Northern California Translators Association (NCTA), a major regional chapter of ATA, had taken action to support the cause. At the conference we circulated a petition on behalf of all "minority" languages requesting positive action by the ATA Board to bring more language pairs into the Accreditation Program. The response from the membership surpassed all expectations: Plenty of signatures came from both majority and minority language translators. Several ATA Board members as well as Steve Sachs, the ATA representative to FIT, assured us we would finally be heard.

It was a stormy conference, leading to a watershed change of ATA officers and directors, a new attitude of openness, and a new Accreditation Committee chair. In a sense, our cause served as a symbol for the changes that were necessary for ATA at this juncture.

9. SUCCESS AT LAST

What followed immediately after the conference was a complete documentation of our case to Maria Brown, Chair of the Accreditation Steering Committee, then some further correspondence, and positive feedback from the Board mainly through President Edith Losa and President-Elect Peter Krawutschke. Yet even this climate of receptiveness could not assure speedy resolution of the problem. There was miscommunication, illness of a key individual, and papers lost. Consequently, the first Board action was negative. All the while we managed to maintain our high hopes, but did not want to take any chances. Therefore we proceeded with preparations for the second FATA accreditation exam, this time at FinnFest 1994 in DeKalb, Illinois. Once again the examiner would be Dr. Merne, his travel from Finland funded this time by a grant from the Finlandia Foundation. Just before the July exam, the ATA Board, on the recommendation of the Accreditation Steering Committee, approved both this and the previous FATA exam as valid ATA accreditation exams.

Success was at hand, but there was a last wrinkle to be ironed out. Two of the four FATA members who had passed the exam had let their ATA membership lapse out of disillusionment with the process. The solution was a one-time waiver of the 60-day waiting period between joining ATA and taking the exam. The ATA Accreditation Committee also made good on its earlier promise to publish in the *Chronicle* their guidelines for the adoption of any new language combination into the program.[4]

At the 1994 ATA Annual Conference in Austin, there was jubilation and congratulations from colleagues who had been following our saga and supporting our cause. At the conference accreditation exam sitting, three candidates were translating into Finnish. For the first time, a Finnish Language Chair and two graders of Finnish exams participated in the Graders' meeting. From our more experienced colleagues in other languages, we learned much about the passage selection process and consistency in grading.

Even though the number of U.S.-based translators of Finnish is currently small, and likely to remain so for some time to come, there is a good deal of work in the language, which since the beginning of this year is one of the official languages of the European Union. In this work, translators continually need to establish and maintain professional standards. At present, as far as we are concerned, the ATA Accreditation Program provides the best vehicle for this purpose. We are proud to be part of the program and committed to its goals. We heartily recommend it to our colleagues working in other "minority" languages.

Finally, we hope that the foregoing proves that persistence -- or *sisu*, as we call it in Finnish -- does pay off!

10. REFERENCES

[1] The contributions of fellow panelist Douglas Robinson to this paper are gratefully acknowledged.

[2] Finnish American Translators Association. *Articles of Association.* 1989, 1.

[3] *The ATA Chronicle.* Vol. XXII, September 1993, 26.

[4] Implementing new language accreditation exams. *The ATA Chronicle.* Vol. XXIII, June 1994, 11 and Vol. XXIV, July 1995, 34.

PUBLIC PERCEPTION OF TRANSLATION WORLDWIDE

János Samu
East-West Concepts, Inc.

Keywords: Licensing translators, interpreters, work requirements, income, expenses, translator's name in books, royalty, teaching foreign languages, career value, public acceptance of translators.

Abstract: An overview is given about the various factors constituting the basis for public acceptance and recognition of the profession of translators and interpreters in 13 countries. In the countries examined the name of translators and interpreters rarely appears in daily newspapers and periodicals. In most countries anyone can call himself a translator or interpreter and work without a license as a freelancer. Companies hiring staff (in-house) translators are more likely to require a diploma or a degree in foreign languages. Very few countries have licnesing requirements even for court interpreters, but even courts bypass the degree or licensing requirements.

The income of a freelance translator or interpreter in many countries is lower than that of a plumber or electrician. Current earnings of translators and interpreters are tabulated and presented with per capita income data of the countries surveyed.

In some countries the name of the translator rarely appears along with the author's in a book that he/she has translated. There is, however, a growing trend to have it included as the translators' self-awareness is growing. The royalty issue is also discussed.

When on an assignment away from their home, translators and interpreters are being treated differently. The payment of travel time, meals, mileage, etc. is commonly accepted in some countries, while in English speaking countries it depends mostly on the interpreter's negotiating skills.

To determine translation as a career choice our profession was compared to 17 other professions in the examined countries. According to the survey conducted from among the examined professions, translators rank the highest in Albania and the lowest in Greece.

1. THE SCOPE OF RESEARCH

We wanted to see globally how we are doing. Therefore we surveyed translators and interpreters in Albania, Austria, Colombia, Germany, Greece, Hungary, Portugal, Russia, South Africa, Spain, Sweden, United Kingdom and the United States. According to a preliminary study 55 per cent of the translators do interpreting too. Considering this as the majority, we have decided to include them as well. Some information came from individual translators, some from translators associations, from publications and some from government sources. Considering the small size of the groups and control groups, this study can only be considered as a "pilot test", and possibly the foundation of a more extensive survey. The expected cost of a more extensive and more scientific test at this point was prohibitive to us. The results, however, can be considered representative for the various countries.

2. MEASURING PUBLIC PERCEPTION

This task was rather difficult, because the values and even the criteria could not be clearly defined. The following items were selected as values:

- Professional licensing requirement
- University or college degree requirement
- Income and compensation for services
- Being treated as a professional
- Translators' name in print
- The profession of translator as a career choice

The last value selected was the most difficult to measure, because it is very subjective. In order to receive usable data we surveyed only those, who are the most likely users of translations, that is people, who had at least high school diploma.

3. LICENSING REQUIREMENTS

It can be accepted as a general rule, based on the results obtained from all 13 countries that anybody may become a freelance translator or interpreter without any license. There are only a few restrictions or exceptions.

In Colombia the Ministry of Justice tests and certifies those who want to become "official translators". The test consists of a three-hour written and oral exam and if the translator passes it, he/she will be entitled to stamp his seal on all his "official translations".

In Spain an official appointment followed by an examination at the Oficina de Interpretación de Lenguas is required for sworn (official) translators and interpreters (traductores-intérpretes jurados). The requirements for court interpreting are rather confusing, because Article 231.5 of the Ley Orgánica del Poder Judicial states that in oral proceedings the Judge or the Court may authorize "any person who is acquainted with the language used" to act as an interpreter, after being duly sworn. Articles 601 and 657 of the Ley de Enjuiciamiento Civil, Articles 398, 440, 441, 442 and 520.2 of the Ley de Enjuiciamiento Criminal contradict this. In short, anybody with some knowledge of a foreign language can act as a translator or interpreter before the courts as long as the intervention of the Oficina de Interpretación de Lenguas is not required.

In Austria both conference and court interpreters, but not the translators, have to be licensed.

Hungary also has an interpreting license requirement for persons interpreting in state administrative matters and in courts. The law is quite vague stating that "...in state administrative matters only persons holding a valid Interpreter's License are permitted to interpret". Since this is the maximum definition, government officials accept many kind of documents. These are sometimes issued by universities, sometimes by the Foreign Language Testing Center, sometimes by foreign organizations, even by Town Clerks who do not speak a foreign language. They are accepted as long as they look official and bear the "Interpreter's License" designation in Hungarian.

In Russia the situation is similar to the economy - chaotic. According to the Constitution everyone has the right to testify in court in his native language (that is in any of the more than 100 languages of the former Soviet Union), but no court interpreting process has ever been devised and none is in existence. There is a draft of a bill today on the desk of the Minister of Justice, but the professionals do not have high hopes that it will ever be discussed amidst the thousand of other matters waiting to be handled by the Russian legislation. Currently the parties in the case find their own interpreters, agree on a fee and sign a contract. This frequently results, as one may expect, in unprofessional interpreting.

Germany has a unique requirement - court interpreters must obtain a clearance from the police (Certificate of Clean Records) before accepted for the case.

There is one more documentary requirement, namely proof of citizenship and residency permit in some of the countries. In Germany, Russia, Sweden and the United Kingdom the interpreters with assignment from military units or installations must be the nationals of these countries. The German and British government offices and courts require a residency permit from non-citizen interpreters.

In the federal departments of the United States, the Secretaries and their deputies use only the services of interpreters, who are U.S. citizens.

4. UNIVERSITY OR COLLEGE DEGREE REQUIREMENTS

Employers represent the major category where university or college degrees are required. Since most of them are not able to verify the proficiency of the applicant, they rely on documentary evidence. This is generally a college or university degree in one or more foreign languages. In these countries the majority of employers require a degree from in-house translators:

Spain	University Licenciado degree, although not necessarily in translation.
Greece	University degree or Certificate of Proficiency in the language, issued by the Ministry of Education - Foreign Language Department
Germany	University degrees in translation, technical translation or interpreting issued by any of the 8 German universities offering the subjects of translation or interpreting as a major.
Russia	Degree from an institution of higher education in translation or in a foreign language. At private companies there is a trend, however, to give preference to those, who have work experience in foreign countries.
Portugal	Some kind of academic degree.
Albania	Degree in a foreign language from Tirana University.

5. INCOME AND COMPENSATION FOR SERVICES

5.1. Translators

Translators living in different countries with different economic systems and under different taxation rules derive different incomes. The income figures are presented in Table 1. along with the latest available per capita income statistics (Ref. 1., 2., 3. and 5.) to give a more reliable comparison.

In several cases we obtained monthly income ranges from our sources, and the a mathematical mean of those was used for inclusion in the table. Monthly wages did not prove to be the best for comparison because in some countries, like in Germany and Austria the employees get salaries for 14 months (one month salary as Christmas bonus and the other month extra to enjoy their vacation). Thus annual incomes proved to be the best working numbers. To further simplify the task of comparison we converted the income figures expressed in the various national currencies into U.S. dollars using the official exchange rate of September 14, 1995.

It is interesting to note that in Russia most translations are sent to Moscow, or the Moscow translation companies will send translators and interpreters to the customers' sites even at large distances, like the Sakhalinsk Island in the Far East, Vorkuta or Norilsk in the Far North. The income of those who accept the assignment under the harsh climatic conditions is significantly higher than of those who do the translation in their cozy Moscow home or office. They are paid incentives and supplementary allowances.

Germany does not suffer from this centralization problem, but the unification of the country has its mark on the income of people. Translators in the Eastern part of the country, similarly to other employees or intellectual workers may earn 30 to 50 percent less than their counterparts in Western Germany.

The relatively high income of German translators can be contributed to two factors. The old tradition of translator training at nine (9) German universities and the acceptance of the profession by the Government and by the public. The work of the BdÜ is also commendable in gaining this public and official acceptance for the German colleagues.

When comparing the income of translators in Albania, Germany, South Africa and Hungary to the per capita national income, it shows that translators in these countries can afford a much higher standard of living in their home country than the average translator in the United States. It was interesting to find out that our colleagues in Albania, Austria, Germany, Hungary, Portugal, Russia, Spain, Sweden and the United Kingdom are very busy and they can hardly handle the amount of work. The economic transition of the former Soviet Union and of Eastern Europe and the formation of the European Union have a lot to do with this.

5.2. Interpreters

Due to the nature of their business their daily rates appeared to be more reliable in the evaluation process than any other value. This is what we presented for five categories of interpreters in Table 2. The highest rates are paid generally to conference or simultaneous interpreters, although there are exceptions. Court interpreters, probably because they are paid from the almost always tight government budgets, usually earn much less than consecutive or simultaneous interpreters. Looking at this from a professional point of view, the lower rates are not justified at all, because in court they actually do the same job as consecutive and sometimes simultaneous interpreters. On the other hand when the court system allows practically anyone to act as a court interpreter, like in Russia and South Africa, it is understandable that courts are hesitant to pay out more money for services rendered by non-professionals. Unfortunately due to the lack of standards,

Country	ALBANIA	AUSTRIA	COLOMBIA	GERMANY	GREECE	HUNGARY	PORTUGAL	RUSSIA	SOUTH AFRICA	SPAIN	SWEDEN	UK	USA
In-house translator	3,600	47,115	4,800	47,297	5,862	6,363	NO DATA	4,500	13,079	21,825	33,613	23,250	30,600
Freelance translator	9,200	33,000	5,400	84,459	4,187	19,696	1168	11,400	13,079	31,746	67,226	55,800	36,000
Per capita income	930	20,200	1,844	17,085	8,930	5,420	9,030	2,100	2,850	13,267	17,030	14,990	24,305

Table 1.: Annual income of translators compared to the per capita national income (US$).

Country/ type of interpreting	ALBANIA	AUSTRIA	COLOMBIA	GERMANY	GREECE	HUNGARY	PORTUGAL	RUSSIA	SOUTH AFRICA	SPAIN	SWEDEN	UK	USA
Conse-cutive	60 - 100	300 - 450	240	945 - 1,250	NO DATA	80 - 200	390	90 - 120	70 - 435	500	143	NO DATA	150 - 400
Conference	110 - 180	500 - 800	290	675 - 1,250	NO DATA	250 - 400	390	150 - 200	225 - 500	410	710	NO DATA	250 - 500
Escort	40 - 45	300 - 350	100	470 - 740	NO DATA	80 - 120	25 - 50	20 - 30	80	290	110 - 150	NO DATA	100 - 250
Protocol	120	1,000-1,200	380	490 - 680	NO DATA	250 - 400	NO DATA	200	80 - 444	560	710	NO DATA	250 - 450
Court	60	NO DATA	100	540	NO DATA	60 - 80	270	10 - 12	33	520	210	NO DATA	200 - 450

Table 2.: Current compensation level for interpreting assignments per day (US$)

Country Rating	*ALBANIA*	*AUSTRIA*	*COLOMBIA*	*GERMANY*	*GREECE*	*HUNGARY*	*PORTUGAL*	*RUSSIA*	*SOUTH AFRICA*	*SPAIN*	*SWEDEN*	*UK*	*USA*
1.	Lawyer	Lawyer	Physician	Nurse	Physician	Lawyer	Govt. off.	Govt. off.	Physician	Physician	Physician	Nurse	Comp. eng.
2.	Physician	Govt. off.	Lawyer	Physician	Lawyer	Physician	Biologist	Lawyer	Priest	Lawyer	Lawyer	Teacher	Physician
3.	Translator	Physician	Comp. eng.	Translator	Milit. off.	Comp. eng.	Lawyer	Physician	Lawyer	Artist	Comp. eng.	Physician	Lawyer
4.	Teacher	Teacher	Biologist	Carpenter	Govt. off.	Electrician	Physician	Comp. eng.	Comp. eng.	Teacher	Electrician	Priest	Milit. off.
5.	Biologist	Comp. eng.	Govt. off.	Teacher	Teacher	Translator	Artist	Translator	Govt. off.	Milit. off.	Teacher	Lawyer	Plumber
6.	Comp. eng.	Translator	Carpenter	Biologist	Biologist	Beautician	Comp. eng.	Milit. off.	Biologist	Govt. off.	Priest	Driver	Electrician
7.	Govt. off.	Carpenter	Teacher	Lawyer	Priest	Artist	Milit. off.	Beautician	Teacher	Priest	Biologist	Electrician	Carpenter
8.	Nurse	Biologist	Translator	Priest	Comp. eng.	Carpenter	Teacher	Plumber	Milit. off.	Biologist	Salesman	Govt. off.	Govt. off.
9.	Artist	Artist	Priest	Salesman	Electrician	Plumber	Translator	Nurse	Plumber	Translator	Translator	Plumber	Cook
10.	Electrician	Plumber	Salesman	Electrician	Plumber	Biologist	Nurse	Electrician	Electrician	Comp. eng.	Nurse	Milit. off.	Beautician
11.	Carpenter	Priest	Nurse	Plumber	Nurse	Govt. off.	Salesman	Salesman	Translator	Nurse	Cook	Biologist	Salesman.
12.	Milit. off.	Electrician	Artist	Milit. off.	Carpenter	Teacher	Electrician	Teacher	Nurse	Electrician	Carpenter	Carpenter	Translator
13.	Driver	Beautician	Cook	Comp. eng.	Beautician	Milit. off.	Cook	Carpenter	Artist	Plumber	Govt. off.	Translator	Priest
14.	Plumber	Milit. off.	Electrician	Govt. off.	Driver	Salesman	Priest	Driver	Salesman	Carpenter	Milit. off.	Cook	Teacher
15..	Beautician	Salesman	Driver	Driver	Salesman	Nurse	Carpenter	Biologist	Beautician	Driver	Beautician	Artist	Nurse
16.	Cook	Nurse	Plumber	Cook	Artist	Priest	Beautician	Cook	Cook	Cook	Plumber	Comp. eng.	Driver
17.	Salesman	Cook	Beautician	Beautician	Cook	Cook	Plumber	Artist	Carpenter	Beautician	Driver	Salesman	Biologist
18.	Priest	Driver	Milit. off.	Artist	Translator	Driver	Driver	Priest	Driver	Salesman	Artist	Beautician	Artist

Table 3. TRANSLATION AS A CAREER CHOICE IN 1995

professionals accepting court interpreting assignments also fall into the same basket and get paid accordingly. There are plans for certification of court interpreters and restructuring the system in several of these countries.

4. BEING TREATED AS A PROFESSIONAL

Interpreters are always on the go, because they have to render their services at different locations Sometimes also translators have to travel to the customer's site, similarly to the interpreters, therefore in this comparison we treated them as interpreters. They are entitled to certain extras in addition to the agreed-upon fee. Some extras are almost automatically given, but some of them require the negotiating skill of the interpreter/translator.

When the interpreter has his meals with the client or with the client's guests his tab will be picked up by the client. This has become general business practice. It is, however, not common yet to get paid for preparation time, travel time and being reimbursed for mileage. Safe-arrival phone calls, not to mention the first class train or the business class air trave appear to beong into the luxury category for many interpreters. I have personal experiences with companies that treat the interpreter the same way as they treat their guests when it comes to extras. On the other hand I have examples of companies, where the interpreter had to negotiate to have water on the table when he was interpreting. The treatment of interpreters and translators is a good measure of the degree of the respect and also sometimes of the disrespect of our profession. Table 3. gives an overview of this treatment.

5. TEACHING FOREIGN LANGUAGES

We can all agree that speaking a foreign language does not make one an interpreter or translator. The same applies to teachers too. A good translator or interpreter is not necessarily a good teacher of the language. Still we found that except South Africa and Spain interpreters and translators are teaching foreign languages. Our research revealed that in the majority of cases teaching a foreign language is a supplementary activity primarily for those, who start their career as a freelance translator and interpreter. Until they are established and have a steady flow of orders they supplement their income by working as teachers at language schools, continuing education courses and community colleges. Later they drop teaching from their work schedule. This is valid mostly for the United Kingdom and the United States.

In countries with strong traditions of foreign language teaching, significantly larger number of translators and interpreters are involved in this activity. In Albania, after several decades of isolation there is a tremendous interest in doing business with foreign countries. This resulted in an increased demand for teachers or persons capable of teaching and tutoring a foreign language. Thus 63% of the translators teach or tutor English, Italian, German and Greek. In Russia the demand is also very high and 51 % of the translators and interpreters are involved in teaching to supplement their income. Due to the uniqueness of the Hungarian language Hungarians always had a strong emphasis on learning foreign languages, so the recent change of the political system has not significantly increased the demand and the translators' involvement in teaching has remained practically unchanged.

Coun-try	ALB.	AUS.	COL..	GER.	GREE.	HUNG	PORT.	RUSS.	S. AFR..	SPAIN	SWE.	UK	USA
%	63	33	16	22	4	53	28	43	0	0	11	21	9

Table 4.: Percentage of translators and interpreters involved in teaching or tutoring foreign languages in 1995

Country / Extras	ALBANIA	AUSTRIA	COLOMBIA	GERMANY	GREECE	HUNGARY	PORTUGAL	RUSSIA	SOUTH AFRICA	SPAIN	SWEDEN	UK	USA
TRAVEL TIME	NO	YES	DOMESTIC: NO FOREIGN: YES, 50% OF FEE	FLAT FEE OR NEGO-TIATED % OF FEE	NO	YES, IF OVER 4 HOURS, CONSEC. RATE	NO	YES, IF OVER 4 HOURS, CONSEC. RATE	NO	YES, IF OVER 2 OR 3 HOURS; THEN 2/3 OF FEE	AIIC RATES, FOR EARLY MORNING OR LATE NIGHT TRAVEL	NEGOTIABLE	NEGO-TIABLE
MILEAGE OR KM FEE	NO	30 US CENTS PER KM	NO	36 US CENTS PER KM	NO	20 US CENTS PER KM	29 US CENTS PER KM	NEGO-TIABLE	NEGO-TIABLE	24 US CENTS PER KM	STANDARD DE-DUCTIBLE		28 US CENTS PER MILE
PARKING & TOLLS	NO	YES	NO	YES	NO	YES	NO	NO	YES	NO	NO		YES
TRAIN CLASS	SECOND	SECOND	NEGO-TIABLE	FIRST	SECOND	FIRST	SECOND	SECOND	SECOND	FIRST	FIRST	SECOND	COACH
PLANE CLASS	TOURIST OR ECONOMY	TOURIST OR ECONOMY	NEGO-TIABLE	BUSINESS	TOURIST OR ECONOMY	TOURIST OR ECONOMY	TOURIST OR ECONOMY	TOURIST OR ECONOMY	TOURIST OR ECONOMY	TOURIST OR FIRST	BUSINESS	TOURIST OR ECONOMY	TOURIST OR ECONOMY
TYPE OF HOTEL ROOM	SWB	SWB	SWB	SWB	SWB	SWB	SWB	SWB	SWB OR SHARE	SWB	SWB	SWB	SWB
PHONE	NO	YES	NO	NO	NO	NO	NO	NO	NO	NO	NO	NO	YES
MEALS & BEVERAGE	YES	YES	USUALLY	FLAT FEE ONLY	NO	YES	FLAT FEE ONLY	NO	YES	YES	PER DIEM SEK 1,500	YES	YES
OTHER EXPENSES	NONE	NONE	NONE	PREPA-RATION TIME	NONE	NONE	NONE	NONE	NEGO-TIABLE	ESP 7,000 PER DAY	NONE	NEGO-TIABLE	NONE

Table 5. TRAVEL RELATED EXTRAS FOR INTERPRETERS

6. NAMES IN PRINT

It is not fair to any of us if our name is excluded from the books we translate. The efforts required for a good translation frequently exceed the efforts of the author. Still publishers in some countries tend to forget about this simple recognition of the translator. Luckily there is a growing trend especially in Europe to protect the copyright of the translations as well. Any translator negotiating the translation of a book should demand the inclusion of his/her name and his/her decent royalty share in the contract. The table below shows the situation in the surveyed countries.

COUNTRY	TRANSLATOR'S NAME IN PRINT - ROYALTY STATUS	REMARK
ALBANIA	Appears in every book translated. Receives 1-3 % royalty.	New achievement
AUSTRIA	Always appears in every book translated. Royalty negotiable.	
COLOMBIA	Rarely printed in the books. Royalty rarely paid. Max. 1/2%.	
GERMANY	Always appears in every book translated. Low royalty is possible.	
GREECE	NO DATA	
HUNGARY	Always appears in every book translated along with the editor(s) of the book. The translation is copyrighted to the translator and the editors. Royalty negotiable. 1 to 2 % are usual.	Copyrighting to the translator and editor is a new rule.
PORTUGAL	Appears in every book translated. Royalty paid only to famous writers if they acts as translators.	
RUSSIA	Always appears in every book translated along with the editor(s). Very low royalty.	
SOUTH AFRICA	Both the inclusion of the translator's name and the royalty are negotiable.	
SPAIN	Always appears in every book translated. Royalty is 3 %.	
SWEDEN	Usually appears in the books translated. Royalty only in special cases.	
UNITED KINGDOM	Usually appears in the books translated. Royalty is negotiable.	
UNITED STATES	In literary translations it appears in the books, but from other publications it is frequently omitted. Royalty is subject to negotiation, but it is still a white elephant.	

Table 6.: The translator's name in the books translated and royalty to the translators

7. TRANSLATION AS A CAREER

How desirable is the profession of translators in the surveyed countries? How much recognition does this profession get? We have surveyed 318 people in 13 countries to see where would they place the profession of translators on the current scale of importance. It is important to stress the currency of importance, because political and economic changes affect the demand for certain professions. Nowadays, for example, economists and financial experts are in great demand in Eastern Europe. There were 28 translators among those surveyed and the rest came from all walks of life. The only condition was to have at least a high school diploma. Apart from the fact that the rating was somewhat subjective, the way the test designed, gave exactly what we expected, the *general* perception of our profession. Various other professions could be added to the list to manipulate the current rating of the translators, but we are not convinced that it would have been more representative than the current survey. Table 5. shows where we stand now.

8. CONCLUSION

The profession of translators is not among the most popular professions according to the result of the survey, especially in English speaking countries. (Ref. 4.) We need better public relations, and we need to learn from our German and Swedish colleagues. Maybe we do not have to start with an expensive TV commercial, but rather with talking to the guidance counsellor of the local high school. We need to educate children and adults about our profession. But first of all we must deliver an excellent product to the client to be able to charge a relatively high rate.

9. ACKNOWLEDGEMENTS

I wish to thank all those colleagues and others who have sacrificed their weekends, days or nights to answer my questions and to participate in the survey, frequently on a very short notice and sometimes without knowing me at all.

10. REFERENCES

1 *World Almanac*. (A Pharos Book, 1995 - New York)

2 Falus Szikra, Katalin. *The Position and Conditions of Intellectuals in Hungary* (Akadémiai Kiadó, Budapest 1995)

3 *KSH Statisztikai Közlemények* (1995) (Monthly Statistical Bulletin) 5

4 Bacon, Josephine. "*The Status of Interpreters in English-Speaking Countries: A Challenge for the Future.*" in VISTAS, ATA Proceedings 35th Annual Conference, Learned Information, Medford, NJ. p. 101-104.

5 *Changes Around the World.* Asiaweek, 38., 1995, p. 57.

MARKETING THE MEMBERSHIP: REGIONAL GROUP DIRECTORIES

Caitilin Walsh
Northwest Translators and Interpreters Society (NOTIS)

Keywords: Marketing, Regional Groups, ATA Chapters, Cooperating Groups, Membership Directories, Directories.

Abstract: This session will be a continuation of the group discussion on Marketing the Membership presented at the 1994 ATA Conference in Austin. The issues relating to the decision to publish a membership directory are similar to those relating to referral services and marketing the membership in general. Since nearly all Chapters and cooperating groups of the ATA (as well as the ATA itself) publish a membership services directory, we have much to learn from each other. This panel discussion will concentrate on the philosophical and practical aspects of their distribution and presentation.

The panel will be moderated by Caitilin Walsh, Vice President of the Northwest Translators and Interpreters Society (NOTIS): the other panel members are Tony Roder, President of the Northern California Translators Association (NCTA) and Ann Sherwin, Administrative Manager of the Carolina Association of Translators and Interpreters (CATI).

INTRODUCTION

A general discussion about Membership Services Directories will be followed by an in-depth look at the next generation of information distribution. The general discussion will focus on the following areas:

- Overall philosophy and purpose
- Distribution and financial considerations
- Structure and content

Overall Philosophy and Purpose

There are translators who don't know how to pay the rent. There are also customers who don't know how to find a translator or how to find the appropriate translator. In addition, there is a definite need for a networking tool for members.

Which of these needs should a directory try to meet? Should a directory market its members, presenting them in a professional light? Should it be considered a community service? Should a directory be a fund-raiser (as in the case of the ATA TSD)? For whom is the directory intended, and for what purpose? Is its main purpose to be a tool for the potential translation/interpretation user, or should it double as a networking and referral tool for members? The answers to these questions relate back to a group's decisions about its role as a marketing partner to its membership.

Distribution and Financial Considerations

For many groups, producing a membership services directory is the single largest expense incurred each year. Groups need to keep costs within their means, taking into account distribution policies and advertising. Some groups give the directory to their members, but sell it to businesses: other groups sell it to their members and give it to businesses: and still others either sell or give it away to everyone. What are the reasons behind these decisions?

Even the cost of mailing a directory can be prohibitive. First class can cost as much as the directory, but bulk mailing can be a headache and book rate slow. Should recipients remain on the mailing list indefinitely, or should they have to renew their request annually? How many members want a copy just to see how their own name looks in print?

Another way some regional groups defray costs is to carry advertising. Not all directories do this. How do groups that do this handle courting advertisers and decide on rates? Does it have an impact on a group's non-profit status?

Structure and Content

To be useful to both the client and the networking translator, a directory must contain two elements: a mechanism for finding a translator or interpreter meeting certain criteria, and pertinent information about that person. Many directories add supporting material concerning the profession and practical information on using the directory and translators, as well as information about the group itself.

Directories are generally indexed by language direction tables referring the user to member profiles. We will have a look at different approaches to indexing, member presentation, and making the directory user-friendly. We will look at different groups' directories, comparing approaches and key elements, including:

- Indexing: language pairs, specialty areas, name, zip code
- Presentation of the individual translator: information provided, use of space
- Client education: how to use the directory, how to use a translator
- Information about the regional group: mission, by-laws, other information
- Advertising: included or not, sizes, target audience

ELECTRONIC DIRECTORIES

Almost all chapters and cooperating groups of the ATA publish a directory of its members on an annual basis in a traditional printed booklet. Sometimes it consists of a simple typewritten copy photocopied and stapled together, at other times considerable creative effort goes into professional design and presentation. Whatever the format, all printed material suffers from the modern axiom that information is out of date as soon as it is printed. For this and other reasons, many groups dream of an electronic version of their directory.

Though seemingly situated at opposite ends of the spectrum, print and electronic directories are closely related. The greatest difference between them is one of form, and not content. One is a familiar object held in your hands, the other a fleeting series of strange electronic switches, disappearing as quickly as it appeared. It is impossible to expect many people to be as comfortable with a computer program as a printed book, with its centuries of tradition. That noted, the same inception and design considerations must be made for both print directories and their electronic cousins:

- *Ease of use.* If it is easy to use, it *will* be used, which is better for those being marketed.

- *Speed.* This ties in with ease of use: if it takes too long to obtain results, it will not be used. Two factors contribute to speed: design and size.

- It should be relatively *easy to update*; automating the process makes updating easier, which in turn makes the directory a more valuable tool.

- It should be relatively *inexpensive to design, manage and distribute*.

- *Equipment requirements.* The equipment needed for both the design and management should be accessible, both technically and financially.

There are currently two approaches to electronic directories: On-line, searchable databases and standalone published data. We will look at four different electronic databases existing today: The Dutch on-line database; the French Télétel, also on-line; and two standalone files: the FLEFO Translator's Database; and the *NOTIS InfoBase*.

The On-line Systems

While the dream may focus on a dial-up database, groups in the US are not as well placed as many of their foreign colleagues to exploit this option. Two factors contribute to this: several other countries have videotext systems installed. With BTX, Minitel and others, the infrastructure is ready and waiting for the service. Secondly, national professional groups, backed by government regulation of their professional status, are more likely to have the financial means to pursue their use. In these countries, on-line makes sense.

NGVGIDS

Fully described in an article in the *ATA Sci-Tech Newsletter*[1], the Dutch Translators Association (Nederlands Genootschap van Vertalers) has the oldest on-line system. The potential translation user dials up, searches a current database of all 1,100 NGV members, and retrieves information on the translator of their choice. The NGVBBS consists of a rented space on the country's largest and most sophisticated BBS, NEABBS. The interface is text-based, and all searching must be done while connected.

Since May 1991, the NGV member directory can be publicly accessed in the Netherlands and Belgium via the Dutch Videotext network. The Videotext system attracts more than 400,000 callers a month (mainly for the on-line national telephone, fax and postal code directories). Interested callers are free to use the database.

The NGV member directory was designed for use without manuals and includes seven help screens. The Dutch Videotext system takes care of many of the variables. It provides a bullet-proof environment, and updating becomes a matter of keeping the database current. The end user's platform is also not a concern on an on-line service. However, since the database runs on a high-end machine, some programming and financing were necessary to get the directory up and running. In addition, rented space means ongoing maintenance costs. For those interested in this model, the NGV has proposed it as a possible worldwide standard and made the structure available through FIT.

The French Minitel System

The Société Française des Traducteurs makes use of the French videotext system Télétel. The potential client uses his Minitel or Minitel emulation software to dial up (3615 SFT) and search a database. All searching must be done while connected, and is limited to people with access to the Télétel system. The service is not free, carrying a surcharge of FF 2.23/min. Access is widespread, with an installed base of 6.5 million Minitel terminals and 600,000 personal computers with emulation software in France alone. Connections are available via the Internet through different providers.[2]

Like the Dutch Videotext system, Télétel provides a familiar and very stable interface (in this case, graphical). The end user's platform is not a concern to the group placing the directory on-line, and emulation software for the Macintosh, Windows and UNIX is widely available as freeware. The same programming and monetary considerations as the Dutch solution come into play, although updating would be as automatic as maintaining the data.

Downloadable Electronic Directories

In the US, the telecommunications situation differs radically from most European nations. With no telephone monopoly, we have traded standardization for lower prices. Since there is no national videotext system in place, an on-line database would have many technological hurdles to overcome before it came to fruition.

In addition to the technical considerations, there are more human reasons to put a translator's database on the user's machine. This is a nation of convenience seekers. Just as a reference book on their shelf would be more likely to be consulted than a copy in a library, so would an icon on their desktop be more accessible than having to use their modem to dial up.

There are some caveats to this approach: a file that takes too long to download discourages people from acquiring or updating it. A large standalone file would also use valuable hard disk space and would be more likely to be deleted. Because the directory must be readable on the user's computer platform, a widely distributable format is needed.

The main advantage of this strategy is cost: data publishing generally requires less skill than database programming, putting it more within the reach of volunteers. A deregulated telecommunications system and profusion of Internet providers mean that there is almost certainly an inexpensive or even free space on a server for making the directory available for downloading.

FLEFO's TBD File

The Translators Section on FLEFO on CompuServe is host to a text-based translators' list, a cooperative venture of FLEFO and the American Translators Association, listing hundreds of translators the world over. The potential client downloads the file and uses their word processor to search the file. The 170+ page file is updated every few weeks by adding translator-written profiles submitted as text files. It is only accessible to CompuServe members. CompuServe is a commercial information service with currently over 2.5 million subscribers worldwide.

There are many advantages to this system: the file is easy and inexpensive to generate and to update. The FLEFO file is a compilation of many smaller texts, but could just as easily be generated from a database. Since the user uses his or her own Word Processor, he or she is (hopefully) familiar with the tool used to access it. Because it is a simple text file, cross-platform compatibility is assured, with the usual caveats associated with character set substitution. The file is also relatively small (in comparison to a database), though it could become unwieldy as it grows.

The principle drawback to a text file as a directory is its lack of sophistication. The file is only as effective as the user's word processor search function. Searches for a combination of language pair and specialty area are generally not possible in this format. This, combined with the difficulties of controlling the quality and format of the listings, makes it a less than ideal solution, though some would argue, better than nothing.

The NOTIS InfoBase

The Northwest Translators and Interpreters Society (NOTIS) publishes a quarterly electronic directory, the *NOTIS InfoBase* (affectionately known as NIB). The InfoBase is a help file using the Microsoft Windows Help Engine. The potential user obtains the file, either from the NOTIS BBS, CompuServe, the NOTIS WWW home page or by requesting a diskette, and consults the file on their computer. The InfoBase is limited to computers running Windows and to Macintosh computers with Microsoft Word installed.

The main advantage to this format is its ease of use. The Windows Help Engine was designed for beginners, and is familiar to most people running Windows. The look and feel of the InfoBase is similar to the print directory, which should make the print directory user feel more at home. The Windows Help compiler creates compact files: over 1MB of database information compresses to a 250K small file.

Cost was not a great issue in this system, since volunteers have done all the design and implementation. Updating is largely automated using non-programmer tools such as database queries, mail merge, and word processor macros, and could be taken over by a less-skilled volunteer. Since NOTIS has its own BBS, a distribution mechanism was already in place. Further distribution on CompuServe and the World Wide Web are donations.

There are still some drawbacks to this method: the data published is still static, but quarterly updates mean the information is still more up-to-date than the printed directory. The main disadvantage to this solution is that not all potential translation users are using the Microsoft Windows platform.

CONCLUSION

In spite of seemingly enthusiastic acceptance, electronic directories still remain a bit of a curiosity. Diskettes with the *NOTIS InfoBase* were sent with the NOTIS Print Directory in early 1995, along with return postcards asking for the user's preference for future updates' format. Out of 550 mailings, 33 postcards were returned, 11 requesting only the print directory, the other 22 requesting NIB in addition to the print directory. It appears the print directory is here to stay.

If we assume the NOTIS InfoBase experience is typical, it would seem most people are unwilling to give up on paper. For most regional groups, the traditional printed directory will remain the mode of choice for publishing information about their member's services. With time, the market should become more open to less traditional electronic media and technology will be more accessible. For now, insuring that a Directory will be used is of far greater service to the groups' members.

NOTES

[1] Bouthillier, Christoph. "Good News from the Netherlands," *ATA Sci-Tech Newsletter,* No. 3, Year 8, July 1992.
[2] For more information on Minitel access via the Internet, point your Web browser to: http://www.minitel.fr/.

Connections

Spanish

TRANSLATING THE ENGLISH PASSIVE VOICE INTO SPANISH

Marian B. Labrum
Brigham Young University

Keywords: English<=> Spanish Translation, Translator/Interpreter Training, English <=> Spanish Translation Theory and Practice.

Abstract: The so called passive voice found in English presents a challenge to translators who wish to translate into Spanish. This presentation proposes to examine the "passive voice" in English, as well as in Spanish. Once the grammatical functions have been discussed, source text problems and solutions will be explored. Special emphasis will be placed in explaining the appropriate options that the Spanish language presents to the translator.

1. INTRODUCTION

The so called passive voice found in English presents a challenge to translators who wish to translate into Spanish. Spanish prefers an active transitive verbal modality, and therefore favors the active voice. Although certain types of discourse in Spanish seem to employ a true passive construction, such as technical-scientific and newspaper writing, it is still frowned upon, as it will be discussed in this paper. Spanish offers several other options for translating the English passive voice.

2. WHAT IS THE PASSIVE VOICE

The term *passive* "has been used to cover a wide variety of constructions in many different languages" (Siewierska 2). Plainly speaking, the passive voice is a construction in which the following three elements must be present: 1) A *"to be"* verb, 2) a *"past participle"*, and 3) an agent preceded by the preposition *"by."*

A typical English, passive voice construction may look like this: "**The story *was written* *by John*.**" ***Was*** is the *"to be"* verb in the past tense, ***written*** is the "*past participle*" of the verb ***to write***, and ***John*** is the agent preceded by the preposition ***"by."***

In typical Spanish passive voice construction may look like this: "**El libro *fue escrito por* *Juan***," in which ***fue*** is the *"to be"* verb, ***escrito*** is the *"past participle"* of the verb ***escribir***, and the agent is ***Juan***, preceded by the preposition ***"por."*** This one-to-one correspondence between the English construction and the Spanish construction seems to indicate that the English passive voice may be rendered directly into the Spanish passive voice without much challenge. However, nothing can be farther from the truth. The English passive voice is a grammatical phenomenon to which linguists have devoted enormous time and interest. This is because; "we do not yet have a generally accepted, cross-linguistically valid definition of the [*English*] passive at our disposal that does justice to the relevant phenomena attested in the languages of the world." (Andersen 1, [italics added for clarity]).

The passive voice in Spanish is used much less frequently than the passive voice in English (Tarr, Centeno and Lloyd 168, Cantelli Dominicis and Reynolds 293). The Royal Academy of the Spanish Language states that: "nuestro idioma tiene marcada preferencia por la construcción activa" ["*our language shows a noticeable preference for the active construction*"](Vázquez-Ayora 107-108). In fact, speakers of Spanish and English both "have at their disposal two voices or ways to indicate the relation of the subject of the verb to the action expressed by the verb" (Cantelli Dominicis and Reynolds 292). These two ways are:

1) True passive voice and,

2) Active voice

In a sentence like, **The book was written by John**, a Spanish speaker would have the choice of translating the idea as:

1) "*El libro fue escrito por Juan*" - a true passive construction, or

2) "*Juan escribió el libro*," which is an active construction.

Spanish prefers the active voice over the passive voice, and the overuse of the passive voice in Spanish is regarded as a stylistic defect (Cantelli Dominicis and Reynolds 293). This is because the excessive use of the passive voice in Spanish sounds foreign. English and Spanish both construct the passive voice in the same manner. However, while the passive sentence could be constructed as an **active** construction in both Spanish and English, Spanish prefers the active construction over the passive construction. English does not use the active voice nearly as often as Spanish (Cantelli Dominicis and Reynolds 293).

In this paper, we will limit our discussion to the two most prevalent types of English passive construction and their translation into Spanish. The two English constructions we will look at are:

1) The Personal Passive,

2) Impersonal Passive

The first English construction, the **Personal Passive**, is represented by a sentence like:

"The book was written by Cervantes."

The second English construction, the **Impersonal Passive,** is represented by a sentence like:

"The book was written in fewer than ten day."

Let us look at the challenge presented by these two constructions when translating English into Spanish. The first sentence to be translated from English into Spanish is "**The book was written by Cervantes.**" A translator working into Spanish would have two choices: the passive voice or the active voice. If he chooses to translate the idea in the ***passive voice***, a good translation would be "**El libro fue escrito por Cervantes.**" The choice of the passive construction is of some concern because it lends a "foreign" flavor to the sentence. This is because Spanish construction prefers an active transitive view. Thus, the translator may chose to translate the same idea as "**Cervantes escribió el libro.**", which is an ***active voice*** construction. Choosing the active voice would eliminate introducing a structure that sounds "foreign" in Spanish, and it would render the same meaning.

Not all usage of the English passive sentences are so clear cut. Let's look at the sentence corresponding to the second category, the ***impersonal passive***, as in the sentence, "**The book was written in fewer than ten days.**" The agent or "doer" of the action, (**by Cervantes**), is no longer expressed. This statement made outside of the context of the surrounding text, does not tell us who the agent is. Who wrote the book in fewer than ten days?

This type of construction is called an ***agentless passive construction*** (Decles, Guentcheva, Shaumyan 11), and is usually referred to as simply the ***impersonal passive*** (Siewierska 93). This type of construction could be translated into Spanish as, "**El libro fue escrito en menos de diez días**," but Spanish has a more regular construction for rendering the passive voice, when the agent has been deleted: the ***se pasivo*** form. Using the ***se pasivo*** form, our translated sentence becomes, "**El libro se escribió en menos de diez días.**"

Stockwell, Bowen and Martin state that:

> "Although the passive in English is frequently and highly productive (specially in more formal style and in writing), the passive in Spanish is relatively rare and less productive for . . . the fact that the sense of the passive is very regularly conveyed by a sentence type that results from agent deletion. . ." (234).

The sentence "**The book was written in fewer than ten days**" can also be translated into Spanish as "**Escribieron el libro en menos de diez días.**" This Spanish construction, known as the "***third person plural*** " or "***they construction***" is also used to translate an idea when the agent has been deleted. "**El libro se escribió en menos de diez días**", and/or "**Escribieron el libro en menos de diez días**", are legitimate translation choices in Spanish, and are preferable over the Spanish passive construction, "**El libro fue escrito en menos de diez días**". Even though this construction does exist in Spanish, Spanish prefers the ***"third person plural"*** **[tercera persona plural]** and the "***se***" construction over an agentless passive voice.

In other words: a translator of English into Spanish needs to look at the English passive voice with great care. First of all, a decision must be made whether to convey the meaning using a passive voice in Spanish, and run the risk of sounding foreign, or use an active voice. This choice will be made on the basis of:

(1) whether an agent is present in the English construction;

(2) whether the nature of the text being translated is one that sanctions the use of the passive voice in Spanish; and

(3) whether the Spanish passive voice has already been used frequently in the target text.

If the agent in the English passive voice has been omitted, the translator must translate the idea using either the third person plural (**Escribieron**, etc.) or the "**se pasivo**" construction, rather than employ a true Spanish passive construction.

Now, let's look at several Spanish syntactic translations of the English sentence "**The child was bitten by a dog.**" This is the English passive voice with agent construction presented in category one.

*Spanish translation using passive construc*tion	**El niño fue mordido por un perro.**
Spanish translation using active voice	**Un perro mordió al niño.**
English passive without an agent	**The child was bitten.**
Spanish translation with "they"	**Mordieron al niño.**
Spanish translation with "se pasivo"	**Se mordió el niño. ***
Spanish translation with "se pasivo" and personal "a"	**Se mordió al niño. ***

The last two constructions, marked with an asterisk, correspond to two serious mistakes that a translator could easily make, while applying the Spanish "**se pasivo**" option.

In the Spanish sentence "**Se mordió el niño**", we are actually saying *The child bit himself.* In "**Se mordió al niño**", we are saying that *The child was bitten* or *Someone bit the child.* The sentence "**Se mordió el niño**" is clearly an error in meaning even though the construction is correct.

On the other hand, "**Se mordió al niño**" can be misunderstood because it is ambiguous. The choice of the correct construction to be used in translating English into Spanish, where the passive voice shows agent deletion, will depend, in part, on the context of the surrounding source text. The "***se pasivo***" can become a reflexive "***se***" and the prepositional "***a***" may render the idea ambiguous.

There is another peculiarity that must be given consideration when translating the English passive voice with agent deleted into Spanish. Let's look at the following pairs of original English text and its corresponding Spanish translation.

"The books were written in fewer than ten days."

"Los libros se escribieron en menos de diez días."

and

"The prisoners were killed at dawn."

"Se mató a los prisioneros al amanecer."

In the first pair, the Spanish translation of the English sentence follows the dictated construction: "Se pasivo" + verb in third person plural ***escribieron***, which agrees in number with ***los libros***. Because of the construction chosen to translate the English passive voice, the Spanish sentence can mean "***The books were written in fewer than ten days***" or "***The books wrote themselves (or on themselves) in fewer than ten days.***" However, the choice of a meaning-based translation rather than a construction based translation would be made on the basis of the context of the document.

In the second pair the "***se pasivo***" is followed by a verb in third person singular ***mató***, even though it does not seem to agree in number with ***los prisioneros***. This construction appears to contradict the "***se***" construction used in the first example. Nevertheless, "**Se mató a los prisioneros al amanecer**" is correct both in meaning and form. Had we said "**Se mataron los prisioneros al amanecer**", we would have said "**The prisoners killed each other (or themselves) at dawn**", rather than, "**The prisoners were killed at dawn**." The main difference between the first and second pair of examples, is that in the first pair we are dealing with ***objects*** (books) while in the second one we are dealing with ***people*** (prisoners). Spanish requires a personal "***a***" when dealing with people. If the personal "***a***" is left out, the people take an inanimate characteristic and become "***things***". When using "***a***" with people the Spanish "***se pasivo***" construction will always be third person singular, and it will be followed by a personalizing "***a***", unless the meaning of the text indicates a dehumanizing intention, in which case the "***a***" is omitted and the verb agrees in number.

3. SUMMARY

A translator working from English into Spanish must carefully analyze the two English passive construction and determine the best option possible to present the same idea or feeling in Spanish. Blindly applying the correct construction may result in an ambiguous and/or erroneous statement. **Not** paying attention to grammatical peculiarities of the Spanish language (target language), or **not** being intimately aware of the different nuances in the grammatical constructions of the source and target languages, may result in an **inaccurate** translation. Translating the English passive voice into Spanish using the same grammatical components may not be appropriate. The purpose and/or meaning of the surrounding English text may not warrant the use of the passive voice in Spanish.

The nature of the Spanish language prefers the active voice over the passive voice, and provides several grammatical constructions to replace the different English passive forms.

The translation of the English passive voice into Spanish should be approached with care. Grammar rules should not be applied without careful consideration to meaning and intent. Although the passive voice exists in Spanish, it is used with much less frequency than in English. Spanish is a transitive oriented language, and the English passive voice construction is best translated with several other options rather than using a true Spanish passive voice.

4. REFERENCES

Andersen, Paul Kent. A New Look at the Passive. Frankfurt am Main; New York: P. Lang, c1991.

Canteli Dominicis, M. and Reynolds John J. Repase y escriba. New York; Chichester; Brisbane; Toronto; Singapore: John Wiley & Sons, 1987.

Desclés, J. P. Theoretical Aspects of Passivization in the Framework of Applicative Grammar. Amsterdam; Philadelphia: J. Benjamins Pub. Co., 1985.

Siewierska, Anna. The Passive: A Comparative Linguistic Analysis / Anna Siewierska. London; Dover; N.H. : Croom Helm, c1984.

Stockwell, Robert P., Bowen, J. Donald and Martin, John W. The Grammatical Structure of English and Spanish. Chicago: University of Chicago Press, 1965.

Tarr, F.C., Centeno, A. and Lloyd, P. A Graded Spanish Review Grammar. EnglewoodCliffs, New Jersey: Prentice Hall, 1991.

Vázquez-Ayora, Gerardo. Introducción a la traductología. Washington, D.C.: Gerorgetown University Press, 1977.

LEGAL TRANSLATION FROM SPANISH: A LAWYER'S PERSPECTIVE

Thomas L. West
Atlanta Translation, Inc.

Keywords: Spanish, Legal Translation, False Cognates

Abstract: Translators who translate contracts and other corporate documents from Spanish into English should have an idea of how such documents sound when they are drafted by English-speaking attorneys. Unfortunately, however, most resources for legal translators focus only on the litigation side of law practice. The purpose of this presentation is to familiarize translators with legal English in the context of contracts and corporate documents, so that they can use this kind of language when they translate Spanish contracts into English. In Part One, we will examine common provisions of a contract governed by Georgia law. In Part Two, we will consider some provisions of contracts in Spanish that are commonly mistranslated. In particular, we will focus on false cognates, the mistranslation of which could seriously mislead or confuse an American lawyer.

1. COMMON PROVISIONS OF A CONTRACT IN THE UNITED STATES

1.1 Typical Opening Sentence of a Contract

This agreement[1] (the "Agreement")[2] is made and entered into[3] as of this 25th day of August, 1995, by and between[4] ABC, Inc. ("ABC"), a Georgia corporation,[5] and XYZ, Inc. ("XYZ"), a Delaware corporation.

1) The words "contract" and "agreement" mean the same thing, but lawyers tend to use "agreement" to refer to a specific arrangement between parties: a "loan agreement" or a "license agreement."

2) American lawyers often define terms in a contract by beginning words with capital letters. Thus, we would write: John Doe (the "Purchaser") agrees to purchase from Fulano de Tal (the "Seller") the following goods (the "Goods").

3) The contract will be "by and <u>between</u>" the parties if only two parties are involved. It will be "by and <u>among</u>" if more than two are involved. Note that one word --<u>entre</u> -- is used to translate both <u>between</u> and <u>among</u> in Spanish.

4) "Made and entered to" is basically a doublet, <u>i.e.</u>, saying the same thing twice. Note that "celebrar un contrato" in American English is to "enter into" a contract, NOT to "conclude" a contract, as many dictionaries would have you believe.

5) "A Georgia corporation" does NOT mean "a corporation located in Georgia," but "a corporation incorporated under the laws of the State of Georgia." Corporations in the United States are incorporated under state law; there is no nationwide statute corresponding to the *Ley de Sociedades Anónimas* in Spain or the *Lev General de Sociedades Mercantiles* in Mexico. This means that it is possible for a corporation to be incorporated under the laws of one state, but have its headquarters in a different state.

1.2 Preamble. This part of the contract gives the background ("antecedentes") of the transaction.

WHEREAS,[1] Diego desires to disclose to Fulano certain information relating to the design and operation of a cellular mobile communications system in Spain, some of which information Diego deems to be confidential;

NOW, THEREFORE,[2] in consideration[3] of the foregoing premises and the mutual covenants and promises herein contained, the parties hereto[4] agree as follows.

1) "Whereas" is "por cuanto." The entire clause might be translated as follows: *Por cuanto, Diego desea desvelar a Fulano determinada información relativa al diseño y explotación en España de un sistema de comunicaciones móbiles celulares, parte de cuya información Diego considera confidencial.*

2) "Now therefore" is "por tanto." The entire clause might be translated as follows: *Por tanto, en contraprestación de las anteriores consideraciones y de los compromisos y promesas recíprocos formulados en el presente documento, las partes contratantes convienen en lo siguiente.*

3) Consideration, "contraprestación," is a term of art. To constitute consideration, a performance or a return promise must be bargained for. A performance or return promise is bargained for if it is sought by the promisor in exchange for his promise and is given by the promisee in exchange for that promise. Restatement (Second) of Contracts § 71.

4) Note that in English, it's the parties <u>to</u> the contract, whereas in Spanish, it's "las partes <u>del</u> contrato."

1.3 Representations and Warranties. A party affirms or guarantees that a certain state of affairs exists or will exist. If he is lying, he can be sued for breach of contract.

Seller represents[1] that the gross weekly receipts of the business averaged $100 per week, and hereby warrants that the gross receipts for the period from August 31, 1996 to September 30, 1996 will aggregate at least $400.[2]

1) "Representations" are "declaraciones." "To represent" is thus "declarar" or "manifestar." "To warrant" is "garantizar." Lawyers refer to this section of a contract colloquially as the "reps and warranties."

2) The entire sentence might be translated as follows: *El Vendedor declara que el promedio de las entradas brutas semanales fue de $100, y por el presente garantiza que las entradas brutas del período que abarca desde el 31 de agosto de 1996 hasta el 30 de setiembre*

de 1996 ascenderán a un total de $400 como mínimo.

1.4 Covenants. There are two types of covenants: affirmative covenants, which are promises to do something, and negative covenants, which are promises not to do something.

Affirmative covenant: Party A covenants to return the Property upon termination hereof in good condition. *"La parte A se compromete a devolver el Bien al finalizar la vigencia del contrato en buen estado."* Note that it is best to use the word "covenant" to translate "comprometerse a." By all means stay away from "is obliged to," which would never appear in a contract in English.

Negative Covenant. Examples would be a covenant not to compete or a covenant not to transfer shares (*compromiso de no transmisión de las acciones de la sociedad*).

1.5 Indemnification.

Seller[1] shall indemnify Buyer for all damage that Buyer or third parties, including but not limited to Buyer's employees, may incur as a result of work performed or material delivered under this Agreement.

1) Although it sounds strange, contracts often admit the word "the" before defined terms.

2) This sentence might be translated as follows: *El Vendedor mantendrá indemne al Comprador de todo daño que pudiese causar al Comprador o a terceros, incluyendo pero sin limitarse a los empleados del Comprador, como consecuencia de trabajos realizados o material entregado en virtud del Contrato.*

1.6 Miscellaneous. This section may be called "generalidades" in Spanish, but do not translate it as "generalities."

Governing Law. *This contract shall be governed by and construed in accordance with the laws of the State of Georgia,[1] except for the choice of law provisions thereof.[2]*

1) The entire sentence might be translated as follows: *Este contrato se interpretará conforme a las leyes del Estado de Georgia y estará sujeto a dichas leyes sin tomar en consideración las disposiciones de las mismas relativas al conflicto de leyes.*

2) The laws of each state contain "choice of law" provisions (also called "conflict of laws" provisions) that determine which state's law governs if, for example, a resident of Tennessee runs a red light in Georgia and hits a resident of Kentucky. By excluding those provisions, the parties to the contract are agreeing that Georgia law will govern even if its choice of law provisions would make another state's law applicable to the contract.

Severability.[1] *If any of the clauses of this Agreement are declared null and void, the remaining clauses shall continue to be effective as if the Agreement had been executed wtihout the invalid portion.*

1) This means that the contract can be "severed" or "separated." If any term is declared invalid, it will be severed, and the rest of the contract will remain valid.

2) The entire sentence might be translated as follows: *Si cualquiera de las cláusulas de este contrato fuera declarada nula o inválida, la validez de las cláusulas restantes quedará en vigor como si el contrato hubiera sido suscrito sin la porción inválida.*

Merger[1] Clause[2]. *All prior contracts or agreements, whether written or oral,*[3] *between the parties hereto are hereby merged into this Agreement, and this Agreement is the sole, complete and entire agreement between said parties.*[4]

1) Merger here does not refer to the "merger of two companies" (fusión), but to the fact that all prior agreements are "merged" into the contract (quedan integrados al contrato).

2) The "merger clause" is sometimes labeled "Entire Agreement."

3) Note that in American law, we refer to "prior oral agreements," not "verbal" agreements.

4) The entire sentence might be translated as follows: *Todo contrato o acuerdo que existiere entre las partes, ya sea verbal o por escrito, queda integrado a este Contrato, que pasa a ser el único acuerdo vigente entre las partes, completo y cabal.*

Assignment. *This Agreement shall not be assigned, in whole or in part,*[1] *without the prior written consent of the other party.*[2]

1) "In whole or in part" is often used in American contracts and is the correct translation of "total o parcialmente."

2) The entire sentence might be translated as follows: *Este Contrato no podrá ser cedido, total ni parcialmente, sin el previo consentimiento por escrito de la otra parte contratante.*

Waiver.[1] The failure by either party to insist on strict performance of any covenant or condition hereof, or to exercise any option herein contained, shall not be construed as a waiver of such covenant, condition, or option in any other instance.[2]

1) Note that "to waive" is "renunciar," which must not be translated as "renounce." It's waiver of a right in English, but renuncia a un derecho in Spanish.

2) The entire sentence might be translated as follows: *El hecho de que cualquiera de las partes no exigiere el cumplimiento específico de cualquier obligación o condición contractual, o no ejerciere ninguna opción previsto en el contrato, en ningún caso será interpretado como una renuncia a dicha obligación, condición u opción.*

Counterparts.[1] *This Agreement may be executed in one or more counterparts, each of which shall be deemed to be an original, but all of which together shall constitute one and the same instrument.*

1) Many dictionaries translate "counterpart" as "contraparte" or " contrapartida." That is not what is meant here. Rather, the word simply means "copy," and should be translated by "ejemplar."

1.7 Closing. *IN WITNESS WHEREOF,*[1] *the parties have executed*[2] *this Agreement as of the date first above written.*[3]

1) "In witness whereof" is "en fe de lo cual" or "en testimonio de lo cual"

2) Note the use of the word "executed," which simply means "signed" (firmado).

3) The entire sentence might be translated as follows: *En fe de lo cual, las partes firman el presente contrato en el día de la fecha que consta en su encabezamiento.*

2. TRANSLATION OF CONTRACTS FROM SPANISH INTO ENGLISH

A good translation of a contract from Spanish into English will take into account the discussion of contracts written in American English set forth above. For example, the American lawyer will expect to see "representations," and therefore "declaraciones" is best translated as "representations" and not as "declarations."

2.1 False Cognates. The biggest trap here is the need to avoid false cognates. While most of these would simply confuse the American attorney, some of them would actually mislead him. The following Spanish sentences will be discussed in the seminar, and a proposed translation for each of them will be provided.

1. incorporación. *Todos socios podrán decidir por unanimidad la incorporación de nuevos socios.*

2. ejecución. *Las partes renuncian a cualquier fuero que pudiera corresponderles y acuerdan someter cualquier cuestión vinculada a la interpretación, cumplimiento o ejecución del contrato a los Tribunales Ordinarios de la Capital Federal.*

3. contrato de adhesión. *El nuevo socio deberá firmar un contrato de adhesión al "Joint Venture Agreement."*

4. personas morales. *El objeto de la sociedad es controlar, administrar, organizar, y crear toda clase de personas morales, ya sea formando parte de las mismas como socio o accionista o simplemente participando en su planeación y organización.*

5. acciones nominativas. *Las acciones representativas del capital social serán nominativas, siempre liberadas e integramente suscritas y pagadas.*

6. conveniente. *El Consejo de Administración se reunirá cuando el Presidente lo considere conveniente.*

7. renunciar. *Cuando dicha entrega no se realice en el plazo convenido, el Comprador podrá rescindir el contrato sin necesidad de intimación previa, sin que ello implique una renuncia a los demás derechos que le correspondan.*

8. compensar. *El Comprador podrá compensar automáticamente y sin necesidad de preaviso cualquier suma que deba pagar a El Vendedor con cualquier suma que El Vendedor le adeude, cualquiera que sea la causa de tal deuda.*

9. objeto del contrato. *El envío de cualquier material o el comienzo de ejecución de cualquier trabajo objeto de la Orden de Compra constituirá la aceptación de todos los términos y condiciones del contrato.*

10. crédito. *El Vendedor no podrá ceder, total o parcialmente, el Contrato ni los créditos a su favor que surjan del mismo.*

11. verbal. *La voluntad de las partes expresada en las cláusulas de este contrato deja sin efecto cualquier acuerdo previo, verbal o escrito, o propuesta unilateral anterior que se le oponga.*

12. cesión. *El Vendedor no podrá ceder, total o parcialmente, el contrato ni los créditos a su favor que surjan del mismo.*

13. propietario. *El Consejo de Administración estará integrado por un número variable de Consejeros Propietarios, que nunca será menor de tres, ni excederá de siete. Podrán elegirse Consejeros Suplentes, que podrán asistir regularmente a las sesiones del Consejo con derecho a voz, pero sólo tendrán derecho a voto en ausencia de los Consejeros que subrogan.*

14. corresponder. *La administración de la empresa corresponderá al Presidente Ejecutivo.*

15. disposición. *Sin perjuicio de las diposiciones del Artículo 9, se podrán establecer normas aplicables a las inversiones contempladas por esta ley.*

2.2 Obscure Constructions. The following three sentences contain language (not entirely contractual) that is difficult to find in bilingual legal dictionaries and therefore merit special consideration.

1. *Las partes se someten a la jurisdicción de los tribunales de la Capital Federal, prorrogándola de conformidad.*

2. *Por todo lo dicho solicito que me tenga por presentado, por parte en el carácter invocado, y que se imprima a la presente demanda el carácter de ordinaria, ordenándose el traslado de ley.*

3. *Si los términos del contrato son claros y no dejan duda sobre la intención de los contratantes, se estará al sentido literal de sus cláusulas.*

3. ADVERTISING

When I was practicing law, I often received brochures from translation agencies offering their services. I was struck by the number of times these brochures contained errors themselves, making me very sceptical whether they would translate our documents accurately. Here are two examples:

3.1 Faulty Terminology. "As a leader in its field, Translation Co. has been providing translation and interpretation services for some of the most prominent international corporations in the U.S. such as Arnold & Porter, Hogan & Hartson, etc." Arnold & Porter and Hogan & Hartson are well-known *law firms*, not *international corporations*. If Translation Co. is not aware of this distinction, will they translate our legal documents accurately?

3.2 Poor English. "Translation Co. has demonstrated to be an invaluable business tool for hundreds of law firms throughout the United States." The sentence should read " has proven to be" or "has demonstrated that it is." If Translation Co. makes mistakes in English in promotional literature, will they translate our legal documents into good English?

APPENDIX I

BIBLIOGRAPHY OF SPANISH LEGAL DICTIONARIES

1. Alacaraz Varó, Enrique and Brian Hughes, *Diccionario de términos jurídicos* (Spa⇔Eng). Barcelona: Ariel Derecho, 1993. ISBN 84-344-1579-8. This dictionary is particularly useful for documents from Spain, and gives examples of the words in context.

2. Cabanellas de las Cuevas, Guillermo and Eleanor Hogue. *Butterworths Spanish-English Legal Dictionary* (Spa⇔Eng). 2 volumes, 1991. ISBN 0-40925-699-2. Can be ordered from the publisher by calling 1-800-562-1197. Cost $90 for the two volumes. This dictionary is especially thorough, and is thus indispensable if you translate a lot of legal documents.

3. *Dahl's Law Dictionary* (Spa⇔Eng). 1992. ISBN 0-89941-807-4. A glance at a random page of this dictionary reveals many words that you could find in your desktop dictionary: *expropiado*, person whose property is condemned; *extender*, to extend; *extensible*, extendible; *extinguir*, to extinguish; *extradición*, extradition.

4. Díaz, Luis Miguel and Ben Lenhart. *Diccionario de términos jurídicos* (Spa⇔Eng). Mexico City: Themis, 1992. ISBN 968-454-306-9. Although very short, this small dictionary is useful because it contains a long list of Mexican abbreviations and acronyms and defines words that are specific to Mexico, such as "ejido" and "amparo."

5. Kaplan, Steven M. *Wiley's English/Spanish and Spanish/English Legal Dictionary* (Spa⇔Eng). New York: John Wiley & Sons, 1993. ISBN 0-471-57678-6. This book is fairly complete, but it is more of a glossary than a dictionary. No context is given for the words.

6. Mazzucco, Patricia Olga and Alejandra Hebe Maranghello, *Diccionario bilingüe de terminología jurídica* (Spa⇔Eng). Buenos Aires: Abeledo-Perrot, 1991. [Argentina] ISBN 950-20-0634-8. This book is indispensable for translating documents from Argentina. The translator's notes are particularly useful.

7. Orellana, Marina. *Glosario internacional para el traductor* (Spa⇔Eng). Santiago de Chile: Editorial Universitaria, 1990. Although not strictly a legal dictionary, this book contains many words in context that you won't find in any other dictionary. Don't miss the list of "common stock phrases" at the back of the book.

8. Solís, Gerardo and Raúl A. Gasteazoro, Jr. *West's Spanish-English, English-Spanish Law Dictionary* (Spa⇔Eng). West Publishing Co., 1992. ISBN 0-314-00846-2. $49 from West Publishing Co. Like the Dahl dictionary, this book contains too many filler words: *política económica*, economic policy; *ponerse de acuerdo*, to agree; *por mayor*, wholesale; *porcentaje*, percentage; *porción*, portion; *postdatar*, to postdate; *pornografía*, pornography.

9. Vanson, George N. and Marilyn R. Tayler. *Spanish-English Legal Terminology* (Spa⇒Eng). Glenview, IL: Scott Foresman Co., 1985. ISBN 0-673-19262-8. This book, though short, is excellent because it uses each word in context.

Connections

Terminology

STERIC ASPECTS OF CHEMISTRY. CONCEPTS, TERMINOLOGY AND NOMENCLATURE

S. Edmund Berger
Independent Technical Translator

Keywords: Stereochemistry, drug receptors, stereoregular polymers, terminology

Abstract: This paper deals with selected aspects of stereochemistry and related terminology of potential interest to translators. It describes the asymmetric carbon atom, optical isomerism, the rotation of plane-polarized light and the nomenclature of optically active compounds. It also deals with biological interactions, chemical receptors, the perception of sweetness and chiral drugs. Finally, a brief review of stereoregular polymers is presented with emphasis on isotactic polypropylene.

1. INTRODUCTION

The chemical reactivity and the properties of organic compounds are often determined only or primarily by the relative position of atoms or atomic groups in the molecule. This presentation deals with some of these steric situations from a conceptual and terminological standpoint and should be of interest to translators working in the chemical, pharmaceutical and related fields.

One example of the type of chemistry that will be discussed in the following is steric hindrance. This well-known phenomenon occurs when a large, bulky atomic group is located close to a reactive group A and by its bulk hinders the approach of another reactive group B to reactive group A. Because, organic groups before they can react must first be able to approach each other in space, the reaction of A and B will be hindered. For example, a bulky tertiary butyl group located in the ortho position to a phenolic -OH group will prevent many reactions which the -OH group would otherwise undergo.

Maleic acid and fumaric acid have the same formula, but their two carboxyl (-COOH) groups are differently oriented in space. In the maleic acid molecule, these two groups are located on the same side of the plane passing through the double bond which both compounds have, whereas in fumaric acid these groups are on opposite sides of this plane. When maleic acid is heated to a very high temperature, it eliminates elements of water forming maleic anhydride. Fumaric acid cannot undergo such a reaction, because the two -COOH groups cannot approach each other in space sufficiently.

Organic chemistry is replete with situations involving such steric factors. This presentation will deal with a few selected concepts of this kind of chemistry: molecular asymmetry and optical isomerism as they apply to sugars, amino acids and drugs followed by some aspects of stereospecific polymerization and stereoregular polymers.

2. OPTICAL ISOMERISM

Isomers are compounds having the same general or empirical formula but different structures and, hence, different chemical properties (example: n-propanol, $CH_3CH_2CH_2OH$, and isopropanol, $CH_3CHOHCH_3$; empirical formula of both: C_3H_7OH). Optical isomerism is a branch of stereochemistry or chemistry as it is related to the spatial or steric arrangement of atoms in a molecule. It is of great interest because it is involved in the biological activity of many compounds, such as drugs, herbicides, insecticides, sweeteners, hormones etc.

Let us now talk about the asymmetric carbon atom. The carbon atom is tetravalent, namely it can combine with four other atoms or atomic groups. When all atoms or groups attached to a carbon atom are different, the carbon atom is referred to as an asymmetric carbon atom. Such carbon atoms are capable of conferring optical activity to the molecule and are also known as asymmetric or chiral centers.

A compound is optically active if its solution can rotate the plane of polarized light either to the right (dextrorotatory) or to the left (levorotatory). Ordinary white light is made up of rays of different wavelengths vibrating in many different planes. Monochromatic light (single wavelength), for example the light emitted by a sodium lamp, also consists of waves vibrating in many planes. If such light is made to pass through a crystal of Iceland spar (transparent variety of the mineral calcite, pure calcium carbonate), the emerging rays vibrate in a single plane determined by the orientation of the crystal. Such light is known as plane-polarized light. Molecules with an asymmetric or chiral center can interact with such light and rotate the plane of the polarized light that passes through the solution of such a compound. The phenomenon is called optical rotation.

Optical rotation can be measured accurately in an instrument called polarimeter. The result is reported in terms of the specific rotation $[\alpha]$, defined as the rotation in degrees brought about by a solution containing 1 g of test substance in 1 mL of solution and examined in a 10-cm polarimeter tube. The result is reported with a notation of the temperature of measurement and the nature or wavelength of the light source. Thus $[\alpha]_D^{25^\circ C}$ means that the measurement was made with sodium light (D line of the spectrum) at 25 °C.

We know that carbon is tetravalent, i.e., capable of forming four bonds with other atoms or groups. The four bonds (or valences) are directed to the corners of a tetrahedron at the center of which the carbon atom is situated. If four different atoms or groups are attached at the four corners, then such a molecule is asymmetric and capable of existing in two forms. These forms cannot be superimposed on each other and thus are different even though they represent the same structural arrangement of atoms. They bear a relationship to each other which corresponds to that existing between an object and its mirror image, or between the left and right hand. These two spatial arrangements result in the existence of two isomers identical in all respects except optical properties: one isomer will be dextrorotatory (right-handed) and the other will be levorotatory (left-handed) by the same number of degrees. Such isomers are called enantiomers or optical antipodes. A mixture of equal amounts of two optically active enantiomers is obviously optically inactive (no rotation measurable in the polarimeter) and is called a racemic mixture or racemate. In such a mixture, the individual molecules are optically active, but because the number of levorotatory and dextrorotatory molecules is the same, the net result is a lack of optical rotation. To indicate that an isomer is dextrorotatory, the prefix "d-" or "(+)" is used. Similarly, "-l" and "(-)-" refer to the levorotatory enantiomer. "d,l-" denotes the racemate.

Many organic molecules have more than one chiral center. For example, tartaric acid has two. In this case, we have four stereoisomers: the l-, d-, d,l- and a fourth called meso-tartaric acid. The meso isomer, like the racemate, is optically inactive, but the reason is internal compensation. In this case, the upper half of the molecule is a mirror image of the lower half which means that any tendency of the upper asymmetric carbon to rotate the plane of polarized light to the right will be exactly compensated by the tendency of the lower carbon to rotate this plane to the left. Result: optical inactivity.

As the number of chiral centers in a molecule grows, the number of possible stereoisomers also grows and so does the difficulty of unambiguously indicating all spatial relationships within such molecules. Several different conventions exist, therefore, to express these relationships. One such convention concerns dextrorotatory glyceraldehyde. All organic compounds, for example sugars, that have the same absolute configuration as glyceraldehyde will have the prefix "D-" regardless of whether they are dextrorotatory or levorotatory.

D-Glyceraldehyde was chosen as reference standard because it is the lowest member of the sugar series and corresponds in configuration to one of four asymmetric carbon atoms in the key sugar glucose. In the series D-(+)-glyceraldehyde, D-(+)-erythrose, D-(-)-arabinose, D-(+)-glucose, the asymmetric carbon next to the primary alcohol group in each of these sugar molecules has the same configuration as D-glyceraldehyde and, hence, all belong to the D-series, even though arabinose is levorotatory and the others are dextrorotatory. This D,L nomenclature is known as the Fischer-Rosanoff nomenclature. No simple relationship exists between the sign of rotation (d or l) and the absolute configuration (D or L).

Because of ambiguities involved in this nomenclature, the Cahn, Ingold, Prelog sequence rule was introduced and is widely used to designate absolute configuration. This convention considers that the groups are attached to the asymmetric atom according to a sequence that enumerates functions according to decreasing numbers of atoms. Configurations are indicated by **R** (rectus or right) if this sequence follows a clockwise direction and **S** (sinister or left) if it is counterclockwise. There are many other such conventions. Optically active compounds are of great theoretical and practical interest, because living systems are chiral. For example, all amino acids obtained by hydrolysis of human protein with the exception of glycine which has no chiral center are levorotatory.

3. BIOLOGICAL INTERACTIONS

Interactions between the human body and drugs, flavorants, food constituents and other chemicals occur at the cellular level. Specific cell sites, called receptors, of unique chemical and steric architecture combine exclusively in lock-and-key fashion with appropriate complementary groups of a chemical. In Dorland's Illustrated Medical Dictionary, a receptor is defined as a "molecular structure within a cell or on the surface characterized by (1) selective binding of a specific substance and (2) a specific physiologic effect that accompanies the binding". The response of receptors to chemicals is usually a highly stereospecific phenomenon with different enantiomers of a given chemical eliciting different receptor responses. Proteins, nucleic acids and polysaccharides possess characteristic chiral structures that are closely related to their functions. Because of chirality, living organisms usually show different biological responses to one or the other of a pair of enantiomers. For example, sodium L-(+)-glutamate tastes good, whereas its mirror image, D-(-)-glutamate is bitter or flat. In the case of the once widely used racemic hypnotic agent thalidomide, its teratogenic activity (a side effect!) has been ascribed exclusively to the (S)-(-)-isomer.

3.1 Perception of Sweetness

The physical and chemical structures of foods stimulate receptors in the nose and mouth. Impulses from these receptors are then processed into perceptions of flavor by the brain. The perception of sweetness is an interesting phenomenon, because sugars are chiral compounds, and the enantiomeric forms do not stimulate the receptors equally.

Sweetness is complicated because of the great variety of chemical compounds that are sweet. Two very similar compounds can have different flavor qualities. For example, alpha-D-mannose is sweet and the beta enantiomer is bitter. We know from experience that sugars such as glucose (dextrose), sucrose and fructose (levulose) are sweet, but "milk sugar" or lactose is tasteless. Another example is that of aspartame (L-aspartyl-L-phenylalanine methyl ester). This compound is a potent sweetening agent, but its other isomer is bitter. Many chemical and physical properties have been examined to determine the requirements for a chemical to elicit a particular sensory response. For sweetness, a theory, called the acid hydrogen-base (AH-B) theory, has been developed. According to this theory, a molecule must have a proton attached to an electronegative center located 0.3 nm away from another electronegative center in order to taste sweet. All sweet compounds have an AH-B system, but molecules that do not meet these strict requirements do not. Among the sugars, those that are sweet in one enantiomeric form may not be sweet in the other.

3.2 Chiral Drugs

About half of the more than 2000 drugs in the US Pharmacopeial Dictionary of Drug Names contain at least one asymmetric center. Many of these drugs are racemates. This is easy to understand considering that standard chemical synthesis of compounds with one or more chiral centers leads to racemic mixtures. Hence, a synthetic compound or drug of this kind usually consists of equal amounts of the therapeutically active and <u>inactive</u> compound. Because drug binding in the human body is often a highly stereospecific process, making optically active compounds or resolving racemic mixtures into optical antipodes are areas of major interest to the pharmaceutical industry. Large and small companies are making efforts to develop proprietary resolutions, asymmetric synthesis, enantiomeric intermediates and optically active drugs. Special, rather complicated techniques are required to achieve such goals. These efforts are sometimes referred to as <u>chirotechnology</u>.

An area that could benefit from the use of a single enantiomer rather than a racemate is that of transdermal drug delivery (skin patches). For example, patches containing only (S)-(-)-nicotine (used in nicotine withdrawal therapy) would be much less irritating than those containing the racemate. Another area is that of cardiovascular or hypotensive drugs. Here the other, biologically inactive enantiomer often causes side effects, whereas a drug containing only the biologically active enantiomer would result in a more potent drug being taken less frequently thus enhancing patients' compliance with doctors' orders.

4. STEREOREGULAR POLYMERS

This section deals with stereoregular polymers, namely polymers with a sterically ordered structure. The discussions that follow assume a basic familiarity with polymer chemistry on the part of the reader or audience.

Stereoregular polymers consist of macromolecules whose units follow one another along the chain with configurations ordered according to some rule. We refer to such polymers as linear tactic polymers and this arrangement as their tacticity or orderliness of the succession of configurational repeating units in the main chain of a polymer.

Before we consider stereoregular polymers in greater detail, let us recall some pertinent and basic aspects of polymerization. To form a high-molecular-weight addition polymer, a polymerizable small molecule or monomer with a reactive carbon-carbon double bond, for example ethylene or propylene, is subjected to appropriate reaction conditions which usually include a catalyst. Depending on the type of catalyst, polymers made from the same monomer may be different in some ways and that means that they may have significantly different and potentially useful properties. Well-known polymerization catalysts include compounds which under reaction conditions can generate free radicals, namely special atomic groupings capable of initiating addition polymerization. We refer to such polymerization as free radical polymerization. Peroxides and azo compounds are examples of such catalysts. Other known types of polymerization are cationic polymerization (catalyzed by cationic catalysts, for example boron trifluoride or aluminum chloride) and anionic polymerization (catalyzed by anionic catalysts such as sodium). In the early 1950's, Prof. Ziegler in Germany discovered a different kind of catalyst for the polymerization of ethylene to polyethylene. This work was not highly remarkable except that it enabled an Italian group of chemists led by Prof. Giulio Natta to discover a new world of polymers - the stereoregular polymers. For their work, Ziegler and Natta received the Nobel prize, and the new catalysts - the coordination catalysts - are usually referred to as Ziegler-Natta catalysts. Early catalysts of this kind consisted of titanium tetrachloride and triethylaluminum. Today there are many other, more effective, catalysts of this type.

Ethylene does not have a structure that yields stereoregular polymers. In fact, when the C=C double bond opens during polymerization and combines with another such ethylenic double bond, the resulting chain has no asymmetric character; in the case of propylene, this event yields an asymmetric structure which can exist in two different configurations. For this reason, Prof. Ziegler made no spectacular discoveries beyond his new catalysts. Prof. Natta did. In 1954, Prof. Natta one day polymerized propylene, a compound similar but also significantly different from ethylene. In working up his batch, he noticed that while 99% of the product was soluble in boiling n-heptane 1% was insoluble. Prof. Natta was intrigued! He studied this anomaly and soon found out that he had on his hands a new kind of polypropylene, not the kind known until that time but something different. From Natta's curiosity grew a new generation of polymers.

Let us now return to tactic polymers. The material Natta discovered turned out to be isotactic polypropylene. The term isotactic refers to a polymer characterized by repetition of the configuration of the monomeric units along the polymer chain. Conversely, an atactic polymer is totally devoid of such steric regularity. In a syndiotactic polymer, the configurations of the monomeric units in the polymer alternate. A polymer, for example polypropylene, made up of blocks of monomeric units having one configuration alternating with blocks of such units having the other configuration are known as stereoblock polymers. The different configurations can arise depending on whether the monomer molecules combine by head-to tail or head-to head enchainment. In general, the nature and the yield of the polymers formed depend strongly on the catalytic system used.

Before we leave this subject, let us emphasize that in polymer terminology the terms configuration and conformation have different meanings and should not be used interchangeably. Configuration is the spatial arrangement of atoms in a molecule, especially in one containing one or more asymmetric carbon atoms. Conformation is the overall spatial arrangement of the atoms and groups in a polymer molecule, or the general shape of the molecule. Thus, we talk about two different configurations of the propylene repeating unit in polypropylene, but of the conformation of the polypropylene chain.

It is difficult to polymerize propylene to high-molecular-weight products with conventional free-radical, anionic or cationic catalysts. These catalysts yield amorphous, low-molecular-weight materials which, in general, have not achieved commercial significance. While these products can be used as chemical intermediates, such as detergent alkylate, they are not useful as polymers in their own right. The discovery of isotactic polypropylene changed this. This polymer had outstanding, commercially useful properties (high crystallinity, toughness, insolubility in organic solvents) and was relatively inexpensive. Isotactic polypropylene has found a wide variety of applications including films, fibers, car bumpers, garden furniture and many others. The spectacular growth of this polymer therefore is not surprising. In 1989, the world consumed 11 million tons of this material!

******

The writer hopes that this brief presentation will be useful to at least some of our translators when they need to translate material dealing, for example, with sugars, amino acids, chiral drugs or stereoregular polymers. The task of technical translation is always so much easier when the translator has at least a basic understanding of the subject matter.

FROM CONCEPT TO TERM: A WHOLE-LANGUAGE REVERSE DICTIONARY

Henry G. Burger
The Wordtree

Keywords: Add-on (Binary). Lect. Onomantics. Reverse dictionaries. Technolect. Wordtree.

Abstract: Translation must relate concepts from two different communities. The problem is not spelling but ideation. Yet almost all current word systems are arranged alphabetically. - To equate from concept-phrases in the left-hand column to their single-word equivalents in the right, is the mission of the rarely discussed Reverse Dictionary. We survey the market and find grabbags correlating mere loanwords, eponyms, etc. Many reversers often merely connect two languages' morphemes: *Kidney, diseases of = Nephritis [Greek, 'kidney'].* - A truly reverse dictionary must be *time-dependent,* relating precedents and consequences: *Unfilled* <vacancy> should lead not only to the synonym, *empty,* but also to causes such as *attrition,* and effects like *waste.* That goal requires us to build on the processual / impactive / transitive words. By enthusing 1,000 contributors from verandah to fieldwork, we have arrayed 30% more transitives than OED. Result is a seamless, multi-occupational database of procedural verbs. - Each is the sum of the previous process plus an addendum: *To Develop & to Enlarge = to Grow.* These procedures nest as branches, producing a Hierarchy half for the system. - A computer can then permute the three elements, so that an "E" page, for example, should show: *Enlarge upon Developing = to Grow.* When alphabetized, these form the second half of the system, an Index. - Henceforth the translator can word any component of an idea, then move out its branches for pinpointing. He can trace by concept (in the Hierarchy) or by alphabet (in the Index). Result: an all-concept reverse dictionary named *The Wordtree* ®. - 26 refs.

1. REVERSE DICTIONARIES POINT FROM NONTECHNICAL TO TECHNICAL.

1.1 There Are Dual Ways to Arrange Terms. - A translator must connect two sets of concepts, namely, from two or more different cultures or even subcultures. Being well versed in both of the languages in which he or she is working, his problem is not how to spell a word, but how to find an exact or close concept. Yet the wordbooks and word systems that are available to translators are almost entirely arranged alphabetically, not by concept. A communication crisis is emerging. - To break the gridlock, we must remember that symboling offers at least two major directions. The first leads from a known word to its meaning. That discipline is called *semantics.* The second orientation is far less understood. Leading **from** a meaning **to** its word is the system vaguely known as *nomenclator.* Its discipline is vaguely termed *onomasiology* (ref. 25:s.v.). Fred Riggs has proposed to parallelize its concept-naming function as *onomantics* (cf. ref. 18); I agree. The remainder of this paper will concern such onomantics, which points from a meaning to the single word expressing it. - To match his two sets, the translator essentially imagines a two-columned system. The Occidental tradition moves from left-up to right-down. In any logical system, the left, or beginning, or Question column contains the "referent." The right, or destination, or Answer column, is called the "relatum."

1.2 Reverse Dictionaries Refer from Phrase to Word. - Heretofore, the principal semantic format has been the dictionary. Such a reference book of course essentially alphabetizes each common noun as a "lexeme" in the "left column," where it serves as a referent. In the matching, right-hand column (with or without interruptions such as etymology) is a restatement with a phrase of definition. Example: In the left column of the "P" pages, *Podiatrist.* In the right column as its relatum: *One who treats the human foot in health and disease; a foot doctor.* - Only in 1987 did the Oxford Unabridged begin listing the phrase, *reverse dictionary* (and not as a separate entry but within the adjective, *reverse).* Since then, it is our impression that its frequency of use has rapidly increased. Why? - By this analysis, we hope to extract their useful part and to discard the far greater amount of mediocrity. We have spotchecked a few concordances for serious discussion of Reverse Dictionaries, such as Modern Language Association's *International Bibliography.* Our spotcheck didn't find it, and so we must begin almost "blind." - Oxford Unabridged cites M. Lehnert's *Reverse Dictionary of Present-Day English* (1971) to limit *reverse dictionary* to an arrangement which inverts each letter or each affix *within* a word. That permutation is useful for Scrabble players wondering, for example, what words could be formed with (say) an extra letter "e." That arrangement was termed by Yakov Malkiel as *rücklaüfig*, 'retrograde.' From here on, we will term a reverse dictionary simply as *a reverser.*

1.3 Reverser Has Changed Its Meaning. The reverser obeys market demands, and thus rarely respells a word. Far more often, a reverser reverses the two *columns* so as to "untranslate" the words. Its left column is usually filled with an allegedly plain-language (non-technical) phrase, whose most unique or important word heads the entry. For instance, its "F" pages would say, *Foot doctor....* - The corresponding right-column entry is the (usually) single word or proper-noun technical term claimed to be the professional equivalent, in this case: .... *Podiatrist.* In that way, a reverser moves from the concept and leads to a word, rather than vice versa. - And so there are at least two broad types of reversers. They should be differentiated henceforth: If the reversing occurs within the word, as in its spelling backwards, the Scrabble-game type, it might be termed as an Internal Reconstructor or "Rucklaufig." If it points from concept to technical word, as almost all reversers nowadays do, it could be termed an "Onomantic." We shall discuss only the latter.
But the theory will apply to any language. Example: the Francophone *Dictionnaire Plus* (ref. 21).

1.4 We Focus on Certain Types. - Because of the current lack of standards and labels, there are now many variations of reverser. Some arrange at least one column by graphics. Typical are atlases, and the pictorial portions of of Longman's *Dictionary of Contemporary English* (ca. 1994), and icons like Macmillan's *Visual Dictionary* (1992), and St. Olderr's *Reverse Symbolism Dictionary* (1992). - Other reversers show referents that are non-generic proper nouns, like *Baedeker's* landmarks of particular cities. Another neighboring category offers relata that are essentially numerical -- concordances like Grun's fascinating *Timetables* (1946 ff.).

1.5 Reversers Require "Lect." - The remainder of this paper will focus on systems both of whose "columns" are words. We consider them the truer reversers. They allow the wordsmith a leap from mainstream-to-technical term, because, by definition, a user better knows the nontechnical phrase than the precise word. Within the limits we set above, the market has perhaps four dozen brands claiming to be reverse dictionaries. Analysis of their strengths and weaknesses might lead us to an overall theory, and hence to higher usefulness. - We immediately find that a typical reverser's relata cover nowhere near the full breadth of the language. Instead, they tend to be restricted to one or another group of non-mainstream words, merely a variety of a language. It may be dialect, slang, jargon, brogue, cockney, pidgin, lingo, cant, argot, dialect, technolect, Saussurean *parole,* sociolect, or other specialty sublanguage within a natural language. Fortunately, all these types fit the OED 1987 Supplement's definition of a thing called a *lect*: "a social variety of a language or a dialect." Thus we may say that the reverser correlates only one or a few lects, and not the full <English> language as its title usually boasts.

2. EACH KIND OF "LECT" ENABLES ONE KIND OF REVERSER.

2.1 Synonyms Principle Some Reversers. - The reverser's probably commonest kind of entry is in fact the mere synonym: Some mainstream word or phrase is listed as referent, and its almost equally known equivalent(s) as the relatum. Thus, a "C" page of Theodore Bernstein's reverser (ref. 2) lists the word *counterpart* in the left column, and as relatum, says simply, *obverse*. Therefore this is not a true reverser. It should instead have said, and on the "S" pages, at that: *Side of a token considered to be its front.* - Such instances of mere synonymy abound. Xerox's *Word Nerd* system gave rise to *Word Menu* which helped generate Microlytic's *Inside Information* (1990). One of its four "views" of this electronic instrument is claimed to be a Reverse Dictionary. The user is to keystroke his problem, such as *left side of a ship*, whereupon the device screens *port* as its right-column answer.

To the present speaker, the foregoing examples do not illustrate a Reverse Dictionary, but merely a Synonymy. And synonyms are already far more systematically covered by many reference books, of which the best known is Roget's *Thesaurus*. Kirkpatrick's Roget (1989), for example, has a category 242 which groups *leftness* not only with *port* but also with 33 alternatives. And instead of *counterpart* equalling just one term, *obverse* (as Bernstein claimed above), Kirkpatrick's (1989 #237) groups *counterpart* with fully 158 synonyms! And so, largely synonymous reversers should be perceived as limited, opportunistic, and haphazard. True, the wordsmith Alan Kaye (ref. 11) liked the fact that Bernstein listed the mainstream referents alphabetically, whereas in the alphabetical dictionary, it is the specialty words that are alphabetized. But that hardly makes it a novel of reference work. March's synonymy (ref. 14), for example, did similarly almost a century ago.

2.2 Eponyms Principle Some Reversers. - And what about left-column listings of proper-name-words, or *eponyms*? Thus, "President William Clinton" is not only one person's name, but also a personifier of certain historical events. -- Eponyms are alphabetized in many places, like Ruffner's *Eponyms Dictionary Index* (1977), and D. Ballentyne et al's *Dictionary of Named Effects ... in Chemistry....* (1980). - Now, eponyms are often ethnic-specific, and as a result should particularly interest translators. Therefore it would be of considerable value to offer a reverse arrangement, that is, from the generic characteristic in the left column to the eponym in the right column. - One such eponym market is crossword puzzlers. There we find most present merchandise to be merely an oversimplified short-order, a "McDictionary." Thus, on the "O" pages of Dell's reverser (ref. 17), at the word *opera*, we read: "*Opera* of Puccini ...[is] *Tosca*." (*Tosca* only?!) - A more durable eponymous strategy would be to emphasize only those proper nouns which come to overspread an entire semantic area and thus will parent generics. In this way, Greco-Roman mythology related a story about a beautiful youth named Adonis. And so the M pages of a reverser may show: *Man of great beauty ... Adonis* (in ref. 2). - We cannot, however, find a reverser that right-columns more than a few dozen. But thousands exist. For example, Penelope, the long-suffering wife of wandering warrior Odysseus, is legendary for her faithfulness. And so one would want (on the "F" pages): "Fidelity ... Penelope." Yet many librarians tell me that such a useful reverser dosn't exist.

2.3 Bilingualisms Principle Some Reversers. - Every bilingual dictionary, even if only one-way, is a kind of reverser (such as the English-Spanish Duden [1985, 1992]). Likewise, the "C" pages of ref. 3 differentiate 61 kinds of *Cutting* words in Greek, Latin, and Anglo-Saxon, such as *-tomy*, with their relata, in this instance, "suffix designating dissection."

2.4 Vocations Principle Some Reversers. - Part-of-speech as relata are, therefore, one kind of reverser. Still others reversers cover just one occupation. Consequently, they might be considered Terminologies or even Collocators. Example: Eric Partridge's several excellent dictionaries of (professional criminals') argot. One (ref. 16) typically shows, on the "S" pages: *To sunrise*, intransitive verb, {of a vagabond} = To jail overnight, and then expel. - Perhaps the vocation most utilized by reversers is the technolect of medicine. Thus, physician Jacob Schmidt authored the aptly-named *Reversicon.* It lists items like these on the "K" pages: *(1) Kidney...{Greek:} Nephros. (2) Surgery of kidney...Nephrectomy* (ref. 20).

2.5 Meronymies Principle Some Reversers. - Other Reverse Dictionaries relate wholes to parts. Thus, in the English-German Duden (1960 ¶184), the "E" pages show *Engine,* while its relata are listed as *carburetor, dip-stick, ignition coil,* etc. Now, George A. Miller termed the relation of parts to a whole as *meronymy* (ref. 15:3). Hence we have Meronym Reversers. Kirkpatrick's Roget occasionally does so: At ¶722, *Combatant*, she generically shows *boxer[:] featherweight, ... heavyweight*[, etc.] (ref. 12). - And as for *general* meronymies, one immediately remembers the Dewey Decimal System.

2.6 Nuancers Principle Some Reversers. — Synonymies may group a word-gob and and yet nuance each component. Under *Purity*, for example, March (ref. 14) distinguished 125 noun relata, such as *Agreeableness of sounds = Euphony.* Today's reversers are like space renters. Their barebones format usually resembles that of an alphabetical telephone directory, in that the relatum word is named without *definition or even collocation*! This brusqueness misimplies that the listing is 100% equivalent. Thus Grambs (ref. 24) reports, *"Decorate lavishly, as with jewels = INCRUST...."* An English-as-Second-Language person might misassume that every princess is personally "incrusted"!

2.7 Spurious Hybrids Principle Some Reversers. - In fact, most reversers dabble in several of the above categories. Eponyms, for instance, may be classicals and yet loanwords. Again, there are now regularly editionalized dictionaries whose referents are neologies, and whose relata are the the mainstream words or even lengthy explanations. (For such a list, see ref. 1.) - Typifying the Opportunistic Encyclopedism that we find, the "M" pages of the *Reader's Digest* reverser (ref. 10), as an example, obtrude a 120-word and eight pictures of certain insects' *Metamorphosis.* And so we see again that reverser really means only: Excerpted from an unabridged dictionary by *any* principle, and an unstated principle at that! - More importantly, we perceive that the average reverser is essentially a translation or analogy from one part of a language (a lect) to another part or to the core of that language. Thus we can finally characterize our topic: ***A reverse dictionary is, at present, any Inter-Lect Equivalence Table.***

3. REVERSERS LACK THE PREREQUIRED WORDING SEAMLESSNESS.

3.1 Reversers Always Trail Unabridgeds. - Such atheoreticalness inherently produces great gaps. As an illustration, we find Bernstein's (ref. 2) "M" pages listing under *Machinery* only one of the thousands of types of mechanism: the Calendering (paper-glossing) machine that he probably saw daily at his newspaper workplace. And the chasms continue. Merely cruise through his remaining offerings at "Ma--." You find that he does not list the vast quantity of other "Ma--" transitives of which the present author has evidence: *majorize, major-league, majuscule, Makehamize, makeshift, makeweight, Makutu.* Also unlisted are five *mal-* verbs like *maladapt.* His continuing gap then misses: *malagasy, malarialize, malathion, malinger, mambo, mammograph, manblock, mantilla,* etc. Such spottiness is patchwork, not system.

3.2 Reversers Pick & Choose Opportunistically. — The reverser lists the lect's root word and therefore suggests little of a language's richness. For example, reference 2 lists eight terms under *Court.* But each shows only their judicial sense. None helps us word the behavior of ***courtship***. And it is the more abstract ("conceptual") terms which tend to be omitted. Yet the cover and introduction to the typical reverser give no hint as to its parochialism. For example, Laurence Urdang (ref. 23) studied George Welsh's reverse system: "Trying to be fair, I cooked up a test [such as]...., What is the nautical term for the left and right sides of a boat?... Unfortunately," concluded Urdang, " neither *left* nor *right* (nor *port* nor *starboard*) is in." - Furthermore, reversers are usually published either by small firms or as spin-offs by full-product-line publishers. In neither case is original research expended on them. Yet just such a fetch from technolect is the missing link most needed by the translator.

3.3 Each Reverser Reports on Few Lects. - The overall value of any reverser must therefore depend on how many lects its editors offer, and how thoroughly its editors know and present each lect. Consider merely the lects which are based on the workplace (sometimes called *ergolects*). The potential list is endless: shipping; sports; marketing, etc. In fact, the Census Bureau's *Classified Index of Occupations...* (ref. 22) enumerates no less than 30,000 occupations. - Some future Ph.D. dissertation should quantify the sources ("lects") and the percent of the known wordstock from each, that each brand of reverse dictionary actually contains, like a tabulation in *Consumer Reports* magazine. Until then, the reverser-buyer is merely gambling.

3.4 Theorize or Peripheralize! - If any work would reveal the best traditions of a reverser, it should be a recent one by a major publisher. Just such characteristics apply to Longman's interesting *Language Activator Production Dictionary* (1994?). On the "S" pages at *sad*, for example, are 13 basic meanings. Three levels are devoted to each. For example, *unhappiness* is differentiated from *heartache*. - What principles guided the new system of fitting concepts to words? -- One is astonished to read a staffer's confession that "its conceptual structure is atheoretical" (ref. 19:173). - Such opportunism will never lead lexicography into lexicology. Studies in the history of science suggest that a discipline which does not formulate theory loses control of its own scientific and technical direction, forfeiting control of its destiny.

3.5 Wordsmiths Need Causation and Result. - The basic problem, perhaps never stated clearly heretofore, is that the traditional dictionary tries to commingle two irreconcilable types of thought: statics and dynamics. Statics is the framework of literature, the humanities, and most of academe, whence many word-workers still come. Dynamics is the world of physical and social engineering. Unfortunately, one cannot well intergrade statuses, such as *beauty*, with movements, such as *to smash.*

Look outside the editorial room. For every poet, dozens of people are engaged in the effectuation industries, from nurse to architect. A conceptual dictionary, an onomantic, must therefore add *time*-based change. The word *wick* should lead not only to the structural term, *candle,* but also to the instrumental process of *capillary action. Duality* should lead not only to *twins*, but also to *financial control by escrow*. *Unfilled* should point not only to *emptiness* but also to *attrition*. *Class* should invoke not only *envy* but also *Marxist warfare*. *Unseen* should connect not only *invisible*, but also *surveilled*. Etc. Reversers will continue to be haphazard until dictionary publishers supplement their rhapsodists with engineers (ref. 6).

4. SEAMLESSNESS PREREQUIRES PROCEDURAL GRADATION.

4.1 Reversers Must Befit the Basics of Science. - In critiquing Welsh's reverser, Laurence Urdang (1990:13) commented that "the problem is less one of simple (!) lexicology as one of information theory and psychology. I might add artificial intelligence...." Urdang is certainly correct. If reverse dictionaries and lexicography are to harmonize the historical trend of science, there must be some theory or principle around which their concepts are organized. The wording disciplines must leave the warm nest of humanism and join modern physical and social science.

4.2 Only an Add-On (Complexifying) Word System Can Offer Seamlessness. - To organize verbal completeness in procedural reportage, one must follow the principle of additivity. That is, the word representing every step must be specified. The best-known category of wordbook to do so is the humble *cookbook*. Thus, Susan Stevens's *Delitefully HealthMark* (1989:52) lists nine steps to converting a pound of spinach into Sesame Spinach Salad. - The most obvious kind of additivity would be from simple to complex. Early and current Rogets misassumed that the world grows in some straight kind of line, as from A to Z or Roget's category 001 through 1,000. However, outside of wordsmithing, such ortho-linear beliefs have been doubted for at least three decades.

4.3 Evolution Inherently Diversifies. - One form of additivity is evolution. And the master trend in science has long been evolutionism. It is a biological crisscrossing of hardly-predictable direction toward increased efficiency whence-ever. This is approximately the view of many current anthroplogists that is termed Cultural Materialism (ref. 4). It was forefelt even by the philosopher Herbert Spencer (1896, quoted in ref. 9:173): "Social progress is not linear but divergent... While spreading over the Earth mankind have found environments of various characters, ... [whence] there have arisen genera and species of societies." - Vocabulary develops likewise from unpredictable needs. It is universal at base (whence the idea of "primitives"), but vastly divergent as ethnicities specialize. Likewise, the procedural words of the real world should govern the reverser, not vice versa.

4.4 This Paper Considers Only Transitive Verbs. - Heretofore, wordbooks have commingled concepts and words for structures with those for processes. For important reasons of worldview, concept finders would work far better when structural terms are codified separately from procedural words. I have argued this fundamental point at length elsewhere (ref. 5:C82-C93). The remainder of this paper will discuss only procedural words, English's transitive verbs. (Every language has, by the way, been found to have some way of expressing transitivity.) Additivity in the real world is not instantaneous; it is also *time-dependent* (ref. 7:38-43). We must arrange the entire vocabulary in some evolutionary format that gradates from one sprig of a word branch to another.

4.5 Recent Lexicology Offers Theory of Seamlessness. - Our research (ref. 26) quickly found that impactive verbs, which English manifests as transitives, do exist to differentiate every ethnically-recognized gradation of meaning. But there is no one single forward-and-upward line connecting them all. Rather, the appropriate analogy is to a *tree* of words that accommodates all. And it has varying amd endless lengths of branch. The lengths are based on specific ethnic needs -- are based on zeitgeist.

But how can those needs and resultant words be learned?

4.6 We Went into the Workplaces. — Samuel Johnson once sneered that *his* dictionary (1755) omits words of technique becaue he would not "visit caverns to learn the miner's language...." We think this attitude crystallized a basic flaw in all dictionaries authored by humanists: Techniques are the essence of effectuation and therefore needed for conceptual seamlessness. - By contrast, your present speaker is an anthropologist, minored in engineering. Such an approach heartily favors the effectual occupations that constitute most of a culture. We have therefore haphazardly assembled an informal world-wide network of about a thousand sporadic contributors of "technolect." After all, a language is a combination of all its lects. What better way is there to say, for example, *to list (something) without evaluating it*? Answer, in some groups of people, is: *to laundry-list* it! What word for, *to give (an answer) quickly and with minimal brain input*? - Answer: *to knee-jerk* the reply! And thousands of other realia that we've gleaned, but which miss the more bureaucratic publications.

4.7 Binary Reverser Has Hierarchy & Index. - The product is a seamless reverse dictionary termed *The Wordtree* ®. (The definitions we cite in this paper are the newest we have developed, and not necessarily those of the already-published edition.) It has two parts, the Hierarchy and the Index. By add-on, or binary, analysis of every transitive verb into its two components, the Hierarchy seamlessly moves from mechanical actions, through the subhuman, to even human behaviors (ref. 8). Thus, *to electrify* something and to *switch-activate* it are *to sensitize* it. A typical string itemizes the microactions from the procedure of *Fingering* to the procedure of *Handling*. How the Hierarchy half of an add-on system shows it, appears in Figure 1.

To touch
> To finger
To direct
> To paw
To hold
> To hand
To thumb
> To handle
To grasp

FIGURE 1. HOW A CONCEPTUAL DICTIONARY HOMES-IN ON THE PRECISE TERM: A SUCCESSIVE, TWO-COMPONENT (BINARY) ANALYSIS OF THE TRANSITIVE ENGLISH VERBS FROM TOUCHING THROUGH HANDLING. (EACH NODE PROBABLY ALSO LEADS TO JUNCTIONS WITH MANY OTHER VERBS, HERE OMITTED FOR CLARITY.)

4.8 Conceptual Dictionary Would Offer Benefits to the Translator. - If such a "word tree" were elaborated for the entire transitive vocabulary, a translator could skim any component or neighbor of any idea, then move up or down, trunk-ward or fringe-ward, to find the exact word that connotes his concept. Now, we earlier noted that the traditional alpha dictionaries and even reverse dictionaries cannot afford to differentiate clusters. But add-on, processually graded system, each action fully encompasses one or more simpler processes (down to the basic primitives). They nest. A process-term is inherently an accumulator of a chain of many simpler concepts. Thus, *to bonsai* (a young bush) fully includes the concepts of two verbs, *to bind* (it) and *to stunt* (it). - Consequently, a gradated arrangement of processes need only show each addendum separating each sequential pair of actions. Almost no space is required. The definition is inherent at every site. (But any reverser, including our own, works best with the side-by-side readiness of an unabridged dictionary, preferably the OED.) The two-component, or binary reverser that we are now describing can afford to include 30% more transitive verbs than the 4,000-page Oxford Unabridged. And it does so in just 381 oversized pages!

5. PROCEDURAL GRADATION HAS PRODUCED REVERSABILITY.

5.1 *Wordtree* Shows All Parts of Speech. A binary reverser like *The Wordtree* is centered on transitive verbs. Yet it is not limited to them. Virtually any concept can be considered the cause or the effect of something else. Thus, even a seemingly static noun like *peace* can be viewed as "result of pacification" or "cause of commerce" or "instrument of prosperity," etc. Hence the noun *peace* can be shown in the "P" pages of our system with their transitive-verb relata of *pacifying effect, commercializing cause, prospering instrument,* etc. The first edition of our *Wordtree* lists a quarter-million entries, and we hope that that is a mere beginning.

5.2 Cross-Grid Enables User to Begin with Any Part of an Idea. - Imagine yourself as the captain of a vessel seeking a certain tiny island. If your route is linear and "pointy," you may circumnavigate the world many times before you find that golden isle. But if you have a cross-grid -- both latitude and longitude -- you will know exactly where to find your targeted island on your first visit to the region! - Similarly with words. Many a translation problem hinges on a nuance. For example, there are many words that mean, *To carry.* But suppose you need a term that means 'to carry *with dislike or boredom*"? No wordbook routinely provides such information.

5.3 Listing All Lects Produces a Cross-Grid. - Through its incorporation of technolect, Burger's *Wordtree* has becomes relatively seamless. (And each day's correspondence with neology-reporters fills more gaps.) And when we say time-dependent, we mean that the reverser must pinpoint a term by showing its "befores" and "afters." On its "G" pages at the word *Grow*, for example, should appear: *Develop & enlarge = to Grow. Grow & store = to Gestate. Grow & tend = to Foster.* (And dozens of others.) - The two parts that enter into any procedural word can then be permutated so that all three terms are listed alphabetically. On the "D" pages, for example, we should find: *Develop & enlarge = to Grow.* And on the "E" pages: *Enlarge upon Developing = to Grow*, etc. - The translator turns to the alphabetical Index portion of such a cross-grid system. He can look up any component of any impactive word. He would most economically begin by seeking the rarest or clearest component of the concept. In the present problem of *carrying*, that might be *dislike.* He therefore flips to the "D" pages. On finding *dislike,* he looks beneath for all its derivatives. There he would see: *Dislike upon carrying ... To schlepp.* That's it!

Or suppose he had first looked up the *carrying* component. On the C pages he would see the computerized permutation: *Carry and dislike.... To schlepp.* Either way, he has successfully moved from concept to word. - A single glance shows the user what are the possible terms. What, then, would our Index section relate one verb, let us say, *to fasten*, to the others in the vocabulary? - See Figure 2.

To FASTEN = to HOLD (something) & to STAY (it)...12510.

FASTEN upon BINDING = LASH (14788).
FASTEN upon BOLTING = SNIB (15217).
FASTEN upon BUNDLING = NITCH (15313.
FASTEN upon CONTRACTING = SWIG (12926).
FASTEN upon ENCIRCLING = HANK (12940).
FASTEN upon ENTERING = LATCH (12912).
FASTEN upon FILLING = COTTER (12922).
FASTEN upon GROUNDING = STAKE (12935).

FASTEN upon HALF-EMBRACING = EMBRACE (12947).
FASTEN upon INTERSTRATIFYING = INTERDIGITATE (12930).
FASTEN upon INTRUDING = WEDGE (20259).
FASTEN upon JOINING = ATTACH (12914).
FASTEN upon LINKING = CHAIN (15297).
FASTEN upon MACHINING = DOG (12919).
FASTEN upon METALLICIZING = YOTE (12942).
FASTEN upon MIRING = BEMIRE (15259).
FASTEN upon PIERCING = BRAD (12939).
FASTEN uipon PILLARING = PILLORY (12933).
FASTEN upon POSITIONING = MOOR (12924).
FASTEN upon RAVELING= ENRAVEL (12949).
FASTEN upon RINGING = PARREL (12944).
FASTEN upon SCREWING = SCREWFASTEN (19475).
FASTEN upon SEATING = BARNACLE (12921).
FASTEN upon SECURING= SLOUR (12928).
FASTEN upon SOCKETING = DOP (12945).
FASTEN upon STACKING = LOCKSTACK (12937).
FASTEN upon STRAPPING = THONG (15352).
FASTEN upon STRETCHING = RACK (12917).
FASTEN upon SURROUNDING = CLASP (12943).
FASTEN upon TAUTENING = DRUMSTRETCH (12941).
FASTEN upon TRANSVERSING = TOGGLE (12925).
FASTEN upon TRAVERSING = BOLT (12916).
FASTEN upon TURNING = BELAY (12915).
FASTEN upon UPROLLING = FURL (12931).
FASTEN upon WINDING = BITT (16253).
FASTEN & BECRIPPLE = HEADFAST (12929).
FASTEN & ENCLOSE = COPE (12923).
FASTEN & FEED = MANGER (12950).
FASTEN & FIRM = FIX (12938).
FASTEN & LANGLE = SIDELANGLE (12946).
FASTEN & ORNAMENT = BROOCH (14962).
FASTEN & PATCHWORK= SPOTFASTEN (16708).
FASTEN & RECYCLE = DOUBLEFASTEN (14001).

FASTENING Nuances.... Bitt, Direct, Enfasten, Impute, Lock, Secure, Tie.
FASTENING CAUSES...@ Connect, Control, Possess, Safen, Secure, Shut.
FASTENING EFFECTS...@Exclude, Join, Snug.
FASTENING INSTRUMENTS...@ Band, Bolt, Button, Cleat, Hookandeye, Hole, Indent, Knot, Pin, Seal, Sticky, Stitch, Thread, Tie.
FASTENING PREVENTATIVES...@ Enlarge, Impermeabilize, Loose, Slick, Stray, Wild.

FIGURE 2. FROM THE "F" SECTION OF THE INDEX, SHOWING HOW A PROCEDURAL ADD-ON RELATES ONE VERB (FASTEN) TO THE OTHERS.

5.4 An All-Lect Reverser Succeeds. - To study our all-lect reverser, *The Wordtree,* the American Translators Association called on its Dictionary Review Committee. Chairman Jacopo Màdaro (ref. 13) published a report of almost a thousand words, including: "This work allows a two-way research from the simple to the complex and vice versa (as well as from the known to the unknown)..., all of it in two words! ... *The Wordtree* is the most updated thesaurus money can buy, and much more."

This binary system has even attracted media concerned only with enduring linguistics. Thus, the *Cambridge Encyclopedia of the English Language* explained the add-on reverser as a "new way of studying English: ... Following the natural history model, in which organisms are defined binarily into genus and modifier, each verb is defined in terms of ... a general activity and a qualifying notion." And that Cambridge University encyclopedia devoted almost an entire page to *The Wordtree.* Also published have been eighty-five (85) other lengthy reviews. - The scheme is, hopefully, an ongoing project, indeed, a perpetual one. It is presently available to the public in hardcover. If the reader knows of a transitive absent from wordbooks, he should register it with us. Meanwhile, we gladly provide any inquirer with word-solving tutorial materials.

The so-called Reverse Dictionary thus joins the pantheon of basic wordworks like the alphabetical dictionary and the synonymy. But they are semantic, while it is an onomantic. - Furthermore, each type of reverse dictionary should declare itself: Eponym Reverser, Single-Technolect Reverser, Meronym Reverser, or otherwise. But seamlessness appears only in the Binary Reverser (like *The Wordtree*). Its revolutionary paradigm is slowly being understood. Even in this 12th year after publication of the first edition, we are still receiving, and honoring, book-review requests.

And so there is a new channel to ease the growing pressure on translators. In the past, America has been meager in intellectual materials. You live in the country where $4.3 billion is spent annually on libraries. But (as the American Library Association has advised the present author), that is less than America spends in the same time on *potato chips!* Indeed, translators need all the brainy material they can get. Therefore, let us welcome a new station on the Information Highway -- the ***binary reverser.***

6. REFERENCES CITED.

TO ECONOMIZE SPACE, WE LIST ONLY THE LESS TRACEABLE OF OUR FOREGOING CITATIONS.

1 Algeo, John. "A Short History of New-Word Study." *Dictionaries* 16:3-15, 1995.

2 Bernstein, Theodore M. *Bernstein's Reverse Dictionary.* N. Y.: Quadrangle Books, 1975.

3 Brown, Roland W. *Composition of Scientific Words.* Distr. by Random House, N.Y., 1956.

4 Burger, Henry G. "Cultural Materialism: Efficiencies, Not Descriptions." *General Systems* 20:107-119, 1975.

5 " . *The Wordtree.* Merriam, KS: The Wordtree, 1984. [For current availabilities, see *Books in Print.*]

6 ". "Implications of Social Science for Vocabulary Architecture." *Computational Linguistics* 14(3):104-107, 1988.

7 ". "Roget, Darwin, & Bohr." *International Journal of Lexicography* 4:23-67, 1991.

8 ". Ms. a. "Binary Gradation of Verbs Evolves Causes & Effects." *Submitted ms.*

9 Harris, Marvin. *The Rise of Anthropological Theory.* New York: Crowell, 1968.

10 Kahn, J. E.,ed. *Reader's Digest Illustrated Reverse Dictionary.* Pleasantville, NY: Reader's Digest, 1990.

11 Kaye, Alan S."Critique of *Bernstein's Reverse Dictionary* [2nd ed.]" *Verbatim*, 6ff., spg. 1990.

12 Kirkpatrick, Betty. *Roget's Thesaurus of English Words and Phrases,* New Edition. Harlow, England: Longman, 1989.

13 Màdaro, Jacopo. Critique of *The Wordtree. Chronicle, Amer. Translators Assn.* 19(5):22-23.

14 March, Francis. *March's Thesaurus Dictionary*, 1902. - New ed.: Garden City, NY: Hanover House, 1958.

15 Miller, George A. *WordNet*. Rpt 1, rev. Princeton, NJ: Cognitive Sci.Lab,Princeton Univ.1988.

16 Partridge, Eric.*A Dictionary of the Underworld.* 2d ed.London: Routledge & Kegan Paul,1964.

17 Rafferty, Kathleen, ed. *Crossword Dictionary.* New York: Dell Publishing Co., 1965.

18 Riggs, Fred. "Nomenclators." *International Classification* 18:110-121, 1991.

19 Rundell, Michael, and Nick Ham. "A New Conceptual Map of English." Pp. 172-180 *in* Willy Martin et al., *Euralex 1994 Proceedings.* [Amsterdam:] European Assn. for Lexicography, 1994.

20 Schmidt, Jacob E. *Reversicon.* Springfield, IL: Chas. C. Thomas, 1958.

21 Siebenaler,Anne-Marie. *Dictionnaire Plus.* Mamer, Luxem.: Infor.Tech.Consultants, 1994.

22 U.S.A., Census Bureau.Classified Index of Industries & Occupations. Washington: GPO, 1992.

23 Urdang, Laurence. "Critique of Geo.Welsh, *What's the Word That Means...?" Verbatim,* p. 6, spring, 1990.

24 Grambs, David., ed. 2nd ed. of [Ref. 2]. New York: Times Books, 1988.

25 Gove, P. B., ed. Merriam-Webster's Third New International Dictionary. Springfield, MA: G. & C. Merriam Co., [1966]1967.

26 Burger, Henry G. "Procedure Gradation." *General Systems* 7:293-306, 1962. #

CONCEPTUAL AND TERMINOLOGICAL ASPECTS OF ELECTRONIC ENGINEERING

Roberto Crivello
Freelance Technical Translator

Keywords: Electronic Engineering, Electronics Terminology Italian Localization, Oscilloscope Concepts and Terminology, Video Devices Concepts and Terminology, American Translators Association - Science & Technology Division.

Abstract: Advanced conceptual and terminological aspects of electronic measurements are described from the perspective of an electronic engineer/technical translator. Features, application examples, and specifications of an oscilloscope/digital multimeter are illustrated. This includes autoranging, floating measurements, pulse and video triggering. A partial glossary and some basic references are listed at the end of the paper. Several more localization examples, a longer list of references and a wider glossary of advanced terms, in both English and Italian, will be provided and discussed during the presentation of this paper.

1. OVERVIEW

Due to the prominent position that American companies hold in the electronic industry worldwide, technical documentation of electronic instruments often needs to be translated. The translation of such materials requires a good understanding of basic principles of electronics and the use of related terminology. The purpose of this session is to acquaint translators with concepts and terminology related to the operating basics and specifications sections of a typical user manual. The instrument described in this manual combines an oscilloscope and a digital multimeter (DMM), so features of two devices will be described. These complex and widely used instruments contain several components that allow the presentation of fundamental concepts and terminology in the electronic field. This can be of interest to any technical translator. At the same time, the presentation will show how the concepts and terms introduced should be translated into Italian.

2. OSCILLOSCOPE - DESCRIPTION

The cathode-ray oscilloscope is an electronic device that displays the waveforms of electric signals. This highly flexible instrument is actually an extremely fast X-Y plotter capable of plotting an input signal versus another signal or versus time, as required.

The oscilloscope operates by causing an electron beam to sweep rapidly from left to right across the phosphor screen of a cathode-ray tube. The X-axis (horizontal) input is usually an internally generated linear ramp voltage; its sweep rate is variable over a wide range, and it is specified in seconds, millisecond or microseconds per division. (A division is approximately 1 cm, as marked

on the screen.)
The Y-axis (vertical) signal is applied to the vertical deflection plates of the cathode-ray tube, causing the electron beam to move up and down as the signal voltage varies. The vertical sensitivity of the oscilloscope is adjustable from a few volts per division down to a few millivolts or even microvolts per division. If the signal being examined is repetitive (periodic) at a fast enough rate, the display appears to stand still, thus showing graphically the signal waveform (amplitude and frequency).
From what we have said, it appears that an oscilloscope operates on voltages. Using transducers it is possible, however, to convert current, pressure, acceleration, strains and other physical quantities into voltages, and thus to display graphically a wide variety of phenomena.

3. DIGITAL MULTIMETER (DMM) - DESCRIPTION

The DMM is a versatile electronic instrument used in a wide range of measurements; it is capable of measuring basic electric quantities and displaying them as numbers. DMM are used in calibration applications, to test electronic devices and in general laboratory measurements. They are either portable instrument battery-operated or bench instruments that can operate through batteries or line power. The five basic measurements DMM are capable of are:

1) DC Voltage
2) AC Voltage
3) Resistance
4) Direct Current
5) Alternating Current

DMM operate by first converting AC signals into DC signals, and then employing Analog/Digital (A/D) Converters to measure with high accuracy the DC voltage. The measurement is displayed on a Liquid Crystal Display (LCD).

4. INSTRUMENT USER MANUAL - LOCALIZATION EXAMPLES

4.1 Getting Started Chapter - Section "Oscilloscopes Features"

Example 1 of Oscilloscope Features:

Source Sentence:
Autoranging for quick setup and hands-free operation

Localization of the above sentence:
Modalità Autorange, per una rapida impostazione e un utilizzo a mani libere.

Reasons:

Autorange is an oscilloscope feature that automatically produces a stable waveform of usable size. Autorange continues to change the instrument settings to track additional signal changes and automatically select the voltage and frequency range (see "Range" in the **Glossary**) of the oscilloscope.
I have left Autorange in English because a) on the instrument panel there is an Autorange button,

and b) the translation of Autoranging would be too long (e.g., Impostazione automatica del campo di visualizzazione delle forme d'onda).

Example 2 of Oscilloscope Features:

Source Sentence:
Independently isolated channels to allow improved safety for measurements to 1000 V_{RMS} while floating up to 600 V_{RMS} using XX6175 probes.

Localization of the above sentence:
Canali isolati indipendentemente, per aumentare la sicurezza di misurazioni a 1000 V efficaci mentre si misurano fino a 600 V efficaci, con massa flottante, utilizzando sonde XX6175.

Reasons:
"Floating" here means "taking a floating measurement" (see "Floating Measurements" in the **Glossary**).

4.2 Getting Started Chapter - Section "DMM Features"

Example 1 of Meter Features:

Source Sentence:
True RMS VAC, VDC, Ω, continuity and diode-check functions

Localization of the above sentence:
Misurazione del vero valore efficace di tensioni in c.a. e in c.c., misure di resistenza, verifica di continuità e di diodi.

Reasons:
1) RMS = Root Mean Square. This is a common way to express the effective value of an alternating-current waveform. "True" means the measurement is independent of the type of waveform.
2) V is the symbol of volts, the unit of measure of voltages, so here it means "Voltages".
3) AC means Alternating Current, i.e., the electrons in the conductors flow in both directions.
4) Ω is the symbol of ohms, the unit of measure of resistance, so here it means "Resistance".

Note:
1) I do not abbreviate RMS, and I use the translations of Voltage and Resistance, not their symbols.
2) For the alternating current I use the standard Italian symbol c.a.; you should never use CA or ca.
Also note that "cc" in Italian is the symbol for cubic centimeters, so it is important to use the periods: c.c.

Example 2 of Meter Features:

Source Sentence:
Autoranging or manual ranging

Localization of the above sentence:
Impostazione manuale o automatica della portata.

Reasons:
Autorange is a feature of a DMM that automatically sets the range to the optimum setting to measure an input signal. Autorange continues to change the instrument settings to track additional signal changes and automatically select the voltage range of the DMM (see "Range" in the **Glossary**).

4.3 Getting Started Chapter - Section "Replacing the battery pack"

Source Sentence:
Replacing the Battery Pack:
For portable operation, use the rechargeable battery pack.

Localization of the above sentence:
Sostituzione della batteria
Per il funzionamento senza alimentazione di rete, utilizzare la batteria ricaricabile.

Reasons:
I have often seen "battery pack" translated as "pacco batterie".
What is a battery? A battery consists of a group of cells (a cell is a voltage-producing device, usually operating through a chemical reaction) connected in series to obtain a greater voltage. Cells are often (erroneously) called batteries. Some batteries are rechargeable and some are not. Thus, when the writer wrote "battery pack" he/she meant "battery" or "battery cells". The correct localization of "battery pack" is thus "batteria".

4.4 Operating Basics Chapter - Section "Functional Overview"

Using Scope Mode Section

Source Sentence:
Press AUTORANGE to set the vertical, horizontal, and trigger automatically for a usable display.

Localization of the above sentence:
Premere AUTORANGE per impostare automaticamente la deflessione verticale, la velocità di scansione e i circuiti di trigger per ottenere una traccia utilizzabile.

Reasons:
1) What does the writer mean with "vertical " and "horizontal"?

Two set of plates are present in the CRT: the horizontal deflection plates and the vertical deflection plates. A sawtooth type voltage is applied to the horizontal plates, causing the beam to sweep horizontally. The input signal voltage is applied to the vertical plates, at the same time causing the beam to move vertically.
Thus, "vertical" means "the signal (vertical) deflection system", and "horizontal" means "the beam (horizontal) deflection system".
In Italian the correct term to use are "circuiti di deflessione verticale" and "circuiti di deflessione orizzontale".

2) How do we localize correctly the term "trigger"?
Triggering is quite a complicated concept. Basically, the horizontal (sweep) signal must be synchronized with the vertical (input) signal, otherwise the displayed waveform will drift across the screen randomly.
The trigger system includes:
1) An input amplifier, which consists of a differential amplifier and of the controls Trigger Slope and Trigger Level.
2) A Schmitt trigger (a bistable circuit which produces a rectangular wave, regardless of the input waveform)
3) A derivative circuit, which converts the output rectangular wave of the Schmitt trigger in a series of positive and negative pulses
4) A limiter (basically a diode circuit) which outputs only the positive pulses, used to control the horizontal sweep generator (which produces a sawtooth waveform).
In Italian you can use the term "circuiti di trigger" to summarize all of the above. "Trigger" is an established term in the electronic terminology, so generally we *do not* translate it; however, we have to use it appropriately.

4.5 Operating Basics Chapter - Section "Application Examples"

Source Sentence:
Triggering on a Missing Data Pulse:
A positive-going TTL data pulse, 20 μs wide, should occur at least once every millisecond. The circuit is not working correctly and you suspect an occasional missing pulse. Set up the instrument to find the missing pulse.

Localization of the above sentence:
Localizzazione di un impulso di dati mancante
Un impulso di dati di un circuito TTL di ampiezza positiva, durata 20 μs, deve presentarsi almeno una volta ogni millisecondo. Il circuito non funziona correttamente e si sospetta che sporadicamente venga perso un impulso. Impostare lo strumento in modo da localizzare l'impulso mancante.

Reasons:
I see two point of concern here: 1) What does it mean "a positive-going TTL..." and 2) How to translate correctly the header "Triggering on...".

1) TTL stands for "transistor-transistor logic", a logic circuit design in which transistors act on logic signals, which are signals having one of two voltage levels. In Italian we say "famiglia

logica TTL", but here it is also correct to say "circuiti TTL". These are bipolar circuits, i.e., pulses are either positive or negative. Thus, "positive-going" means "di ampiezza positiva".
2) The basic concept here is that we use the triggering action of the oscilloscope to locate the missing pulse. The setup must be such that the instrument triggers if the signal remains in the low state long enough (more than 1 ms). If it does, we have found an occurrence of a missing pulse. A correct translation of "Triggering..." would be "Attivazione dei circuiti di trigger..." or (better) "Comando del trigger...", but then the header would get quite heavy and too literal: "Attivazione dei circuiti di trigger per localizzare un impulso di dati mancanti" or "Comando del trigger per localizzare un impulso di dati mancanti". I think the translation provided is more appropriate.

Note: Since the setup instructions that follow say that we must set the trigger appropriately and how to do it, no clarity is lost if we do not mention the word "trigger" in this introductory sentence.

4.6 Operating Basics Chapter - Section "Application Examples"

Source Sentence:
Triggering at a Specific Motor RPM Section:
You need to measure the start-up current of a 3600-RPM motor at several specific speeds. A tachometer, attached to the motor, outputs a low-voltage square wave with 100 pulses per revolution. Set up the instrument to trigger at 1200 RPM so you can measure the current at that speed.

Localization of the above sentence:
Trigger ad una specifica velocità di rotazione di un motore
È necessario misurare, a varie velocità, la corrente di spunto di un motore funzionante a 3600 giri/min. Un tachimetro, collegato al motore, fornisce un'onda quadra a bassa tensione con 100 impulsi per giro. Impostare lo strumento in modo che si abbia il trigger a 1200 giri/min., cosicché sia possibile misurare la corrente a questa velocità.

Reasons:
We are dealing here with an electric motor, and very likely with a cage motor, i.e. an asynchronous motor in which the rotor (the secondary circuit) consists of a cage winding. Why a cage motor? Because the writer is concerned with the start-up current, and cage motors at start-up absorb a current which can be 7-8 times the steady state current.
I am not going to describe any further the electric motors; the point here is that to correctly translate "start-up current" (the writer could have used the expression "Inrush Current"), you need to know them somewhat. You could be tempted to say "corrente di avviamento"; however, the standard, technical term used is "corrente di spunto".

4.7 Specifications Chapter

First of all, we need to translate correctly the title of the chapter.

Localization of the term *Specifications*:
Translate it with "Dati tecnici" if the list consists mainly of numerical values (i.e., voltage, current, frequency, start-up current, etc.); translate it with "Caratteristiche tecniche" or

"Caratteristiche generali" if the list consists mainly of general descriptions (e.g., location of fuses and buttons, etc.).

Reasons:
"Specifications" is a detailed description of the characteristics of a device, machine, product or process.
This term is often translated as "Specifiche". In technical Italian this term means "A detailed description of the *required* characteristics of a device, machine, product or process".
When a device, product, etc. is the design or production stages, its characteristics can still be changed as needed, to meet specific requirements, so we have "Specifiche" (i.e., "Specifiche di progetto"). However, when a product is completed and available to the end user, its characteristics can not be changed anymore, so they are not "Specifiche", they are either "Technical Data" or "Technical Characteristics".

4.8 Specifications Chapter - Section "Trigger"

This section is introductory to the next one, which deals with "Burst", but it is also contains some useful concepts and terms.

Sensitivity, Video Trigger Type, typical:
Composite video signal with negative sync pulse amplitude from 0,6 to 2,5 divisions
Signal Formats and Field Rates, Video Trigger Type

Broadcast Systems:	Supports NTSC, PAL, and SECAM
Interlaced:	Field 1 or field 2
Non-interlaced:	Any field or any line
Line Rates:	15 kHz to 65 kHz, in five ranges

Holdoff Range:
500 ns to 10 s

Localization of the above sentences:

Sensibilità, trigger video, valore tipico
Segnale video composito con ampiezza dell'impulso di sincronizzazione negativo da 0,6 a 2,5 divisioni.
Caratteristiche dei campi e dei tipi di segnale, trigger video

Sistemi di diffusione:	supporta NTSC, PAL e SECAM.
Scansione interallacciata:	campo 1 o 2.
Scansione non interallacciata:	qualsiasi campo o riga.
Frequenza di riga:	da 15 kHz a 65 kHz, in cinque intervalli

Intervallo di holdoff
Da 500 ns a 10 s.

A quick explanation of the above sentences and their translation:
A "composite video signal" is the standard television signal. This signal is a complex mixture of video information (luminance and chrominance) plus burst and synchronization pulses (more later on "burst"). So the first sentence tells us that the oscilloscope can trigger on the sync pulses of a video signal. Its translation is quite straightforward.

The next sentence is also quite simple, knowing that we are dealing with a video signal: "interlaced" means "interlaced scanning", and "rate" means "frequency" (not "velocità" or "tasso"!) since we have the standard unit of frequency, hertz.

Note: when you write the name of an unit of measure, *almost always* you write it all lowercase; so: hertz, *not* Hertz; ohm, *not* Ohm; volt, *not* Volt *or* volts; ampere, *not* Ampere; watt, *not* Watt or watts; etc. (Ref. 4).

Holdoff refers to the holdoff circuit; this circuit is controlled by the horizontal sweep generator, and it produces a signal such that the trigger circuit will not accept another trigger signal before a specified amount of time has elapsed. The holdoff helps ensure a stable display.
So, "range" here is "time width". We could translate "holdoff" with "ritenuta", however, anyone who knows how the oscilloscope works knows what holdoff is, and an important part of the localization process is knowing what your audience takes for granted. So it is perfectly OK to leave it in English.

4.9 Specifications Chapter - Section "Measurements"

The following section relates to the burst signal and is based on the previous one.

Source Sentence:
Automated Measurements:
Amplitude, Burst Width, ..., Negative Duty Cycle, ..., Pk-Pk, ...

Localization of the above sentence:
Misure automatizzate
Ampiezza, durata di burst, ..., duty cycle negativo, ..., picco-picco, ...

Reasons:
1) The burst signal is that portion of the composite color television signal consisting of a few cycles of a sine wave of chrominance subcarrier frequency. To simplify, the burst signal carries the chrominance information. I have seen "burst" translated with "raffica", because this is what dictionaries usually have. If you dig a little bit more, you could find "impulso di sincronizzazione di sottoportante", which is the translation of the basic definition. Neither the first translation nor the second one is good. Why? The first one is not used in the field. The second one, even though is technically correct, is not useful: if the reader knows what burst is, he/she does not need to read its definition; on the other end, if the reader is not so knowledgeable (for example, "burst" could appear in a glossary at the end of a medical video camera), the definition does not clarify things. In this last case I would say something like "segnale di sincronismo del colore", which in my opinion allows a person with no background in electronics to grasp a little bit the underlying concept. In the case of this manual, however, the reader is surely familiar with the term "burst", he/she will recognize it, so we just say "burst"!
2) The duty cycle is a measure of the proportion of time during which a circuit is operated. It can be "ciclo di funzionamento" or "fattore di utilizzazione", but again, in this case the English term is so widespread that we leave it as it is.
3) Pk-Pk means "from peak to peak", i.e., from the maximum to the minimum value of a quantity (voltage, current, etc.). Do not write "da massimo a minimo"; the standard way in technical materials is to write "picco-picco", or "da picco a picco". I prefer to use "picco-picco".

5. GLOSSARY

The glossary provided here is partial. A much wider glossary of advanced terms, in English and Italian, will be provided and discussed during the presentation of this paper.

Current Flow ***Corrente***

I have inserted here this term because I have seen it translated with "flusso di corrente". This phrase is a (English) misuse. The term should be "flow of charge". Current is defined as the rate of flow of charge, so "current flow" literally means "flow of flow of charge" and thus should be avoided. If you see this term in a document, just translate it with "corrente".

Floating Measurements ***Misurazioni con massa flottante***

Voltage measurements where the reference voltage is not earth ground. This means that an oscilloscope channel, or a DMM input, is isolated from the main chassis of the instrument. Since the chassis is the ground ("massa") of the instrument, the input circuit is not connected to ground at any point and thus the reference voltage of the signal being measured can "float".

Range ***Portata; Campo***

This term is used in a variety of situations in electronics.

Example 1: Useful voltage range of a meter
Example 2: Range over which the value of an electronic variable (ex: frequency) fluctuates
Example 3: The maximum distance from a radar to a target is the range of the radar

Note: Range is not "gamma" ("gamma" is a very generic term). Technical terms to use:

Example 1: Voltage range (in a meter) = Portata di tensione
Example 2: Frequency range (in an oscilloscope) = Campo (di variabilità) di frequenze
Example 3: Radar range = Portata del radar

6. REFERENCES

I have put here only references which have a definition of every term listed; this allows to put the term in context. I think you should never rely only on a English-Italian dictionary which has a list of terms without definition: if by chance you do not know what the English term means, how can you make sure that the translation provided is appropriate for the document you are translating?

1 The Institute of Electrical and Electronics Engineers, Inc. *Dictionary of Electrical and Electronics Terms*.
A basic reference in the field; every term comes with a standard definition.

2 Gibilisco, S.; Sclater, N. *Dizionario di elettronica*. Zanichelli, 1993.
This is the Italian Edition of *Encyclopedia of Electronics, Second Edition*. Tab Books, 1990. It includes a good index (in English and Italian). This reference is valuable because (following the English source) every term comes with a definition/description (in Italian); this helps to see the term in the right context.

3 Mc-Graw-Hill Zanichelli. *Dizionario enciclopedico scientifico e tecnico inglese-italiano italiano-inglese*. Zanichelli 1986.
This is the Italian Edition of *Mc-Graw Hill Dictionary of Scientific and Technical Terms*, 1978. This reference is valuable because every term comes with a definition/description (in English); this helps to see the term in the right context. Some of the translations provided can be inaccurate or too literal.

4 Lesina, R. *Il nuovo manuale di stile*. 2nd Edition. Zanichelli, 1994.
A very valuable reference which covers almost every aspect of style for technical documentation (in Italian). It contains a good glossary, list of references and index.

5 Biondo, G.; Sacchi, E. *Manuale di elettronica e telecomunicazioni*. Hoepli, 1993.
This valuable book contains terms and concepts of analysis and design of electronics (in Italian). It has a good index. To understand it, you must have a solid background in electronics.

6 Angelini, G. et al. *Manuale di telecomunicazioni e trasmissione dati*. Calderini, 1993.
This valuable book contains terms and concepts of analysis and design of telecommunications and data transmission (in Italian). It has a good index. To understand it, you must have a solid background in electrical equipment, electronics and telecommunications.

7 *FE (Fare elettronica)*. Monthly magazine published by Gruppo editoriale Jackson (in Italian).
In this magazine you can read about a wide variety of electronic products (analysis, design and installation) and review the manufacturers ads to keep up to date with the field. To understand it, you must have a good background in electrical/electronics components and equipment.

SPECIALIZED DICTIONARIES FROM THE PUBLISHER'S POINT OF VIEW

Marie-Jeanne Derouin
Verlag Alexandre Hatier

The art of dictionary-making, international specialized terminology

My purpose will be to show that dictionary-making is an art which requires the skillness of many actors: authors, editors, publishers etc. and aims to meet the needs of the translation profession, from the lexicographical and the scientifical point of view.

Nowadays a publishing house for specialized dictionaries is a rare species. What are the reasons for this? They are manifold and not lastly of an economic nature. But the few publishing houses involved in this work will certainly confirm that it is an exciting challenge.

Its main aim is to prepare international specialized terminology for a variety of users. Up to now the book has been the common data carrier. A change has occurred over the past few years and electronic media are becoming increasingly popular. Other developments such as publishing on demand or terminology services will follow which will radically change the profession of the publisher.

Regardless of the form of the manuscript or the data carrier, the content is the result of fruitful cooperation between authors, publishers, experts, publisher's readers and users. A wealth of competence and experience goes into this cooperation as is reflected in the quality. Nonetheless, it remains the work of human-beings and as such is and will continue to be imperfect.

In the short time at my disposal I would like to describe the design and content of our specialized dictionaries and provide some information of how they are produced.

In its publications, a publishing house for specialized dictionaries provides a scientific presentation of specialized terminology.
As it is almost impossible to possess universal specialized knowledge, a publishing house often specializes in certain areas, for example, Hatier in "Engineering and Applied Sciences".

Another decision to be taken is whether to publish mono-, bi- or multilingual dictionaries. Should the decision have been taken to publish multilingual dictionaries, the next consideration is the language combinations, the scope of the entries and the structure of the information provided. These latter factors depend on the respective target groups. Naturally, professional translators and users evaluating specialized literature place different demands on dictionaries and this must be taken into consideration by the publisher.

Every publishing house has its own profile, a fact which certainly does not exclude the possibility of development or novel products.
A range of dictionaries will be produced which while influenced by market demands is the product of the lexicographical considerations of the publishers. By way of example, I would like to present our range of products.

1. Bilingual dictionaries for engineering and applied sciences

These contain basic specialized vocabulary for complex areas and continue on from general dictionaries. The degree to which these are comprehensive depends on the number of terms included. A balance is sought between the various specialized areas although it is legitimate to place a certain emphasis depending on the importance of the specialized area in a particular linguistic region. Teams of authors work on the dictionaries under the guidance of the publisher who is responsible for the final editorial work. The bilingual dictionaries are two-directional, mainly English-German and German-English but also for French, Spanish and Russian. They contain between 100 000 and 150 000 terms.

2. Specialized dictionaries, bilingual

These contain specialized terminology for complex areas and overlap with the dictionaries described above. They include all sub-sections and form the basis for the specialized dictionaries which contain more specific terminology. These dictionaries are compiled by a team of authors who work or teach in the specialized areas. A publisher of responsible author coordinates the work and is in charge of final editing.
Generally these dictionaries contain between 50 000 and 80 000 terms and usually are published in an English-German, German-English language combination.

3. Special dictionaries, bilingual

These contain terminology for highly demarcated areas and contain more specialized terms than those contained in the dictionaries described in category 2. They vary in size but generally do not exceed 30 000 terms as the terminology can be precisely delimited. They are compiled by individual authors. They are published in the language combinations English-German/German-English and depending on the number of terms both language versions may be contained in one volume.

4. Special dictionaries, multilingual

It is imperative that these dictionaries are confined to strictly delimited specialized terminology in order that exact equivalents may be provided in all languages. The main languages are English, French and German. The inclusion of other languages depends on the importance of the specialized area, the market situation and available authors. Generally in size they do not exceed 20 000 terms. On account of its multilingual form, the dictionaries are generally compiled by a team of authors.

In addition to the type of dictionary, the languages included and the size, references and the way in which they are incorporated determine to a large degree the quality of a dictionary.

The basis for any manuscript is lexica, text books, thesauri, specialized periodicals, standards etc. in the source language. Source references must be included.

The uniformity of the lexicographical system used by our publishing house - the result of years of experience in publishing dictionaries - ensures that our dictionaries have a character of their own. We prefer a strict alphabetical classification, in bilingual versions the socalled nesting system and in multilingual dictionaries the natural word sequence.
We attach great importance to strict differentiation of the meaning of the terms in the target language. To do so, references to the specialized areas and other accompanying information is suitable. This is particularly the case for homonyms.

The creation of specialized dictionaries: synergy between a host of competent experts:

Rarely is a particular dictionary a "spontaneous" production which the author offers to the publishing house. Selecting a new title in the series is the result of careful scrutiny of the market, requiring a lively exchange between the publisher and the user. The topics and language versions are decided upon, potential authors selected and contacted.

Valuable information from our users also influences our decisions on the structure of the data and help us to tailor the form and content of the dictionaries to the current needs of the translator and other users.

Some remarks on the dictionaries "Engineering and applied sciences":

Publication of every single specialized dictionary requires an immense amount of work to compile the terms in the foreign language and in German and in selecting terms to take into account latest developments when giving definitions.
Producing a good manuscript is an extremely complicated task requiring a very good knowledge of languages, in-depth knowledge of the specialized area and lexicographical skills for presenting the data. These are high demands placed upon the author of a dictionary.
In the dictionary series "Engineering and Applied Sciences" it is our principle to find an author well-versed in the specialized area covered by the dictionary. As a result a large number of individual manuscripts are produced varying in "style". By "style" we mean:

- that authors include varying quantities of trivial terms, terms which may even be found in more general dictionaries or, on the contrary, submerge themselves in the specialized terminology and become too specific,

- that in as far as possible they confine the definition of terms to the basic elements, leaving it to the user to construct from this his own, often extremely complex, term or to provide the user, on the basis of complex terminological constructions which are not terms in themselves, with an example as a constructive aid for translation,

- that they provide a large number of equivalents in the target language without additional information making it very difficult to select the correct translation or that they confine themselves to the most frequent, most modern characteristic equivalent in the target language while ignoring all additional meanings (which would require additional explanation) and which are extremely important.

It is the task of the publisher to coordinate these "styles". The publisher fulfils a "role" which doubtless is no easy one. All he needs is in-depth knowledge of languages, primarily knowledge of all specialized technological areas, in particular knowledge of the latest trends, not yet commonplace, he must work extremely hard, be particularly meticulous, have a feeling for the errors and mistakes made by authors without exaggerated doubt leading him to deleting not generally known developments, tact and a good judge of character in dealing with authors: all of this he must master in a short period of time - no more than this is required!

Generally, the expert responsible for final appraisal, particularly when working in parallel with the teams of authors in compiling the manuscript of the dictionary plays no unimportant role as far as the quality of the dictionary is concerned. He works in the specialized area and is confronted daily with the material or he is a user. He provides an additional assurance that the manuscript corresponds to current requirements and thus gives the publisher additional assurance.

Where does the publisher's reader intervene in this process? Obviously in organizing the work for the publication, in ensuring contacts between the publisher, authors and experts, in proving the methods for compiling and processing the manuscript and in reading the manuscript when completed. A publisher's reader examines all the manuscripts word for word, primarily, from a lexicographical perspective as is his task. As he has worked on different kinds of dictionaries over the years, he has a good grasp of specialized terminology and can judge which terminology belongs to which specialized area. He has experience in working with lexica and encyclopedias. In addition to dictionary methodology he has a wealth of knowledge in many areas on the basis of which he can make proposals on specific questions to the publisher or the author.

Finally, mention must also be made of the proof-reading usually done by the publisher's reader or sometimes by proof-readers to keep to a minimum the typographical or printing errors.

From this description of these aspects of producing a dictionary it is evident not only that a dictionary is only as good as the publisher, author, expert and publisher's reader involved in its production but also that the quality depends on the extent to which there is genuine cooperation.

CHEMICAL ANALYSIS SPECTROMETRY

Denzel L. Dyer
Dyer Laboratories, Inc.

Keywords: Chemical analysis; Spectrometry; Spectroscopy; Spectrophotometry

Abstract: Analysts use various kinds of interaction between matter and energy to identify elements and compounds, and to determine their quantities. This paper is an introduction to the terminology used.

1. General

Spectrometry: measurement of interaction between matter and electromagnetic radiation. (exception: Mass spectrometry). The electromagnetic radiation includes the spectrum from ultra-high-frequency radio waves to gamma rays (not alpha or beta rays, which are treated as particles). That covers about six orders of magnitude of frequency or wavelength. Very rarely if at all would any one person (or even one laboratory) work across the entire range.

Name	Frequency sec^{-1}	Wavelength	Energy $kcal \cdot mole^{-1}$
Microwave	$3 \cdot 10^{11}$	1 mm	0.03
Infrared (IR)	$3 \cdot 10^{13}$	10 μm	3
Near Infrared (NIR)	$3 \cdot 10^{14}$	1 μm	30
Visible (Vis)	$6 \cdot 10^{14}$ (green)	500 nm	60
Ultraviolet (UV)	$1 \cdot 10^{15}$	300 nm	100
Vacuum ultraviolet	$3 \cdot 10^{15}$	10-100 nm	300
X-ray, gamma ray	$3 \cdot 10^{16}$ - $3 \cdot 10^{17}$	< 1 - 10 Å (0.1 - 1 nm)	3,000-30,000

Table 1. The spectral range of interest in chemistry. The numbers indicate approximate ranges. Kcal/mole is commonly used in chemistry; multiply by 4187 to convert to joules/mole.

With this wide range of frequencies/wavelengths, the equipment needed is very different in the different ranges. While we tend to think of optical equipment, that is useful only in the middle of the range, although that will get most of the attention here.

Closely related terms:

Photometry:	measurement of (visible) light intensity without much regard to color.
Spectro-:	indicates that individual colors/wavelengths/frequencies/energies are considered separately.
Spectroscopy	carries some connotation of using an instrument by looking into it.
Spectrography:	has some connotation of recording on film.
Spectrophotometry:	suggests measurement of visible light, but commonly used also with respect to ultraviolet and infrared radiation.
Spectroradiometry:	usually applied to measurement of intensity of sunlight or other sources.
Pyranometry:	measurement of total radiation from sun and sky.
Pyrheliometry:	measurement of direct radiation from the sun.

Spectrometry includes:

A. Emission spectrometry: radiation emitted by matter (especially hot matter).
B. Absorption spectrometry: radiation absorbed by matter.
C. Fluorescence spectrometry: radiation absorbed, then (in part) emitted again almost immediately, normally at a longer wavelength (lower energy).

Atoms and molecules have only certain distinct (discrete) amounts of energy (energy levels). Relatively small amounts of energy are involved in rotation of molecules (far infrared and microwave). Larger amounts of energy are involved in vibrations between atoms or groups of atoms (infrared). The maximum amounts of energy involved in spectrometry appear in changes of the electronic structure (visible, ultraviolet, X-ray) and nuclear structure (gamma ray).

Spectrometry is primarily measurement of energy absorbed or emitted as atoms or molecules change from one distinct energy level to another. Each such change involves a fixed amount of energy, which occurs in a package called a quantum or photon. While these photons can be considered as elemental particles, it is often convenient (or at least usual) to think of them as having wave motion with specific frequencies and wavelengths which are linked to their energies:

$E = h\nu$ where E = energy (ergs/atom or molecule)
h = Planck's constant ($6.62 \cdot 10^{-27}$ erg sec)
ν = frequency (sec^{-1})

Frequency and wavelength are inversely related:

$$c = \lambda\nu \qquad \lambda = c/\nu \qquad \nu = c/\lambda$$

where c = velocity of light (usually considered as the velocity of light in vacuum, very nearly $3 \cdot 10^{10}$ cm/sec)
λ = wavelength (cm)
ν = frequency (sec^{-1})

Most of these definitions give numbers multiplied by inconveniently large or small powers of ten, so more convenient definitions are commonly used in various spectral ranges; see Table 1. The wavelength is used most commonly. Measurements in UV/Vis typically refer to wavelength in nanometers (nm) (1 nm = 10^{-9} m). The same units were once (and often still are) called millimicrons. The ultraviolet range, for chemical analysis, is usually about 250 - 400 nm. Ångstrom units, 10^{-10} m or 0.1 nm, are used in the ultraviolet and more often for X-rays and gamma rays. Visible light is in the narrow band of 400 - 700 nm, and infrared from 700 nm up to about 50 μm (although most chemical measurements stop at 16 - 25 μm). The μm unit is properly a 'micrometer', but it is very often called by its older name, 'micron'.

Infrared spectrometry in particular uses the **wave number,** $\bar{\nu}$ (cm^{-1}), defined as the number of waves in a distance of one centimeter. $\bar{\nu} = 1/\lambda$ (cm) $= \nu/c = E/hc$ (where h and c are both constant, so $\bar{\nu}$ is directly proportional to the energy per photon).

2. The high-energy end of the spectrum - the particle range

Here we deal with changes in the atomic nucleus, or with changes in electronic energy levels relatively 'deep' within atoms. The radiation is treated more as high-energy particles than as waves. In order to use 'spectro-' methods, there must be a way to distinguish the particles by their energies. While there are instruments which can act like prisms in part of this range, most measurements use a solid-state detector made of silicon doped ('drifted') with lithium. (Germanium is used for particularly high energies (wavelengths < about 0.3 Å)). A voltage is applied to the detector. An X-ray or gamma ray absorbed in the detector produces electrons (and 'holes') which give a pulse of current proportional to the particle energy, and a pulse-height analyzer (PHA) can give a spectrum which is a plot of the number of pulses vs. the energy of the pulses.

Elements that are naturally radioactive, or which have been made radioactive by exposure to neutrons (NAA : Neutron Activation Analysis) have well-known half-lives and emit characteristic gamma rays. Gamma-ray spectra, especially if they are recorded at different times to help distinguish isotopes with different half-lives, identify and quantitate many elements. However, the need for a strong neutron source (such as a nuclear reactor) limits application of the method.

X-rays can be used in emission or fluorescence. X-rays are emitted when atoms are struck by fast-moving electrons, and the X-rays are characteristic of the element (and the voltage used to accelerate the electrons). This process is most often used in the electron microprobe, an extension of the scanning electron microscope. A very finely focused beam of electrons is directed onto a particular point, or scanned across a sample. The electrons generate X-rays, some of which are caught in a solid-state detector and used to be used to produce a spectrum, or to generate an image indicating the distribution of a particular element. In this application, the detector-analyzer combination does XES (X-ray Energy Spectrometry) or EDXRA (Energy-Dispersive X-Ray Analysis). These systems are not sensitive to the very lightest atoms.

In X-ray fluorescence (XRF), a beam of X-rays is directed onto a sample. They should be 'monochromatic'; i. e., all should have the same energy. That can be arranged by selecting the material for the X-ray tube and the tube voltage, and by removing less energetic X-rays by absorbing 'filters'. The X-rays absorbed by the sample raise inner electrons of atoms to higher energy levels, and when they 'relax' back to the 'ground state' they emit other X-rays. Those X-rays yield spectra which are characteristic of the elements and their concentrations.

3. Medium energy — the optical range

Most analytical chemical work is done in this range, with emission, absorption, and fluorescence all in use. The range generally covers the ultraviolet (UV), visible, and infrared (IR) regions, although materials and techniques are different in the infrared. As a rough rule of thumb, ultraviolet and visible radiations are involved in changes of electronic energy levels of molecules, while infrared radiation affects vibrations between atoms and groups. Far infrared can involve rotations of molecules.

Monochromators, which spread out the light into a spectrum, are usually based on either prisms or gratings. Early instruments used prisms, but later production techniques have put gratings in the lead. For either type, incoming light is focused on an entrance slit, dispersed by the prism or grating, and imaged on an exit slit through which the light goes to a detector. The width of the slits establishes the width of the wavelength band measured. This bandpass is constant for a grating monochromator, but varies with wavelength in a prism instrument. The light may pass through the prism or grating, or it may be reflected at the prism or grating. Another system uses a circular interference filter designed so that the pass band changes as the filter is rotated. In any design, much care is taken to assure that essentially only light of the desired wavelength gets through to the exit slit; other light is 'stray light'. It is important to be able to use a very narrow slit so as to measure only a very narrow band of wavelengths. The ultimate, of course, would be infinitely narrow, but at infinitely low energy. In special cases, tunable laser sources offer extremely narrow bands but high energy over limited spectral ranges.

In the following drawings, S indicates the light source, M the monochromator, X the sample, and D the detector.

3.1 Emission spectrometry (especially for metals)

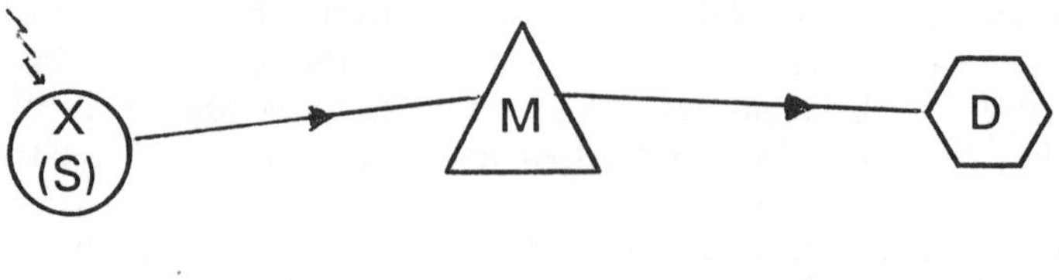

In arc or spark emission, energy is put into the sample by an electrical arc or spark. If the sample is metal, it may be one electrode; or a powdered sample may be placed in a cup in a carbon electrode. Elements in the sample emit light in numerous narrow 'lines' which are separated in a monochromator. The light may be measured by photomultiplier tubes positioned along the spectrum. Alternatively, a photographic film records the light and, after development, the light intensity is determined using a densitometer. Standards of known composition are used to generate spectra for calibration.

In flame emission (atomic emission, flame photometry), solutions of standards and samples are sprayed into a flame. Some elements — especially lithium, sodium, and potassium — are 'excited' by the flame, and emit their characteristic lines, which are measured by a spectrophotometer. As relatively few emission lines are involved here, color filters are sometimes used in place of (or as) the monochromator. Emission intensity is (approximately) directly proportional to concentration, but several standards are normally measured, as the calibration line may not be straight.

Inductively Coupled Plasma (ICP) spectrometry is like flame emission with a much hotter flame. Sample and standard solutions are sprayed into a flow of argon gas. The gas is excited to a plasma (very hot) as it passes through a high-frequency induction coil. The emitted light (many lines) passes through a monochromator and is detected by one or more photomultipliers. In a 'sequential' ICP spectrometer, a single photomultiplier is moved from one position to another along the spectrum to measure certain specified (programmed) lines. A 'simultaneous' instrument has a number of photomultipliers (e. g., 30) preset to measure a number of elements. One advantage of ICP over flame emission is that the high temperature prevents or reduces formation of molecules such as metal oxides or phosphates which do not emit the characteristic lines. As the response is usually linear over a wide concentration range, a single standard can be used.

3.2 Absorption spectrometry

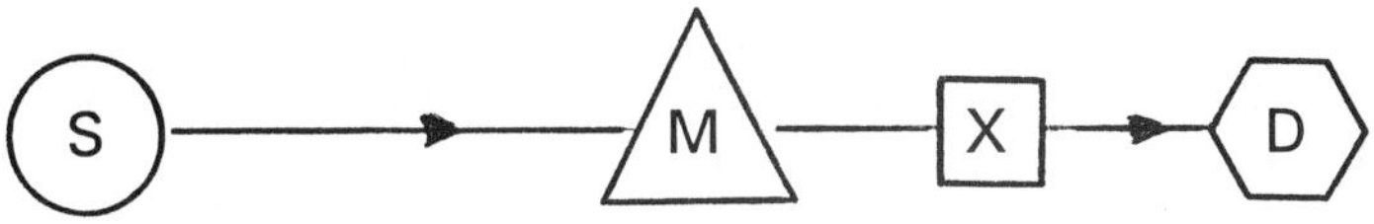

	UV/Visible	Infrared
Light source	tungsten lamp	'Globar'; Nernst glower
Transparent materials	fused quartz; glass (visible only) plastic (visible; aqueous solutions)	salts; silver chloride; sapphire, etc.
Detector	photomultiplier (etc.) diode array	thermal detector (thermocouple)

The diode array, a few hundred photodiodes packed together, can record a UV/Vis spectrum without having to 'scan' the monochromator. It appears to be most useful in a spectrophotometric detector for high-pressure liquid chromatography, where it can generate an absorption spectrum of each component as it elutes from the chromatographic column.

The typical sample holder for UV/visible is a quartz 'cuvet' (or 'cuvette') about 5 cm high, with inside dimensions 1 cm x 1 cm, although other sizes and shapes are used. The light path can be varied from 1 mm or less up to about 100 mm.

Absorption spectra may be plotted either as transmittance (or % transmission) or absorbance vs. wavelength. Plots of log(A) vs. wavelength are occasionally seen. The wavelength scale may or may not be linear.

If I_0 is the intensity of monochromatic light entering the sample (incident light) and I is the intensity of light leaving the sample (transmitted light):

$$Transmittance = T = \frac{I}{I_0} \qquad Percent\ transmission = \frac{I \cdot 100}{I_0} = 100 \cdot T$$

and

$$Absorbance = A = \log_{10}\frac{I_0}{I} = -\log_{10}T = abc$$

The last relation, A = abc, is the Bouguer-Lambert-Beer Law, in which a is the 'absorptivity' (or 'extinction coefficient'), a constant for a given substance and wavelength; b is the thickness of the sample through which the light passes; and c is the concentration of the substance in the sample. Beer stated the relation for c, and 'Beer's Law' is particularly important for chemical analysis: absorbance is directly proportional to concentration. Quantitative measurements normally involve plotting absorbance vs. concentration for several standards, then measuring the sample absorbance and determining concentration from the plot. The plot is expected to be a straight line, at least over a limited concentration range. There are various theoretical and practical reasons why specific systems do not, or do not appear to, follow Beer's Law.

Plastic, glass, and quartz all absorb too strongly for use in the IR, where the sample is typically a thin film between a pair of polished salt (NaCl) plates. Thicker cells with salt plate windows are also used. Solid samples may be ground ('mulled') with mineral oil or a fluorocarbon oil; ground, mixed with potassium bromide (KBr) and pressed into flat disks (pellets); or dissolved in solvents with relatively little IR absorbance (e. g., carbon tetrachloride, CCl_4 and carbon disulfide, CS_2). Large salt prisms have been used in IR monochromators, but gratings are used almost universally now.

Aqueous solutions are bad for IR. Water has several strong absorption bands; worse yet, it dissolves inexpensive salt sample holders. The solubility problem can be avoided by using more expensive materials (e. g., sapphire), or with 'total internal reflection', 'multiple internal reflectance', or 'attenuated total reflectance (ATR)' plates of an IR-transparent material such as germanium. The IR beam enters at an angle and is internally reflected several times at the surfaces. Some of the IR beam ('evanescent wave') actually passes outside the plate and back inside at each reflection. If there is a sample solution in contact with one or both surfaces, absorption of the evanescent wave yields an IR spectrum.

The most severe problem for IR spectrometry has been the lack of sensitive detectors, combined with the fact that there is very little energy in any narrow band of infrared from the monochromator. That is overcome in Fourier-Transform Infrared analysis (FTIR), which uses a fundamentally different system with an interferometer rather than a monochromator:

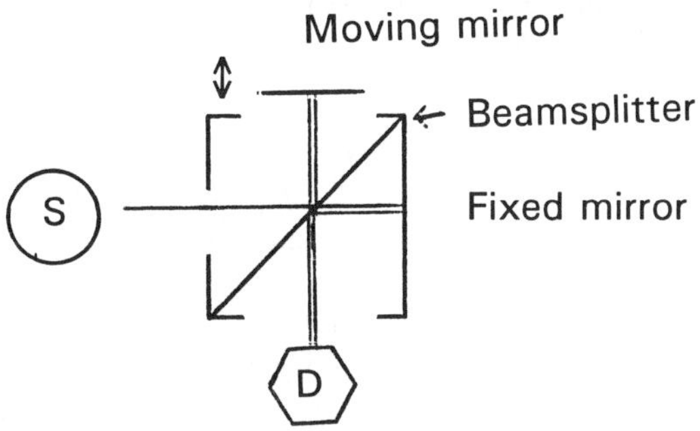

Any single wavelength passing through the interferometer, with no sample present, will generate a sine-wave output of intensity vs. interferometer mirror movement. The FTIR system passes all the IR wavelengths through simultaneously, and some are more or less absorbed. The detector sees a relatively strong but very complex signal which appears meaningless when recorded. A computer with the proper software can perform a 'Fourier Transform' on the signal to regenerate the actual IR spectrum of the sample.

Raman spectrometry, mentioned above, uses an intense monochromatic laser beam (visible or near infrared). Some of the light is absorbed in the sample, and some of that is re-emitted at a different wavelength. The difference in energy of the exciting and emitted light is the same as the energy of some infrared absorption bands. As water does not have a significant Raman spectrum, Raman spectrometry is, among other things, a means of measuring infrared absorbance in aqueous solutions without actually using infrared. However, some bands appear in IR and not in Raman, and some appear in Raman but not in IR.

3.3 Fluorescence spectrometry (Note: the stem is **fluor-**, not flour- !)

In fluorescence, some light is absorbed from the 'excitation' beam and part of that is almost immediately re-emitted, normally at a longer wavelength (lower energy). The fraction that is re-emitted is determined by the 'quantum efficiency'. If the emission is perceptibly delayed, the process is 'phosphorescence', which is much less useful for analysis.

In practice, the exciting light is ultraviolet or blue, and the emitted light is visible or near infrared. The emitted light is usually measured at a right angle to the incident beam, and there will be either a filter or a monochromator for both incident and emitted light. The light source is typically either a mercury arc (which has a strong UV emission band about 365 nm), a xenon arc, or a small UV fluorescent tube. A monochromator in the incident beam allows measurements of the excitation spectrum, and one in the emitted beam measures the fluorescence spectrum. When sensitivity is more important than spectral selectivity, broad band-pass filters are used instead of monochromators.

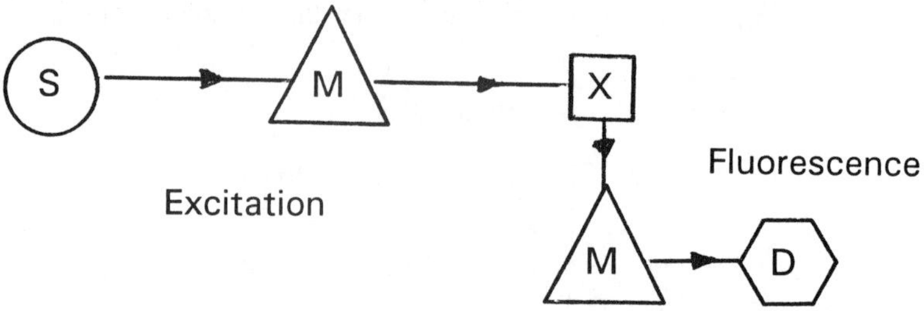

Fluorescence is much more sensitive than absorbance because it measures emitted light against nearly zero background, while absorbance can only measure a small difference between two fairly intense beams. However, relatively few molecules fluoresce, and there are other problems.

There is also some application of 'atomic fluorescence', which is used like atomic absorption to determine metals, except that a strong ultraviolet beam is directed onto the flame, and fluorescence is measured instead of absorbance.

4. Low energy - the radio region

Direct measurements of microwave absorption, measuring rotational energy levels in gases, give some indication of molecular structure. These measurements have little application in chemical analysis.

When a sample is placed in a very strong magnetic field, a few atomic nuclei exhibit 'nuclear spin' (primarily ordinary hydrogen, ^1H (protons); secondarily ^{19}F, ^{31}P and ^{13}C). They can absorb certain microwave wavelengths at certain magnetic field strengths in a phenomenon called 'nuclear magnetic resonance (NMR)'. The absorbed energy is used in making the spin axes of the absorbing nuclei 'flip' into the opposite direction.

A typical NMR spectrometer uses a single microwave frequency (e. g., 90 MHz) with a strong magnetic field (e. g., about 21 kiloGauss) to record proton (^1H) spectra. Because individual protons in a molecule are influenced by adjacent atoms, they absorb the microwave energy at slightly different magnetic fields. The magnetic field strength is varied by careful control of the magnet current to give a record of absorbance vs. magnetic field strength. For the convenience of the chemist, the spectrum is usually presented as absorbance vs. 'chemical shift', which is the difference in field strength from that at which a standard absorbs. The standard is typically tetramethylsilane, $Si(CH_3)_4$, which has 12 identical protons. The chemical shift indicates the environment of the protons inside the molecule, while the intensity of the absorption indicates how many protons are in that environment.

5. Mass spectrometry

Mass spectrometry does not involve the electromagnetic spectrum. It is a 'spectrum' of the relative weights (masses) of fragments of a molecule.

Everything in mass spectrometry takes place in a vacuum. A gaseous sample is introduced and broken down into charged fragments by electron impact or by 'chemical ionization'. The fragments, accelerated by applying a voltage, pass through a mass selector which separates them by their ratio of mass to charge (m/e or m/z). The separate fragments are detected and measured as an ion current. Under constant conditions, a molecule will break up in the same ways, giving a reproducible 'mass spectrum', which is a plot of the relative numbers of fragments vs. their mass/charge ratio. (Most are singly charged, so the plot is effectively relative numbers vs. molecular weight.)

Early mass spectrometers separated fragments as they passed through magnetic and electrostatic fields, one or both of which were varied to direct certain m/e fragments to the detector. Such instruments still have the best resolving power (ability to distinguish slightly different masses), but they require minutes to record a spectrum. Most mass spectrometers in analytical laboratories use a different mass selector. One common selection technique is 'time of flight' (TOF). In these systems, the fragments are accelerated toward the detector by a brief voltage pulse. The lighter fragments are accelerated more and arrive at the detector first. Mass spectrometers are now commonly used as chromatographic detectors, producing mass spectra every tenth of a second.

6. References for terminology

1. Banwell, C. N., *Fundamentals of Molecular Spectroscopy*, McGraw-Hill, New York, 1966. 291 pages covering theory and some practice of microwave, infrared, UV-visible, and magnetic resonance spectrometry. Not current, but still quite informative and much more readable than most such references. Relatively light on details of instruments. Better for general background than for any specific translation work.

2. Christian, Gary D., *Analytical Chemistry*, 5th Edition, John Wiley, New York, 1994. Most college textbooks of analytical chemistry have fairly good presentations of spectrometry. This is one recent example which does quite well.

3. Constantin, E., and Schnell, A, *Mass Spectrometry*, Ellis Horwood, New York, 1990.(Translation of *Spectrométrie de masse*, 1986.) 184 pages. As the authors say, they have "assembled in condensed form the basic essentials and the fields of application". A good introduction with many explanatory drawings.

4. Parker, Sybil S., Ed., *Spectroscopy Source Book*, McGraw-Hill, New York, 1988. This appears to be a compilation of articles from the McGraw-Hill Encyclopedia of Science & Technology. In 288 pages, it covers, briefly, almost all of the theory and practice of spectroscopy. First choice as a broad general reference limited to spectrometry.

5. Willard, H. H., Merritt, L. L., Dean, John A., and Settle, F. A., Jr.: *Instrumental Methods of Analysis*, Wadsworth, Belmont CA, 7th Ed., 1988. Not as complete coverage of spectrometry as [4], but much more attention to actual instruments. Also includes several other techniques. 895 pages. First choice as a general reference in instrumentation with excellent coverage of spectrometry, mass spectrometry, chromatography, and electrometric measurements.

GUNS AND AMMO: TERMINOLOGY

Denzel L. Dyer
Dyer Laboratories, Inc.

Keywords: Ammunition; Cartridge; Pistol; Rifle; Shotgun

Abstract: Terminology of firearms and ammunition is often unclear, imprecise and ambiguous. This paper attempts to introduce some basic technology and some sources.

Anyone accustomed to working in scientific and technical fields appreciates clarity of terminology, precise meanings, and lack of ambiguity. Of course, not all technical fields meet those standards; but the field of firearms and ammunition does not begin to approach any of them. A single term may have two or more meanings, some of them contradictory. Experts disagree over meanings of terms. Writers (especially fiction writers, though the distinction is often uncertain) often use terms extremely casually to make their writing sound authentic. And people in the mass media are far from expert, so if your knowledge of firearms terminology comes from newspapers or television, you are in deep trouble.

This presentation will not produce experts. Its intent is to introduce some of the basic terminology and some sources, and perhaps to help you recognize some of the wilder errors so that you will not introduce any absurdities beyond those in the original text. Experts will consider this terribly oversimplified, and they will argue about several points. Several terms are used without being defined at all. All dates are approximate. Finally, so many different designs have been tried that words like 'always' and 'never' are rarely true.

1. HISTORY

The origins of guns and gunpowder will probably never be known with certainty. Black powder, a 75:25:10 mixture of saltpeter (potassium nitrate, KNO_3), charcoal, and sulfur, may have been discovered in China, Arabia, Germany, or elsewhere. Guns appeared across Europe early in the 14th Century (by 1326 at the latest), and almost every conceivable variation was tried during the next century, although the state of metallurgy and machining was generally inadequate. Terminology was developed, stolen, borrowed, and altered freely. Few people were literate, and standard spelling came centuries later. In any case, literate people (mostly priests) did not associated with gunners, who were thought be be in league with the Devil (a belief that seems still to prevail in some quarters).

2. ARTILLERY

Artillery (used by land forces) is usually divided into guns, howitzers, and mortars. Guns, at least in principle, are used in direct fire; that is, the gunner sees and aims at the target. To fire at a longer range, the gunner points the barrel higher (or uses a heavier powder charge). Bores typically range from about 3 inches (76 mm) to 6 inches (155 mm); naval guns go up to 16 inches (400 mm). Mortars fire at very high angles, typically 60° to 70° above the horizontal. In that situation, aiming higher gives shorter range. Some huge mortars have been made; one had a bore 36 inches in diameter with a ten-foot barrel. In current use, though, mortars are usually relatively light so that they can be moved fairly easily by troops near the enemy. They typically have bores of 3 to 4 inches (81-106 mm). Those mortars are usually not even called artillery, and many are not rifled, depending on fins to stabilize the projectile. Howitzers have an intermediate position. They fire at a lower angle than mortars, but higher than guns, though they may also be capable of direct fire in an emergency.

The smaller guns use fixed ammunition: the powder charge and shell are combined in a metal case so that only one piece has to be handled. Larger guns, howitzers, and mortars are typically separate-loading, so that larger or smaller powder charges can be used as needed to adjust the range. Gun and howitzer gunners add bags of powder; mortar gunners typically remove sheets of propellant ('increments') fastened to the base of the shell.

The old muzzle-loading cannon fired solid shot, originally stone and later cast iron. Later, they were able to fire other projectiles:

- shell: hollow shot filled with explosive, usually with a time fuse to make the shell explode in the air. (German: Granate)
- grape: a set of several smaller cannon balls for use at moderate range (Kartätsche)
- cannister: a thin metal package of musket balls fired at short range; the can burst at the cannon muzzle, so that the cannon acted as a big shotgun (also Kartätsche in German)
- shrapnel, or spherical case: invented by a British officer of that name, this was a shell filled with musket balls and a bursting charge just big enough to scatter the balls a little. It acted like concentrated musket fire, or cannister, but at cannon range (Schrapnell).

Current artillery projectiles do their damage by exploding and scattering fragments. It is important that the explosion occur at the right time. That is accomplished with time fuse for air bursts; VT (variable-time) fuses which are small radar sets that burst the shell at a set distance above the ground; super-quick fuses which fire on the first contact, and delay fuses which fire only after penetration.

Tank guns have become a very special field. At present, they are typically 120 mm (5 inch), and are equipped with some very special instrumentation, including thermal (infrared) sights, laser rangefinding, and a device to measure how much the gun barrel is drooping under its own weight. A typical antitank projectile is a solid rod of a heavy metal (tungsten or uranium) fired at extremely high velocity. The projectile is commonly smaller in diameter than the gun bore, and is carried by a 'sabot' which fills the bore and drops off when it leaves the barrel. The velocity is so great that rifling no longer works, and the projectile is fin-stabilized (APFSDS: armor-piercing fin-stabilized discarding sabot) (Flügelstabilisiertes Treibkäfiggeschoss). On impact, the projectile punches a hole through the tank armor, releasing much heat and many pieces that bounce around inside. Other anti-tank shells include HEAT (high-explosive anti-tank) with a shaped charge that produces a penetrating jet, and HESH (high-explosive squash-head), which is not intended to penetrate but to make a piece of armor flake off (explosively) on the inside of the tank.

3. FIREARMS

Small arms (firearms) terminology is quite fragmented. Basically, though, there are three species (and one or two hybrids), all descended from a common ancestor (hackbut, arquebus) during the last 7 centuries.

Early hand cannons were hard to use from horseback. The rider needed one hand for the reins, at least one hand to hold the gun, and one hand to insert a red-hot wire into the touch-hole. A practical firearm required at least a better means of firing. The first answer was a small pile of powder in a 'pan' around the touch-hole, ignited by lowering a smoldering 'slow match' (a cord impregnated with potassium nitrate) onto it. That development was the matchlock, and it was still used primarily by foot soldiers. It was followed by the wheel-lock, which used a steel wheel spinning against a flint to throw sparks into the powder. The mechanism is used now in disposable lighters, but was very expensive then. Wheelocks were used extensively by mounted troops. They were followed by various types of flintlocks. (Experts and collectors distinguish between the snaphaunce, the miquelet, and the true flintlock.) In the flintlock, the jaws of the 'cock' (or the 'dog') were clamped onto a flint held in a thin sheet of lead. On firing, the cock scraped the flint across a rough steel 'frizzen' to throw sparks into the powder. Sometimes the flash did not propagate through the touch-hole into the chamber, thus the phrase 'a flash in the pan'. The firearms of the American Revolution were flintlocks.

The term 'chamber' (Kammer; Patronenlager) will appear again. It was originally the part of the barrel that held the gunpowder. Now it is the part of the gun that holds the cartridge at the time of firing. The chamber is at the breech end of the barrel, and is enlarged just enough to hold the cartridge (except for revolvers, which have several separate chambers in a 'cylinder').

Early firearms were sensitive to rain and dew because they required powder to be exposed in the pan. That problem was cured with the percussion cap (ca. 1840). That was a thin copper cup containing a small amount of an explosive sensitive to impact (originally mercury fulminate). The cap fit over a matching 'nipple', and the flash was carried through a hole in the nipple to the powder charge. As the projectile was typically a lead ball, these were called 'cap and ball' firearms. They were the principal arms of the American Civil War.

With minor exceptions, generally unsuccessful, firearms (and artillery) up to this time were muzzle-loaders. The advantages of breech loading were known, but there had been no practical means for sealing the breech to prevent escape of powder gas at high pressure in the vicinity of the shooter's face. The brass cartridge case provided such a seal. When fired, the brass expands enough to keep the gas from flowing back around it, while the mechanical parts of the arm support the brass well enough to prevent its bursting. The impact-sensitive explosive (primer) was moved to the inside of the case, so that everything was in a single package. All that happened generally about 1850 - 1880. Around 1900, black powder was replaced with smokeless powder, made of nitrated cellulose. Sometimes nitroglycerin was added to make a 'double base' powder. Aside from being essentially smokeless, these powders are much safer to handle.

'Cartridges' (Patronen) were originally hollow wood tubes with plugs, each holding the powder to fire one shot. Later they were made of paper, and now they are metal (almost always brass, which has better mechanical properties than steel or aluminum). Note that cartridges are very commonly and very incorrectly called 'bullets'. The bullet (Geschoss) is only the projectile.

With breech-loading arms, the breech is opened to load; but it is essential that the breech not open accidentally at the time of firing — chamber pressures are in the range of 30,000 to 55,000 pounds/in^2 (20-40 MPa). Some low-power arms get by with 'blow-back' action (Feder-Mass Verschluss), in which the inertia of a heavy 'breech-block', with some help from a spring, keeps the breech closed until the bullet is out of the muzzle and the pressure has dropped. More powerful firearms must have a locked breech. There are many ways of accomplishing the locking. For rifles, the most common is a 'bolt' (Zylinderverschluss). It is named for the common door bolt operated by raising a handle, sliding the bolt into its locked position, and lowering the handle into a notch. The firearm bolt action differs in that the force is applied to the end of the bolt, and extra 'locking lugs' are provided. The next most common breech-block is a piece that slides forward and back, and is held in the locked position by a 'breech-block lock' that slides or pivots vertically.

While those mechanical developments were going on, the arquebus was evolving into the rifle, pistol, and shotgun.

- Shotguns (Flinten) are not rifled (smoothbores) and usually fire (from the shoulder) a large number (1 - 2 ounces; 30 - 60 g) of small lead pellets (shot). They can also fire single projectiles ('slugs') of various shapes. The diameter of the bore ranges from about 0.41 inch to about 0.73 inch, but it is usually described as the 'gauge' (bore in British English). The gauge is (approximately) the number of lead spheres of bore diameter that would make one pound (except that the 410 gauge is really .41 inch). Shotgun cartridges are called 'shells', and are described by the gauge, the length of the case after firing, the size of the pellets, the total weight of pellets, and the powder charge (often in 'drams equivalent', the amount of black powder needed to give the same effect!). Designations of pellet sizes are not always consistent, but smaller numbers indicate larger pellets. The largest pellets in common use are 00 Buck, about 8 mm diameter, and the smallest common ones are 9 (about 2 mm). In German, a shotgun shell is a "Schrotpatrone" (not a "Granate"), and the "Kaliber" is the gauge, often with the length in millimeters.

  Some shotguns have a single barrel which is opened by 'breaking' (unlocking and tilting the barrel forward and down). A pair of barrels ('double-barrel') is much more common. In the US in particular, repeating shotguns are common. They have a single barrel and a magazine holding 2 - 5 or more shells, and have a slide action operated by a manual push-pull motion on the part of the stock that extends forward under the barrel. These are often called 'pump guns'. In many repeating shotguns, manual operation is replaced by an autoloading ('automatic') action utilizing the force of recoil or of the powder gas to operate the mechanism ('action'). Shotguns generally have no sights, but may have a 'rib' running along the top to help pointing.

- Rifles are designed to be fired from the shoulder. Their barrels have spiral grooves (Züge) on the inside to make the single projectile spin for much better accuracy. The portions of the barrel between the grooves are 'lands' (Felder). The caliber is the bore diameter across the lands. The projectile was originally a lead ball, but now (since about 1870) it is commonly an elongated bullet. Early rifles were very slow and difficult to load, as the bullet had to fit tightly so that the rifling would spin it; thus the ball had to be hammered down the length of the barrel. The Pennsylvania rifle (ca. 1700) (also known as the Kentucky rifle) used a smaller ball wrapped in a greased piece of cloth ('patch'). About 1840 a French Captain Minié invented a bullet which was small enough to load easily but which expanded on firing so that it would fit the rifling.

Metal cartridges were known about the time of the Civil War, but did not predominate until about 1870. Rifles typically have a single barrel, although double rifles have been common for hunting in Africa. There are some single-shots, but nearly all are repeaters, distinguished by the way the mechanism is operated (bolt, lever, or slide action). As for shotguns, manual operation is often replaced by an autoloading mechanism.

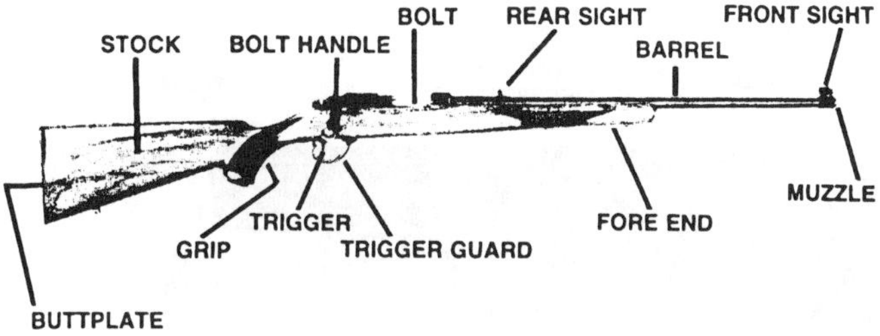

Pistols are designed to be fired from one hand (though both hands can be used). Essentially all are rifled and fire single projectiles. Aside from single-shot pistols, usually used for hunting and long-range target shooting, the major types are revolvers and autoloaders (semiautomatics, often called 'automatics'). A revolver has a revolving 'cylinder' with several 'chambers', each holding one cartridge. When the hammer is cocked, a chamber is rotated into alignment with the barrel and locked in position. There is a slight gap between the cylinder and barrel, and some gas escapes there. An autoloader has a 'magazine' holding several cartridges which are fed, one at at time, into the single 'chamber' at the breech end of the barrel. Note that many people consider that only the autoloaders should be called pistols, and that the general class is 'handgun'. Note also that a pistol may have a cylinder or a magazine, but not both.

'Action' has a special meaning for pistols:

Single action (SA): The hammer can **not** be cocked by pulling the trigger; it must be pulled back in a separate movement.

Double action (DA): The hammer can be cocked **either** by pulling the trigger or by a separate movement.

Double action only: (DAO) The hammer can be cocked **only** by pulling the trigger.

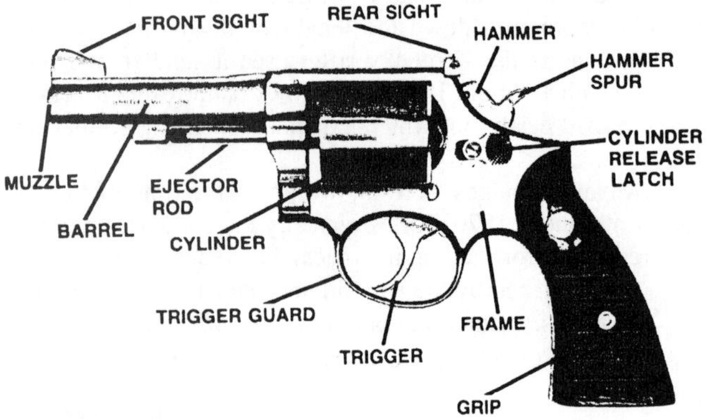

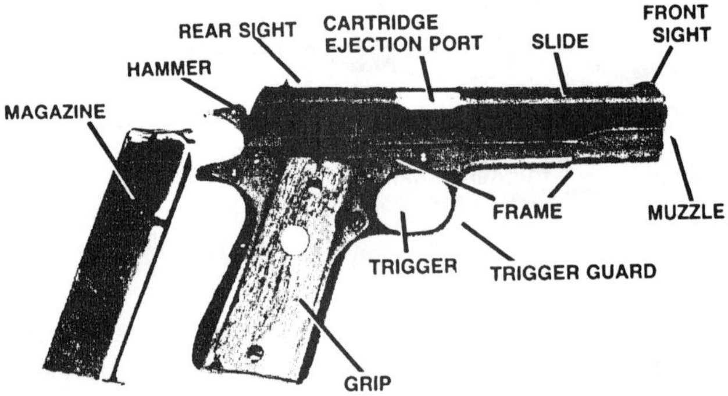

Here we must note a term with contradictory and confusing meanings: "automatic". Technically (and legally, in the US), any firearm that will fire more than one shot if the trigger is pulled and held back is called "automatic". A firearm which uses the force from firing to eject a fired cartridge case and load a fresh cartridge that is only fired after the trigger has been released and pulled again is "semi-automatic" (or an auto-loader). Historically, though, the first firearms that would automatically reload were called "automatic", and the term is very commonly used with that meaning. Because of this contradictory usage, arms which are techically "automatic" may be called "full automatic" to help distinguish them from semi-automatics (auto-loaders). As a terminology rule of thumb, no firearm generally advertised or stocked in a store in the US will really be automatic. Since the National Firearms Act of 1934, possession of an automatic arm requires a license from the Secretary of the Treasury and a special fee.

1. Rifle and Pistol Names, Calibers and Cartridges:

The most important things to remember are that most of the English terms are names which should not be translated, and that the numbers are usually nominal, not precise measurements.

Continental European caliber designations are relatively easy to understand. They indicate the caliber in millimeters (perhaps approximately, usually in the range of 5 to 9) and the length of the cartridge case (without the bullet) in millimeters. The numbers may be followed by an 'R' indicating a rimmed case. Some have names appended, but they are rarely needed to distinguish between cartridges. Some of these cartridges are also used extensively in the US, where they usually retain their original names.

British and US nomenclature is much more complicated. The first number always indicates (approximately) the caliber in inches (usually in the range of .17 to .50 for modern firearms). The caliber is usually followed by some further name or number. The names are usually either the name of the manufacturer or the designer, or a name given by the designer. A second number often indicates a powder charge (in grains), while a third number indicates a bullet weight (in grains). (1 grain ≈ 64.8 mg; 1 mg ≈ 0.0154 grain). Note that in American usage the decimal point is not preceded by a zero, and it is not mentioned when speaking: .22 is spoken as 'twenty-two'. Note also that the powder charge might be either black or smokeless powder (not interchangeable) and that actual powder charges may be different from the number.

Some common examples:

9 x 19 mm (or 9 mm Parabellum, or 9 mm Luger). A very common pistol and submachine gun cartridge. Note that there are several other 9 mm cartridges which are generally not interchangeable.

.257 Roberts A popular US rifle cartridge designed by N. H. Roberts.

.250-3000 Named (1915) for its muzzle velocity of 3,000 feet/second.

.30-30 Winchester ("Thirty-thirty") Another very popular (though obsolescent) US rifle cartridge, originally loaded with 30 grains of smokeless powder.

.30-06 ("Thirty ought six") US military cartridge adopted in 1906 for the 1903 Springfield army rifle; widely used in hunting and target shooting

.22 LR LR : Long Rifle. A very popular low-power rimfire cartridge; US production is said to be about one million per day. Despite the name, this cartridge is very widely used in target pistols. German: .22 Lang für Buchsen.

.45ACP ACP is Automatic Colt Pistol, referring to the semiautomatic pistol adopted by the US in 1911. There is a completely different .45 Colt for revolvers; it is sometimes called the .45 Long Colt as an added distinction.

.45-70-405 .45 caliber, 70 grains of black powder, 405 grain lead bullet, originally used in the US Army 1873 Springfield rifle.

2. Rimmed, rimless, rimfire and center-fire

When a cartridge is loaded into the chamber, something has to keep it from falling on through, or at least from moving in so far that the firing pin cannot reach it. The first cartridges were cylindrical, with the base slightly wider, leaving a rim which held the cartridge in place. When the brass was folded in to leave the wider base, there was a circular groove on the inside which could hold the impact-sensitive priming compound. As the case must be thin and soft enough to be dented by the firing pin, rim-fire works only for low-power cartridges. Center-fire cartridges have a much stronger base, with a separate primer pressed into a 'cup' in the center. There is a 'flash hole' through the cup to the powder charge. When several cartridges are stacked in a magazine, the rims get in the way. As they are no longer needed to hold the priming, cartridges for repeating rifles and autoloading pistols are typically rimless. (They have a groove around the base where they are caught by the 'extractor' which pulls the empty case out of the chamber.) Straight-sided cases (usually for pistols) are held in place because the chamber is slightly larger than the bore for just the length of the case. Bottle-necked cases, common in rifles, are held in place by the 'shoulder' at the base of the 'neck'. There are also 'belted' cases which have a thick rim.

3. Some special terms easily confused:

Assault rifle (Sturmgewehr): Military rifle with selection of semi- or full-automatic fire. It uses a relatively low-power cartridge, and is generally designed for low-cost production, light weight, and use by minimally trained troops. 'Assault' here is the military term: the final advance over the last 50-100 yards to take an objective held by the enemy.

Assault weapon: Political/legal term variously defined in laws, ordinances, and speeches. In general, the term refers to a semiautomatic arm which resembles an assault rifle or submachine gun, but it is also applied to pistols and shotguns.

Browning, bluing: Producing a thin layer of iron oxide to protect against rust. The result may be brown, but is usually blue. Usually done by immersion in hot salt solutions, but there are 'cold bluing' methods. A metal part to be blued later is "in the white". But note that 'Browning' was an American designer and is the name of a current manufacturer.

Caliber: For artillery and naval guns, barrel length measured by bore diameters; e. g., a 5 inch 38 caliber gun would have a barrel 5 x 38 = 190 inches long.

Clip: See 'Magazine', but note that they are different.

Cordite: A British type of smokeless powder made in long thin 'cords' until ca. 1920; often encountered in fiction as the 'smell of cordite'.

Corrosive: Refers to a cartridge with potassium chlorate in the primer.

Dumdum: An early form of expanding bullet made at Dumdum Arsenal, India, in the 1890s, in the attempt to make the small bullets used with smokeless powder as dangerous as the much larger and heavier bullets used with black powder muskets.

Firearm: Generally, an arm carried and used by one person; but note that **US legal** definitions are variable and inconsistent.

Gewehr (German): A gun; usually a rifle. A "Maschinengewehr" is usually a "machine gun", not an "automatic rifle".

Handloading: The process of putting a fresh primer, powder charge, and bullet into a used brass cartridge case (the most expensive part).

Headspace (Verschlussabstand): The space available for the cartridge between the face of the breechblock and the part of the chamber that keeps the cartridge from going forward.

Magazine: The part of a repeating firearm which holds cartridges for use. In the magazine, a spring forces cartridges into position to be fed into the chamber by operation of the action. In many arms, the magazine is an integral part, while in others it can be replaced by separate loaded magazines fitting into the "magazine well". These separate magazines are often incorrectly called "clips". A clip is either (a) a device which holds a set of cartridges and which is inserted as a whole into the magazine, (Patronenrahmen) or (b) a device which holds a set of cartridges which are pushed from the clip ("charger") into the magazine (Streifenlader).

Maschinenpistole (British 'machine pistol'): US: submachine gun, a relatively small full-automatic arm firing pistol cartridges, usually with blow-back action. However, the German MP44 was an assault rifle.

Mil: Angular change in aim that moves the point of impact of an artillery projectile one yard at a range of 1000 yards.

Minute of Angle (MOA): Angular change in aim that moves the point of bullet impact one inch at 100 yards (not exactly a mathematically true minute of angle).

Parabellum: The pistol, and especially the 9 x 19 mm cartridge for it, designed by Georg Luger and adopted by the German army in 1908.

Proof: Used in the German sense of 'test' (but the German word is "Beschuss"). Testing a gun with a particularly heavy charge to make sure that the gun is strong enough.

Sawed-off: a shotgun (rarely a rifle) having a barrel cut very short. The minimum shotgun barrel length under US law is 18 inches (16 inches for rifles).

Schützen (Schuetzen): German: to guard or protect; or to shoot, especially at targets. In American English, a particular kind of rifle target shooting introduced by Americans of German descent.

Trigger weight (Abzugswiderstand): Force which must be applied to the trigger to make a cocked firearm fire; measured by hanging a weight on the trigger. Typically about 4 pounds. About 12-18 pounds must be applied to cock a double-action pistol.

Twist (Drall): Pitch of rifling, expressed as one turn in a distance along the barrel.

Varmint: Slang for 'vermin', small animals considered pests and not protected by game laws. They are generally small and so alert that a hunter cannot get close. A "varmint rifle" may be of small caliber (.22-.25) but must be highly accurate at 200-400 yards. The German "Raubzeug" (predator) is not quite equivalent.

Visierung (German): The sighting system, including both front and rear sights.

Wadcutter: (Flachkopfwettkampfgeschoss) Pistol bullet with a flat nose; used in target shooting to make clean-cut holes which are easier to score.

4. REFERENCE MATERIALS FOR TRANSLATORS AND INTERPRETERS

Barnes, Frank C., *Cartridges of the World*, Digest Books Inc., 3d Ed., revised, 1972. Northfield, IL.

Blair, Claude, Ed., *Pollard's History of Firearms*, Macmillan, New York, 1983. 559 pages, 14 of which are index. Many photographs and drawings. Primarily historical, but covers up to about 1970. Excellent reference on early firearms.

Cormack, A. J. R. *German Small Arms of World War II*, Exeter Books, New York, 1979. English. Limited time period. Includes some drawings with parts indicated, usually with German names. Some information on ammunition.

Drury, Ian, and Gibbons, Tony: *The Civil War Military Machine*, Smithmark, New York, 1993. American Civil War, 1860-65. Particularly good coverage of artillery of the period (ca. 1820-1890); also covers rifles and pistols.

Ezell, Edward C. *Small Arms of the World*, Stackpole Publishing Co., Harrisburg PA. Several editions (e. g., 11th, 1977, 671 pages). Earlier editions were edited by Smith or by Smith and Smith. Descriptions of recent and current military firearms, with nomenclature; some historical background.

Glück, Christa, and Görtz, Joachim.: *Wörterbuch der Waffentechnik*, Ger-Eng, Eng-Ger; E. Schwebend GmbH, Schwäbisch Hall. 5th Ed., 1972. The only G < > E dictionary I have found in the field, and generally good, though not quite current.

Manucy, Albert, *Artillery Through the Ages*, US Government Printing Office, Washington DC, 1949. 92 pages. Black powder/muzzle-loading cannon, carriage, and projectile design and nomenclature. Has 51 figures and a short (2-page) glossary.

Murtz, Harold A., Ed.: *The Gun Digest Book of Exploded Firearms Drawings*; various editions, DBI Books, Northbrook IL. Detailed drawings and parts lists for hundreds of firearms.

Nonte, George C., *Firearms Encyclopedia*, Book Division, Times Mirror Magazine, Inc.,1973. Also Harper & Row, New York. 341 pages, arranged alphabetically; English, but has a 4-page glossary of terms in English, German, Spanish and French. This may be the best available source of English terms, but will be hard to find.

Peterson, Harold R., *Round Shot and Rammers*, Stackpole Books, Harrisburg PA, 1969. Describes muzzle-loading artillery, especially in US service. 128 pages, abundant drawings.

Waffentechnisches Taschenbuch, 8th Ed., 1989, Rheinmetall GmbH, Düsseldorf. 690 pages 4" x 7". **Very** technical. Many photographs and drawings. Emphasis is on artillery, with some coverage of small arms and rockets.

LEXICAL OPERATIONS IN TRANSLATION. SEGMENTATION AND INTEGRATION OF MEANING

Kinga Klaudy
University of Budapest

Keywords: translational operations, mental transformations, semantic changes, typology of operations

Abstract: The paper describes the notion of "translational operations". Translating any sentence from one language into another, even the simplest one, translators carry out a number of operations. Some of these operations can be explained by the differences in the lexical and grammatical structure of the languages - these operations are called "language-specific" operations. Others are explained by the differences of cultures, between generally shared knowledge of the members of different cultural communities - these operations can be called "culture-specific operations". Other operations can be explained neither by structural differences between the languages nor by cultural differences but by the nature of the translation-process itself, that is by the necessity to express ideas in the target language which were originally conceived in the source language. These operations can be called "translation-specific" operations.

The paper explains the advantages and shortcomings of different classifications, and offers an explanation, aiming to avoid mixing the different principles. The classification of operations followed by the author is based exclusively upon the "operational" - that is, "technical" - properties of the operations (not on their reason, purpose etc.) By describing, classifying and explaining translational operations the author hopes to serve one of the main objects of translation studies, the discovery of rules governing the seemingly subjective decisions of translators. The paper will discuss two types of lexical operations characteristic of the process of translation from Hungarian into English, German, French and Russian and vice versa: segmentation and integration of meaning in translation.

In the course of the translation of even the simplest of sentences, translators perform a most complicated sequence of actions, which include the replacement of SL lexical units by TL lexical units, the restructuring of the sentence structure, the changing of the word order, the omission of certain elements and the addition of others etc. Due to the differences between SL and TL lexical systems, even the seemingly simplest action - the replacement of SL lexical units by TL lexical units - can become a complicated task, involving a very complex decision making process to make the right choice from among the different possibilities offered by the TL. The above mentioned actions (omission, addition, replacement, narrowing and widening the meaning etc.) we call translational operations.

1. TRANSFER OF MEANING

Investigating the operational part of translator's activity has become a rather neglected field in translation research, either left out altogether, or sometimes even accused of leading us into the wrong direction, and bringing translation research to a dead end.

Some claim that investigation of the translation operations themselves, takes our attention off meaning. According to them, after having extracted the meaning from the source language form, the translator should immediately forget the SL form, and reformulate the message in his or her target language independently from the source language form. This would mean that the process of translation is nothing but an analysis of the source language followed by synthesis in the target language, or in other words decoding of the source text and encoding the target text, while there is no direct transcoding from the source language form to the target language form. In this interpretation the basis for the transfer is a semantic representation which is independent from languages involved, thus languages do not influence the process of translation at all (Ref. 1).

Empirical studies of target texts translated from different source languages nevertheless reveal striking differences according to different source languages. Translated TL texts possess quantitatively measurable textual properties that differ from those of original TL texts, moreover these properties differ according to the source language of the translated texts (Ref.2)

2. TRANSFORMATION OF MEANING

Hence my conviction that linguistic differences between the SL and the TL can not be overlooked in translation studies. The claim that translation is a meaning-based phenomenon, does not make the differences between languages irrelevant. Meaning is a language-specific phenomenon, and translation operations do consist of different transformations of and alterations to the source language meaning, which are introduced by translators consciously or automatically. The conscious or non-conscious character of these operations can be another interesting topic for further research, but this time we are not going to deal with the question of consciousness.

Aversion against terms like "techniques of translation", "transformations in translation" "operations in translation", "solutions in translation" can be explained by the fact, that these terms suggest that transformations, operations, solutions made by translators always consciously, deliberately. They suggest that translators consciously apply some techniques, operations, solutions to transform the SL form into TL form.

The statement that in the course of translation the SL form is transformed into the TL form is so common in theoretic literature on translation, that we do not feel any longer its absurdity (Ref. 3). Transformation in this sense is nonsense, of course. The translator does not do anything to the SL form. The SL form, the SL text remains unchanged, intact etc.

What really happens, could be best described as the birth of a new entity. The TL text is a new entity, but the circumstances of its birth do leave their mark.

A translated TL text, or in other words, a text conceived in a foreign language and an original TL text, that is a text conceived on TL are two, rather different things. The road, leading from the mind to the linguistic form is never direct, never simple even when we formulate our ideas in our mother tongues. But if the thought takes its origin in an another language the road from the mind to the linguistic form is incomparably more complex.

3. MENTAL TRANSFORMATIONS AND LINGUISTIC OPERATIONS

Transition from the mind to the linguistic form is a mental operation complex enough even within the same language, how much more complex this process becomes if there are two languages involved in it! If the transition from thought to linguistic form requires the working of a mental switch even within one language, the same process between two languages certainly takes a double-switch at least. Double-switch here means transformation, but not of the SL form or the SL text which of course remains unchanged, but rather various mental transformations, which finally result in a TL text based on the SL text.

The existence of these mental transformations is undisputable, I think. The question is whether these mental transformations, mental operations include only SL-analysis and TL-synthesis, that is only the decoding of the SL text (transition from the SL-form to the thought) and the encoding of the TL text (transition from the thought to the TL form), or include at the same time a certain kind of transcoding as well, or in other words, if it is conceivable that there exist more or less direct roads as well, that lead from the SL form to the TL form. This is the most important question to ask, and using the terms "transformations", "operations" "techniques" we suggest that in some cases there is a direct road between two languages -but it leads through the translators mind.

Making use of the terms "transformation of meaning", or particularly "segmentation and integration of meaning" which is the topic of my paper, I mean these mental operations, carried out routinely, systematically and reliably (though not necessarily consciously) by all professional translators, but only randomly and unreliably by beginners.

4. THE HISTORY

The discussion of translation techniques can be traced from Vinay and Darbelnet (Ref. 4) through Eugene Nida (Ref. 5) to works of scholars from the former USSR Barkhudarov (Ref. 6) Komissarov (Ref. 7), Shveitser (Ref. 8) etc.

Vinay and Darbelnet made a list of seven main technical procedures ("les procédés techniques") in translation: emprunt, calque, traduction littérale, transposition, modulation, equivalence, and adaptation. In the glossary of technical terms at the beginning of their book we find several other procedures as well, for example; actualisation, amplification, articulation, compensation, explicitation, implicitation, généralisation, particularisation. Nida uses the term "techniques of adjustment". The main techniques of adjustment in his work are: additions, subtractions and alterations.

Barkhudarov, the well-known Russian translatologist uses the term "transformatsiya". He differentiates four types of transformations in translation: "perestanovka" ('transposition'), "zamena" ('substitution'), "dobavleniye" ('addition'), "opushcheniye" ('omission'). According to him, the most important reason for additions in translation from English into Russian are the elliptic nominal structures in English, that is the omission of certain semantic components in English surface structure which were present in the deep structure. As ellipsis is not characteristic of Russian, the omitted semantic components are reconstructed in the Russian surface structure (***pay claim*** → ***trebovaniye povysit' zarplatu*** 'demand to raise the pay'; ***gun license*** → ***udostovereniye na pravo nosheniya oruzhiya***, 'license for right to carry weapon.

A very detailed typology of lexical and grammatical transformations, including grammatical additions in Bulgarian-Russian and Russian-Bulgarian translation can be found in the work of the Bulgarian scholar Vaseva (Ref. 9). According to her additions are generated by the "linguistic asymmetry", that is by the necessity to express explicitly meanings in the target language that are contained implicitly in the source language. Cases of grammatical additions in Vaseva's work are explained by the so called "missing categories" and categories with different functions: Bulgarian has articles, while Russian has none; the possessive pronoun can be omitted in Russian, unlike in Bulgarian; the copula can be omitted in Russian but not in Bulgarian; the direct object can be - rarely - omitted in Russian, but never in Bulgarian etc. Besides the grammatical additions Vaseva shortly refers to the so called pragmatic additions as well, when concepts generally known for the source language audience may be unfamiliar for the target language audience and therefore they call for explanations in the translation.

The different kinds of lexical and grammatical and even pragmatic transformations in the process of English-Russian, German-Russian, French-Russian, Spanish-Russian translation were very thoroughly described by the translation-specialists of the former USSR (Ref. 6,7,8,10,11). The question however was not raised by anybody, if these transformations have any psychological reality or not. They did not touch on another important question either: what forms the basis for the transfer itself; in other words: is semantic representation in the translator's mind independent from the languages involved in the process of translation or not?

5. LEXICAL OPERATIONS AND SEMANTIC CHANGES

I think, both the question of the psychological reality of transformations and the question of the character of semantic representation is still open and I am afraid it will remain open in the near future. Nevertheless, this shouldn't prevent us from the study of semantic changes in translation.

The next question to be answered is the following: what is the relationship between the concept of "semantic changes in translation" and the concept of "lexical operations or transformations in translations". Many scholars who do not object the concept of "semantic changes", on the other hand reject the concept of "lexical operations or transformations" (Ref 12).

6. THE CONCEPT OF TRANSLATIONAL OPERATIONS

In my paper I will use the term "translational operations" in the sense described above, that is as a complex mental operation, taking place when the road conducting from the mind to the linguistic form is not direct but leads through another language.

The description of translational operations is in the centre of my book published recently under the title "The Theory and Practice of Translation" (Ref 13). It is an attempt to describe a system of operations taking place in the translation of English/German/French/Russian into Hungarian and vice versa. Though Hungarian is not a widely spoken language, looking at English/German/French/Russian from a Hungarian point of view, can provide interesting insights nevertheless. The Hungarian "looking glass" can reveal striking similarities between these otherwise rather different languages. Describing the translational behavior of Hungarian in the process of its translation into English/German/French/Russian, we in fact describe how a Finno-Ugric language works in the process of translation into the Indo-European languages and vice versa.

7. THE TYPOLOGY OF OPERATIONS

Translating any sentence from one language into another, even the simplest one, translators carry out a number of mental operations. These operations can be classified on the basis of different principles, taking as starting-point the reason of the operation, the purpose of the operation, the level of the operation etc. Thus operations can be: obligatory, optional and facultative operation; automatic and non automatic-operations; word-level, phrase-level, sentence-level and text-level operations; lexical, grammatical, stylistic and pragmatic operations etc.

Some of these operations can be explained by the differences in the lexical and grammatical structure of the languages - these operations are called "language-specific" operations. Others are explained by the differences of cultures, between generally shared knowledge of the members of different cultural communities - these operations can be called "culture-specific operations". Other operations can be explained neither by structural differences between the languages nor by cultural differences but by the nature of the translation-process itself, that is by the necessity to express ideas in the target language which were originally conceived in the source language. These operations can be called "translation-specific" operations.

All classifications have their advantages and shortcomings. I would like to offer an explanation, which tries to avoid mixing the different principles. The classification of operations followed here is based exclusively upon the "operational" - that is, "technical" - properties of the operations (and not on their reason, purpose etc.).

Lexical operations:

1. concretization (narrowing) of meaning
2. generalization (widening) of meaning
3. segmentation of meaning
4. integration of meaning

5. lexical omissions
6. lexical additions
7. transposition of meaning
8. substitution of meaning
9. antonymous translation
10. total transformation of meaning
11. compensation for losses in translation.

Grammatical operations:

1. grammatical concretization and generalization
2. grammatical segmentation
3. grammatical integration
4. grammatical additions
5. grammatical omissions
6. grammatical transpositions
7. grammatical substitution

The main types of lexical and grammatical operations, mentioned above, are divided into further subtypes, that is the total number of translational operations illustrated and explained in the book is approximately one hundred. The examples are taken from more than two hundred literary works and their translations, selected from the work of more than two hundred different translators. We are not going to describe unique, especially successful solutions of famous translators, but rather limit our research to the average.

Paper will discuss two types of lexical operations characteristic of the process of translation from Hungarian into English, German, French and Russian and vice versa: segmentation and integration of meaning in translation.

8. SEGMENTATION OF MEANING

Segmentation of meaning is a lexical operation whereby the meaning of a SL unit can only be rendered by two or more TL units. The main kinds of segmentation of meaning in Hungarian → Indo-European translation are the followings:

1. segmentation of inchoative verbs in H→IE translation
2. segmentation of adverbs of manner in H→IE translation
3. segmentation of verbs of saying in H→IE translation
4. segmentation of semantically rich verbs in H→IE translation.
5. segmentation of kinship terms in both direction
6. segmentation of culture-specific words in both direction

We are not going to discuss all the types of segmentation, only four of them: segmentation of inchoative verbs in H→IE translation, segmentation of adverbs of manner in H→IE translation, segmentation of semantically rich verbs in H→IE translation, segmentation of kinship terms in both direction.

Meaning-segmentation of nouns is motivated by the different segmentation of the world (colour terms, kinship terms etc.). Meaning-segmentation of verbs is motivated by the differences between Hungarian and Indo-European langauges in morphology and lexis. Morphology: Hungarian is a dominantly synthetic language while English, German, French and Russian are more or less analytic languages. Hungarian verb has a very complex conjugation, the personal pronoun, the possessive pronoun, the accusative ending and sometimes the auxiliary verb are all included into the Hungarian verb form, while they stand separately in IE languages. Lexis: thanks to the large storehouse of prefixes and suffixes that can be appended to Hungarian verbs, they can carry meaning which English, German French and Russian need several words to express.

8.1. Segmentation of Inchoative Verbs

The first type of segmentation to be discussed is the segmentation of inchoative verbs in translation. This occurs characteristically when translating from Hungarian into Indo-European. The complex meaning of Hungarian inchoative verbs (***elpityeredett***) is rendered by two separate verbs in IE languages: one for expressing the beginning of the action (***start, beginnen***) and another for expressing the action itself (***snivelling, zu flennen***).

H→E Erre aztán Rozsákné is felkelt, felöltözött és ***elpityeredett.***
This was enough to make Mrs. Rozsák get up too, put on her clothes, and ***start snivelling.***

H→G Erre aztán Rozsákné is felkelt, felöltözött és ***elpityeredett.***
Da stand nun auch Frau Rozsák auf, zog sich an und ***begann zu flennen.***

8.2. Segmentation of Adverbs of Manner

The second type of segmentation to be discussed is the segmentation of adverbs of manner. This occurs characteristically when translating from Hungarian into Indo-European. The complex meaning of Hungarian verb including not only the action but also the way or method of doing it (***berontott, belibbent, elcipelt, bevillamosozik***) can only be rendered by two separate words in IE translation: one for expressing the manner, style or way (***hurriedly, hereingetanzelt, de force, tramvayem***), and one for expressing the action itself (***return, kommen, traîner, doehat'***)

H→E ... ***berontott*** az asszisztens.
... the assistant ***returned hurriedly.***

H→G Thalamus, az udvari borbély ***belibbent.***
Thalamus, der Hofbarbier, ***kam hereingetänzelt.***

H→F ***Elcipelt*** egy budai kocsmába.
Il m'entraîna de force dans un cabaret de Buda.

H→R Az lesz a legokosabb, ha ***bevillamosozik*** a belvárosba.
Luchshe vsego ***tramvayem doehat'*** do centra.

8.3. Segmentation of Semantically Rich Verbs

The next type of meaning-segmentation is probably the most frequent one in H-IE translation. Semantically rich Hungarian verbs are very often rendered by an IE verb of general meaning (E: ***take, make, do*** G: ***machen, commen, tun*** F: ***faire, prendre, avoir***, R: ***prinimat', proizvodit', vzyat'***) and one or two nouns of specific meaning.

H→E Mikor a gróf felébredt, ***kikocsizott***, ha ugyan Estella megengedte.
When the count awoke ***he went out for a drive in his coach***, if Estella allowed him to.

H→G Ott az osztály végén parasztgyermekek ***tanyáztak***.
Hier, am Ende der Klasse, hatten die Bauernjungs ihr ***Lager aufgeschlagen***.

H→F Ők is ***cihelődtek***.
Elles aussi ***rassemblaient leurs affaires***.

H→R Nagy csöndben voltak, egyikük sem ***pisszent***.
Vsyo eto delaloc' v glubokom molchanii: nikto ne ***izdval ni zvuka***.

8.4. Segmentation of Kinship Terms

Meaning-segmentation of nouns can be motivated by the different segmentation of the world by different languages. Kinship terms may have more detailed description in one language and a less detailed one in the other. Hungarian for instance has a generic term for ***brother and sister*** while English, German, French and Russian have not. The meaning of the Hungarian word ***testvérek*** can only be rendered by two words in IE languages: ***brother and sister, Brüder und Schwester, frère et soeur, brat i sestra***.

H→E Azután elment, mert már jöttek a varázsló rokonai és ***testvérei***, s azokkal ő nem volt ismerős.
Then she went away, because the ***brothers and sisters*** and aunts and uncles and cousins of the magician were gathering, and she didn't know any of them.

H→F ***Testvéreim*** nincsenek. Rokonom csak egy van, egy gazdag gyáros...
Je n'ai plus ***ni frère, ni soeur***. Mon seul parent est un riche fabricant ...

9. INTEGRATION OF MEANING

Integration of meaning is a lexical operation whereby the meaning of two or more SL units can be rendered by one TL unit. Meaning-integration is also motivated by the different segmentation of the world and by the synthetic or analytic character of languages. The elements of analytic English verb forms can be integrated into one synthetic Hungarian verb. The main kinds of integration of meaning in Hungarian → Indo-European and Indo-European → Hungarian translation are the followings:

1. integration of inchoative verbs in IE→H translation

3. integration of adverbs of manner in IE→H translation
2. integration of semantically week verbs in IE→H translation
4. integration of kinship terms in both direction

We are not going to discuss all types of integration here only two of them: the integration of inchoative verbs in Indo-European → Hungarian translation, the integration semantically weak verbs in Indo-European → Hungarian translation.

9.1. Integration of Inchoative Verbs

The first type of integration to be discussed is the integration of inchoative verbs. This occurs characteristically when translating from Indo-European into Hungarian. Verbs expressing the beginning of action in IE languages (***start, beginnen, commencer, stat'***) can be incorporated into the main verb (***sipping, zu schluchzen, apparaître, dogadyvat'sya***) resulting in a synthetic Hungarian verb (***belekortyolt, felzokogott, feltüntek, megsejtette***).

E→H Billy ***started sipping*** his tea.
Billy ***belekortyolt*** a teába.

G→H ... im Nebenzimmer rechts ***begann*** die Schwester ***zu schluchzen***. ... a jobb oldali szobában húga ***felzokogott***.

F→H ... et des étoiles aussi ***commençaient apparaître*** a l'horizon obscurci, ...
... és az elsötétedett látóhatáron már ***feltüntek*** a csillagok is ...

R→H Po vidimomu s serediny zimy Dymov ***stal dogadyvat'sya*** shto ego obmanyvayut.
Úgy a tél dereka felé Dimov úgy látszik ***megsejtette***, hogy felesége megcsalja.

9.2. Integration of Semantically Weak Verbs

The next type of meaning-integration is probably the most frequent one in IE-H translation. IE verbs of general meaning (***take, tun, prendre, pochustvovat'***) are amalgamated with nouns of specific meaning (***sip, Blick, café, robost'***) resulting in a synthetic Hungarian verb form (***kortyintott, bepillantsak, kávéztunk, megszeppent***).

E→H Franny nodded, and ***took a sip*** of her milk.
Franny bólintott, és ***kortyintott*** a tejből.

G→H ... da doch ein Sinn und ein Verlangen in mir wäre, auch ***ein Blick*** in dies und jenes ***zu tun***, ...
... pedig bennem megvolna a vágy, hogy ***bepillantsak*** ebbe-abba.

F→H Il revint le soir à la même heure que la veille. Nous ***prenions notre café***.
Este megint csak abban az órában érkezett meg, mint előző nap. Éppen ***kávéztunk***.

R→H Akakiy Akakievich uzhe zablagovremenno ***pochuvstvoval*** nadlezhashchuyu ***robost'***.
Akakij Akakijevics már előre kellőképpen ***megszeppent***.

By describing, classifying and explaining translational operations we hope to serve one of the main objects of translation studies, the discovery of rules governing the seemingly subjective decisions of translators.

10. REFERENCES

1. Seleskovitch, D. *Language, langues et mémoire*. Paris: Minard. 1975.

2. Vehmas-Lehto, I. *Quasi-Correctness. A critical study of Finnish translations of Russian Journalistic texts*. Helsinki: Neuvostoliitto Instituutti. 1989.

3. Komissarov, V. N. *Lingvistika perevoda*. Moscow: Mezhdunarodnye otnosheniya. 1980.

4. Vinay, J. P. - Darbelnet, J. *Stilistique comparée du français et de l'anglais*. Paris: Didier. 1958.

5. Nida, E. A. *Toward a Science of Translating*. Leiden: Brill. 1964.

6. Barkhudarov L. S. *Yazik i perevod*. Moscow: Mezhdunarodniye otnosheniya. 1975.

7. Komissarov, V. N. *Slovo o perevode*. Moscow: Mezhdunarodnye otnosheniya. 1973.

8. Shveitser, A. D. *Perevod i lingvistika*. Moscow: Voenizdat. 1973.

9. Vaseva, I. *Teoriya i praktika perevoda*. Sophia: Nauka i isskustvo. 1980.

10. Gak, V. G. *Sravnitelnaya tipologiya frantsuzskovo i russkovo yazykov*. Leningrad: Prosveshcheniye. 1977.

11. Lvovskaya, Z. D. *Teoreticheskie problemy perevoda*. Moscow: Vysshaya shkola. 1985.

12. Englund Dimitrova, B. "Semantic change in Translation - A Cognitive Perspective". In: Y. Gambier and J. Tommola (eds.) *Translation and Knowledge*. Turku: University of Turku. 1993. 285-297.

13. Klaudy, K. *A fordítás elmélete és gyakorlata. Angol, német, francia, orosz fordítástechnikai pédatárral.*(The Theory and Practice of Translation. With illustrations in English, German, French, and Russian) Budapest: Scholastica. 1984.

MEDICAL TERMINOLOGY MANAGEMENT IN A MULTILINGUAL ENVIRONMENT

Helen Knight, Brenda Rudder, and Clove Lynch
Family Health International

Keywords: Translation-oriented Terminology Management, Good Terminology Practice, Medical Event Coding, FHIbase

Abstract: Terminology management facilitates the transfer of information in the biomedical sciences. As mediators of this information, translators are responsible for the accuracy and consistency of the terminology they use. Incorporating terminology management into the translation process can greatly facilitate translators' efforts in this area. Terminology databases are effective tools for conscientious terminology management. To this end, Family Health International (FHI) has developed FHIbase, a specialized, multilingual medical terminology and coding database.

INTRODUCTION

Translators must be accurate and consistent in their use of specialized terminology to effectively meet the demands of an increasing volume of technical information. Incorporating translation-oriented terminology management (TOTM) into the translation process can greatly improve translation quality. A valuable new tool to foster consistent use of accurate terminology is FHIbase, a multilingual medical terminology database developed at Family Health International (FHI).

1 TRANSLATION-ORIENTED TERMINOLOGY MANAGEMENT

1.1 Principles of Translation-Oriented Terminology Management

The need for accurate and consistent use of terminology in technical communication is widely acknowledged, yet it is a difficult goal to achieve. Translators have a vital contribution to make toward this goal. Their efforts can be facilitated by incorporating terminology management into the translation process. To a greater or lesser extent, terminology work is something that translators do whenever they work with specialized texts. TOTM simply codifies that process and makes it more efficient, by using a systematic approach to term documentation that involves researching and recording terms, definitions, contexts, sources, notes etc. Any translator can implement TOTM as a standard part of the translation process.

The basic principle of terminology work involves concepts, rather than words, and the linguistic form of these concepts, expressed in "terms" (Sager 17). Terms must be designated for each concept and clearly defined independently in each language before equivalents can be matched across languages. In this sense, terminologists do not translate terms, much less definitions. This means that "for each language, the methodology consists of identifying the concept (which may differ from one language to another), then proceeding to describe it within each culture's conceptual system" (De Bessé). Thus, *translation-oriented* terminology work is primarily a monolingual activity, focusing on the clear definition of concepts to which terms are assigned (ISO v). Unclearly defined terms are easily misused (Lewalle 3). Precisely defined terms avoid ambiguity and facilitate effective translation. For this reason, terms and their definitions (whenever possible) are the foundation of TOTM.

1.2 Good Terminology Practice

Each particular TOTM environment affects how term information is documented, according to a given terminology management approach. Certain criteria, however, provide basic guidelines for Good Terminology Practice (GTP). For example, when selecting terms, it is helpful to distinguish between an actual term and an ordinary word. A term can be made up of one or more words and has a precise meaning in a specific context (De Bessé). A good definition, besides being precise, is non-circular, concise, and tailored to the users' needs (Landau 124, Opitz 172, De Bessé). Above all, a good definition should refer to the concept denoted by the term, place the concept within a system of concepts, and define the limits of the concept with respect to other concepts (De Bessé, ISO v, Wright 389).

In addition to the basic term and its definition, other elements can be included to provide more thorough term documentation. A good context illustrates the term in use and should be relevant and concise. The best context is extracted from an original language document. A source should be included for contexts and definitions, and may cite an individual (informant), printed resources or electronic media; it should follow bibliographic code conventions, and refer to a complete bibliographic reference entry (ISO 3, 6, Opitz 167, Wright 7, 8, 10). Notes can also be included as a means of flagging discrepancies between concepts from one language to another, or explaining any controversial elements of an entry. Including concept relations can sometimes be useful and may involve graphic representations of hierarchical relationships between concepts. These suggested elements can be elaborated on and added to as necessary.

1.3 Terminology Databases (TDBs)

Many translators already implement such practices to a greater or lesser extent, one example being the compilation of personal glossaries. The terminology database (TDB) takes this idea a step further, offering a means to store vast quantities of data and to manipulate that data in a variety of ways. Many TDB programs are available, some offering more flexibility than others, but all providing a way to efficiently store and retrieve specialized terminology.

2 FHI AND CLINICAL TRIALS

Terminology work plays an increasingly important role in medical information management. The health care industry is overwhelmed by the amount of information generated in the course of research, development and implementation of medicines, procedures and equipment. Inadequate handling of information ultimately results in either poor or expensive service delivery. Rising costs have created a sense of urgency within the industry to streamline the information flow. Translating technical documentation is one aspect of managing this information.

Translation is an essential component of information management at Family Health International, a nonprofit organization that conducts research in reproductive health, family planning and the prevention of sexually transmitted disease, including AIDS.

2.1 CT Translation Team Process

In the Clinical Trials Division (CT) at FHI, clinical study documentation is translated from English into the language spoken at the study site. During a study, data gathered from various countries are generally sent to CT and then translated into English to be analyzed. In addition, all study findings must be translated into English for reporting purposes.

Although much professional terminology work often begins with term selection, the CT translation team is given terms verbatim from study centers and must research the context. The team relies on informants (subject specialists) and monolingual works as primary resources. If terms are transparent, they are translated and sent out for primary review. If significant problems are found, terms are researched until a satisfactory solution is found, at which point they are translated and submitted for primary review. Differences between the translator's and reviewer's versions are resolved in a second review and revision stage. All terms, whether reviewed once or twice, are given another review before being considered final.

3 FHIBASE

FHIbase, FHI's solution to the need for accurate and consistent use of terminology, is a multilingual medical database including three domains (or sections): Procedures and Medications, Medical Events, and Family Planning. FHIbase incorporates a unique coding and classification system, has powerful search functions, and was created using MultiTerm Professional for Windows (a free-format database application produced by TRADOS GmbH).

Information in FHIbase can be displayed in a variety of ways, using any combination of index and target fields. The different fields make sorting in the database flexible. For example, the database can be sorted using index fields, text fields, or term attributes, among others. These criteria can also be used to select output that enables production of a dictionary with exclusively Family Planning terms in Spanish, for example, or with all database domains in all languages. Within each domain, different text fields can be assigned according to the type of information that needs to be documented. Within the entry, term attributes are used to give further information about a term; for example, to designate synonyms, to indicate whether a drug is a brand or generic, or to show the review status of a term.

3.1 Procedures and Medications

The Procedures and Medications domain was developed in response to the demand for accurate and consistent translations of clinical procedures and drugs. Translation of brand names of medications is particularly difficult, even with appropriate resources. FHIbase enables the translator to consult one source that has consolidated term information from a variety of sources. Both proprietary names of medications and clinical procedures have their idiosyncrasies. For example, Figure 1 shows the English term "culdocentesis," the French equivalent of which is an eponym, "ponction du Douglas".

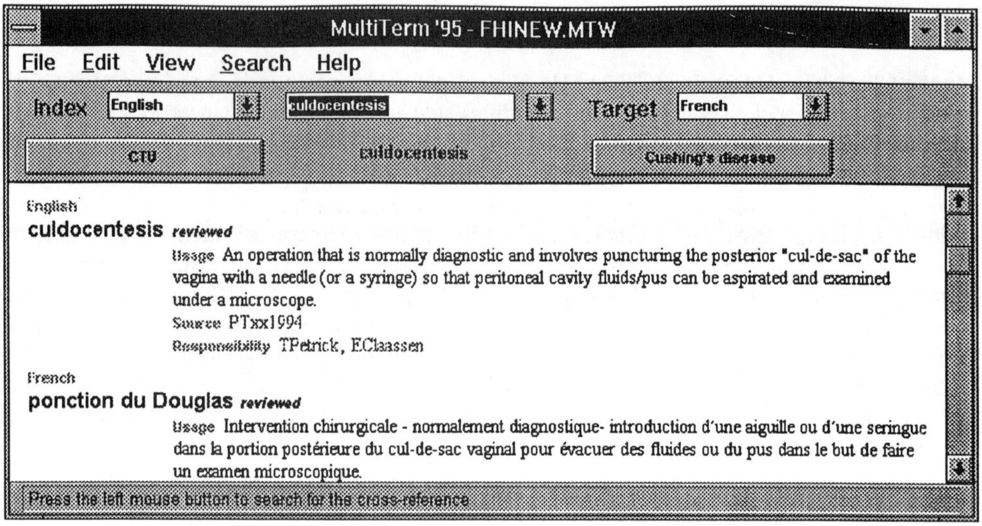

Figure 1

3.2 Medical Events

The Medical Events domain of FHIbase was developed in response to the demand for accurate and consistent translations of adverse events. These terms range from common side effects (e.g., headaches) of certain drugs and/or devices to serious adverse events (e.g., congestive heart failure). Many of the Medical Events terms are multiword terms that are not easily located in dictionaries. Medical Events entries allow the term to be documented as it was recorded in the original document (Specific Term, i.e. STerm), in its physician-preferred form (Preferred Term, i.e PTerm), and as it is most likely to appear in a text. Figure 2 illustrates one search function in FHIbase, where keying in only two words, "abdominal bleeding," brings up the multiword term "abdominal bleeding on steroid contraception."

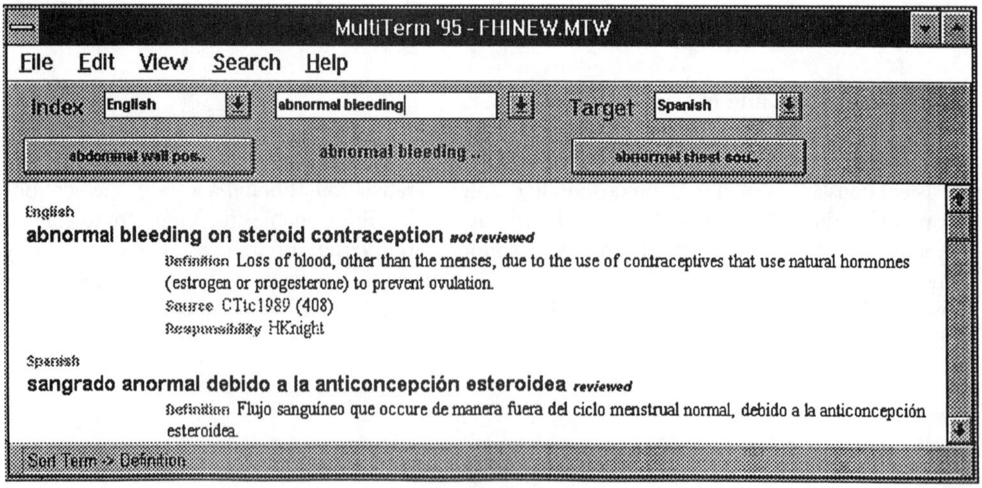

Figure 2

3.3 Family Planning

The Family Planning section was developed in response to the demand for the accurate and consistent translation of family planning terminology used throughout FHI. Figure 3 indicates that in the entry "cesarean section," several possible term choices in English map to only one instance of the term in Spanish.

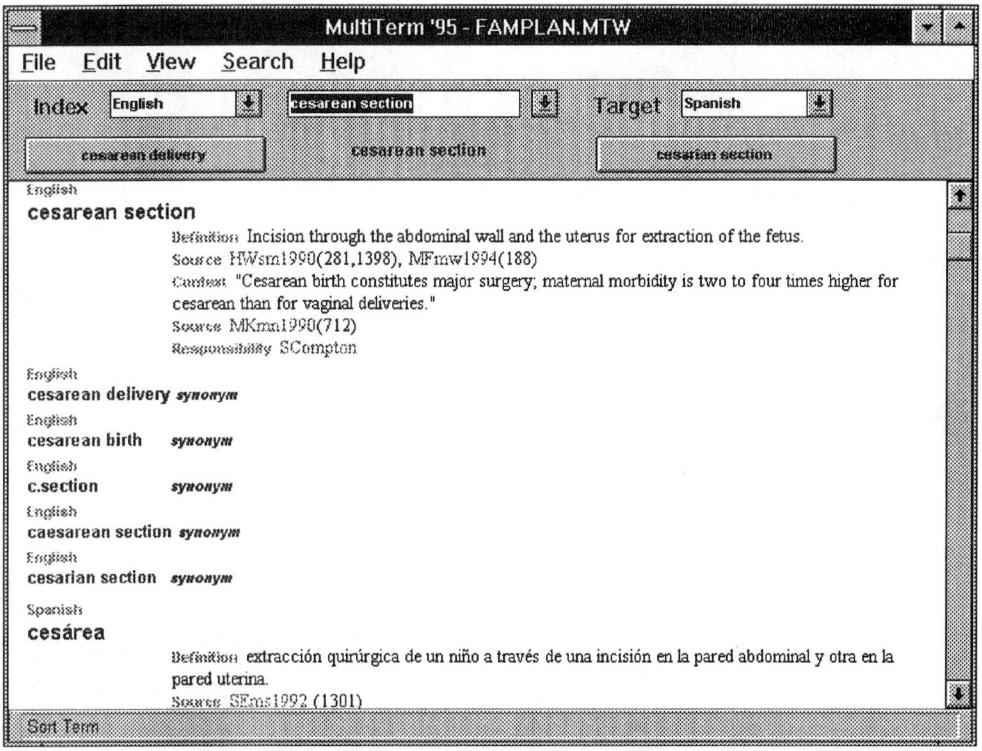

Figure 3

3.4 Coding

Any transfer of scientific and technical information, either intralingual or interlingual, uses terminological organization as the key to effective communication. Medical terminology management involves classifying concepts within and across subject areas as well as across languages. Coding is one response to the question of identifying and classifying medical events reported in the course of study, diagnosis and treatment.

Medical event coding enables a discrete semantic unit (a term) to be uniquely identified within a dynamic and highly structured classification system. The organizing principle is similar to that of postal zip codes, in which each value combines with others to represent particular geographic regions with increasing specificity. In the medical domain, this is accomplished by assigning numeric and/or alphanumeric codes first to categories of medical events (such as diseases or symptoms), then to body systems, then to specific organs within each body system and finally to descriptors such as left, right, etc. The result is a metalanguage that can pinpoint concepts in given domains and subdomains. This is useful in identifying terms that are conceptually related but that may not look alike, for example identifying the term "heart attack" with its more clinical description "myocardial infarction."

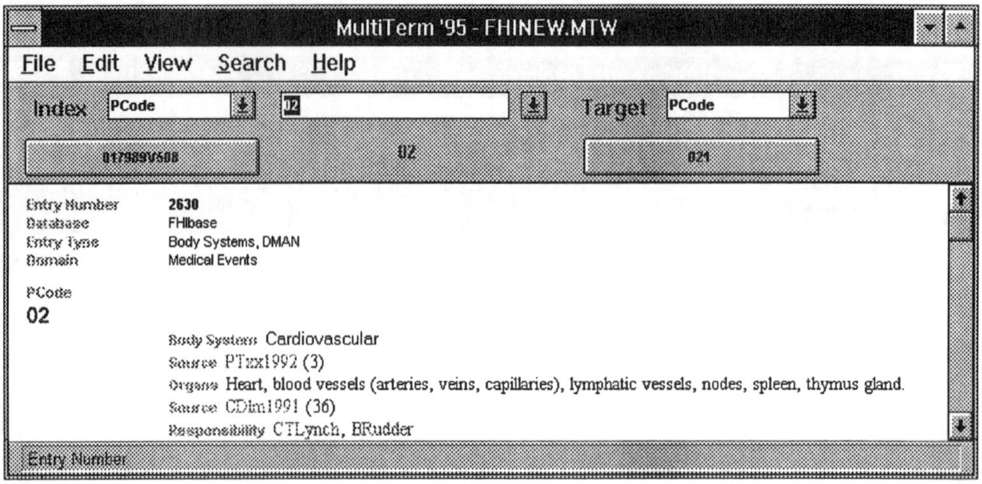

Figure 4

Coding is a way to identify medical events that are often documented in different ways and reported during a study. FHI uses three levels of coding to classify its terminology, all of which combine to form the inherent structure of FHIbase. Codes serve as pointers to term records, each SCode mapping to an STerm, each PCode mapping to a PTerm and various records containing special FDA-preferred codes. As shown in Figure 4, the cardiovascular system has been coded 02. Figure 5 shows that all categories of heart trauma are coded 021.

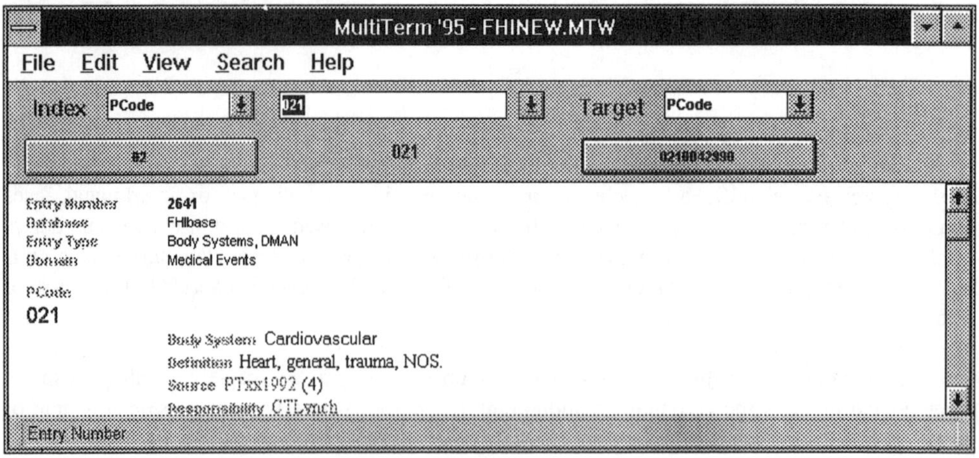

Figure 5

A full code category can be displayed in the database (Figure 6) in the same way that terms are viewed in a hit list (Figure 7).

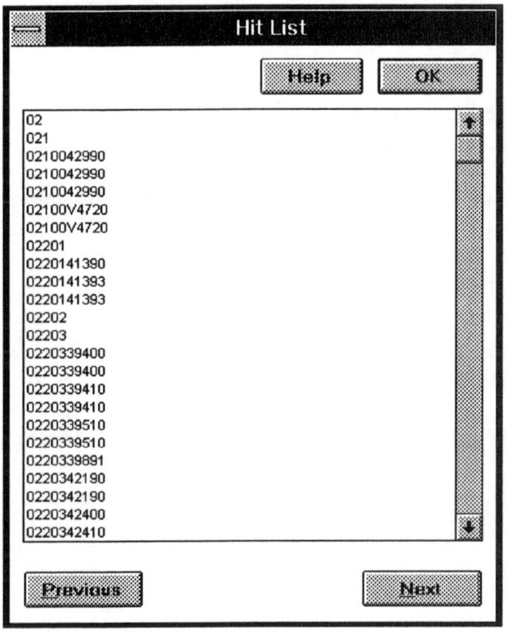

Figure 6

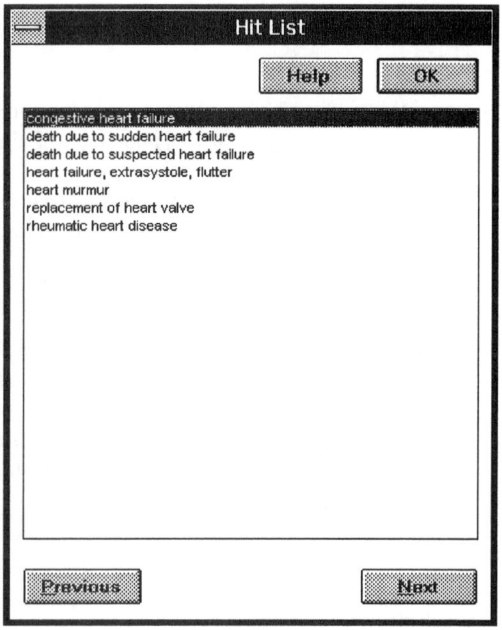

Figure 7

CONCLUSION

Development of a terminology database maximizes the potential to efficiently store, retrieve and process large volumes of data. Translation-oriented terminology management (TOTM) can then be used to improve translation quality by ensuring the accurate and consistent use of specialized terminology. Family Health International has incorporated TOTM into the translation process with the development of FHIbase, a multilingual database that builds on organizational conventions in medical terminology, such as coding.

REFERENCES

De Bessé, B. Terminological Definitions. In: Wright SE, Budin G, editors. Handbook of Terminology Management. Philadelphia: John Benjamins, forthcoming.

International Organization for Standardization, Technical Committee 46. Liaison Statement Concerning ISO/TC37/SC1/WG1 Work Item 18: Documentation for terminology work and terminography. Official communication. Ottawa: National Library of Canada, Office of Library Standards, 1991:3, 6.

International Organization for Standardization, Technical Committee 37. ISO CD 704: Terminology Work - Principles and Methods. Geneva: ISO, 1994:v.

Landau, S. Dictionaries: The Art and Craft of Lexicography. New York: The Scribner Press, 1984.

Lewalle, P. Terminology Work in Health and Medicine. In Wright SE, Budin G, editors. Handbook of Terminology Management. Philadelphia: John Benjamins, forthcoming.

Opitz, K. The Terminological/Standardized Dictionary. In Hartman RRK, editor. Lexicography: Principles and Practice. London: Academic Press, 1983:163-180.

Sager, JC. Translation and Terminology. The ATA Chronicle, 1995 XXIV(6):17-18.

Wright, SE. Term Entry Formatting Conventions. In: Wright SE, editor. Proceedings of the Kent State University Terminology Workshop; 1993 Jun 25-28; Kent (OH). Unpublished:3-26

Wright, SE. Terminology of Terminology. In: Wright SE, editor. Proceedings of the Kent State University Terminology Workshop; 1993 Jun 25-28; Kent (OH). Unpublished:381-392.

Acknowledgements

The authors wish to thank Emily Claassen and Melissa McPheeters of FHI for their valuable contribution to this manuscript.

Support for this work was provided by Family Health International (FHI), although the views expressed in this article do not necessarily reflect those of FHI. FHI is an international nonprofit organization that conducts research and provides technical assistance in reproductive health, family planning, sexually transmitted diseases and AIDS. It is based in Research Triangle Park, North Carolina, USA.

CAR TERMINOLOGY IN THE SPANISH-SPEAKING WORLD

Andre Moskowitz
Portuguese/Spanish, Translator/Interpreter

Abstract

This paper presents regional Spanish terms for eighteen car-related items corresponding to the following US English terms:

1) *car*
2) *bumper*
3) *clutch*
4) *gasoline*
5) *hood*
6) *horn*
7) *hubcap*
8) *inner tube*
9) *jack*
10) *license plate* or *auto tag*
11) *rim*
12) *roof rack*
13) *steering wheel*
14) *tire*
15) *trunk*
16) *speed bump*
17) *stop sign*
18) *dump truck*

Because Spanish is spoken in twenty countries, the study of its regional and national variation is important to translators, terminologists and lexicographers as it bears directly on questions such as, "What term should a translator use when the name of a particular item varies by region?", and "How should a lexicographer define a term that is used differently in different regions?"

The primary goal of this paper, however, is not to indicate the terms that should be used when translating into Spanish. That choice will often depend on the translation's intended audience. The objective is to present dialectological data concerning which terms are actually used in the regions under study.

0 INTRODUCTION

This paper presents a summary of the results of a motor vehicle terminology usage survey carried out on approximately 200 Hispanic informants. Each section is divided into two subsections:

1) Data by Country
2) Observations

0.1 Data by Country

The data by country subsections consist of tables in which the terms used by the informants in each of the twenty Spanish-speaking countries are presented. The countries are given in a geographical rather than alphabetical order:

SPAIN
CHILE
ARGENTINA
URUGUAY
PARAGUAY

BOLIVIA
PERU
ECUADOR
COLOMBIA
VENEZUELA
CUBA
DOMINICAN REPUBLIC
PUERTO RICO
PANAMA
COSTA RICA
NICARAGUA
HONDURAS
EL SALVADOR
GUATEMALA
MEXICO

The countries are presented in this order to highlight what may be called "lexico-geographic blocs", or groups of countries that are in geographic proximity and also share the same lexical usage for a given item.

Following each Spanish word, a numerical proportion is given indicating the number of informants out of the total who gave that particular response. The question was posed as follows:

"¿Cómo se llama el objeto de la foto?", at which time a picture of the object was shown to the informant.

Thus, in the first table,

SPAIN	coche (7/7), auto (2/7).

is to be interpreted as, "Of the seven Spaniards who were shown a picture of a car, all seven gave the term *coche* and two also gave the term *auto*." In a number of cases (such as this one) informants gave several terms as their response, and not all informants answered all questions.

The terms in italics refer to spelling, not pronunciation. Where spelling is not standardized, pronunciation is given in broad Spanish transcription (the International Phonetic Alphabet is not used). When informants offered two similar variants, words or letters appear in parentheses. Thus, where *(camión de) volteo* appears, both *volteo* and *camión de volteo* were given by the informants. Also "o/a" at the end of a word means both masculine and feminine were given. Thus *maletero/a* means that both *maletero* and *maletera* were given as responses by the informants.

0.2 Observations

In these subsections the data collected are summarized and various observations are made regarding how the Spanish-speaking world appears to be divided on the use of automobile terms.

1. CAR

1.1 Data by Country

SPAIN	coche (7/7), auto (2/7).
CHILE	auto (8/8).
ARGENTINA	auto (12/13), coche (10/13).
URUGUAY	auto (5/5), coche (3/5).
PARAGUAY	auto (4/5), coche (4/5).
BOLIVIA	auto (6/7), carro (5/7).
PERU	carro (13/15), auto (11/15).
ECUADOR	carro (12/13), auto (8/13).
COLOMBIA	carro (11/11).
VENEZUELA	carro (5/5).
CUBA	carro (6/9), máquina (6/9).
DOMIN. REP.	carro (6/6).
PUERTO RICO	carro (6/6).
PANAMA	carro (5/5).
COSTA RICA	carro (4/4).
NICARAGUA	carro (4/4).
HONDURAS	carro (5/5).
EL SALVADOR	carro (4/4).
GUATEMALA	carro (6/6).
MEXICO	coche (14/17), carro (11/17).

1.2 Observations

Carro is used in parts of Mexico, all of Central America, the Hispanic Antilles (Cuba, the Dominican Republic and Puerto Rico) and in about half of South America.

Auto and/or *coche* are the most commonly used terms in Spain, the Southern Cone (Chile, Argentina, Uruguay and Paraguay) and much of Mexico.

Auto is used everywhere in technical language and in advertising.

There is evidence that in Mexico *carro* is more common in the north, while *coche* is preferred in the central region, including the *Distrito Federal.*

In Cuba *máquina* appears to be used more in popular and familiar speech; *carro* is more formal.

As you move south in South America, *auto* becomes increasingly more common at the expense of *carro* and in Peru and Bolivia, the two terms appear to be in open competition.

2. BUMPER

2.1 Data by Country

SPAIN	parachoque(s) (5/5).
CHILE	parachoque(s) (7/7).
ARGENTINA	paragolpe(s) (12/12).
URUGUAY	paragolpe(s) (6/6).
PARAGUAY	paragolpe(s) (5/6), parachoque(s) (2/6).
BOLIVIA	parachoque(s) (8/8).
PERU	parachoque(s) (10/10).
ECUADOR	guardachoque(s) (11/12), parachoque(s) (3/12).
COLOMBIA	bómper (11/17), defensa (10-/17), guardachoque(s) (6/17), parachoque(s) (4/17).
VENEZUELA	parachoque(s) (9/9).
CUBA	defensa (8/8).
DOMIN. REP.	bómper (5/7), parachoque(s) (4/7).
PUERTO RICO	bómper (6/6).
PANAMA	bómper (7/7), defensa (5/7).
COSTA RICA	búmper (5/5).
NICARAGUA	bómper (7/7).
HONDURAS	bómper (7/7), guardachoque(s) (2/7).
EL SALVADOR	bómper (5/5).
GUATEMALA	bómper (7/7).
MEXICO	defensa (14/14).

2.2 Observations

Parachoques or *parachoque* is used in Spain and some Spanish American countries, mostly South American.

Bómper is used in Central America and in the Dominican Republic and Puerto Rico.

Defensa is used in Colombia, Panama, Cuba and Mexico.

In most of Central America people say (BOM-per) whereas in Costa Rica many people pronounce the word (BUM-per). This may be due to the fact that a large percentage of Costa Ricans read English.

3. CLUTCH (PEDAL)

3.1 Data by Country

SPAIN	embrague (6/6).
CHILE	embrague (8/8).
ARGENTINA	embrague (11/11).
URUGUAY	embrague (4/4).
PARAGUAY	embrague (4/4).
BOLIVIA	embrague (7/7).
PERU	embrague (11/11).
ECUADOR	embrague (11/11).
COLOMBIA	cloch(e) (10/12), embrague (3-/12).
VENEZUELA	cloch(e) (7/7), croche (7/7).
CUBA	cloch(e) (8/8), embrague (3/8).
DOMIN. REP.	cloch(e) (6/6).
PUERTO RICO	cloch(e) (5/5).
PANAMA	cloch(e) (7/7).
COSTA RICA	cloch(e) (5/5).
NICARAGUA	cloch(e) (4/4).
HONDURAS	cloch(e) (4/4).
EL SALVADOR	cloch(e) (5/5).
GUATEMALA	cloch(e) (7/7).
MEXICO	cloch(e) (9/9).

3.2 Observations

Embrague is used in Spain and in South America from Ecuador to the south. In Colombia, the three informants who gave *embrague* were all from southern Colombia (Valle, Cauca and Nariño Departments).

Cloch(e) is used in Mexico, Central America, the Hispanic Antilles, Colombia and Venezuela. It is pronounced (CLO-che), (CLOCH), (CLOSH) and (CLUCH), and the choice generally depends more on social class and familiarity with English than on region. In Venezuela, however, the pronunciation (CRO-che) is common, especially in popular speech.

4. GASOLINE

4.1 Data by Country

SPAIN	gasolina (5/5).
CHILE	bencina (7/7).
ARGENTINA	nafta (5/5).
URUGUAY	nafta (5/5).
PARAGUAY	nafta (4/4).
BOLIVIA	gasolina (4/4).
PERU	gasolina (7/7).
ECUADOR	gasolina (10/10).
COLOMBIA	gasolina (4/4).
VENEZUELA	gasolina (6/6).
CUBA	gasolina (6/6).
DOMIN. REP.	gasolina (5/5).
PUERTO RICO	gasolina (5/5).
PANAMA	gasolina (3/3).
COSTA RICA	gasolina (3/3).
NICARAGUA	gasolina (3/3).
HONDURAS	gasolina (4/4).
EL SALVADOR	gasolina (3/3).
GUATEMALA	gasolina (3/3).
MEXICO	gasolina (5/5).

4.2 Observations

Gasolina is the most frequently used term everywhere except the Southern Cone.

5. HOOD

5.1 Data by Country

SPAIN	capó (6/6).
CHILE	capó (8/8).
ARGENTINA	capó (14/14).
URUGUAY	capó (5/5).
PARAGUAY	capó (5/5).
BOLIVIA	capó (7/9), capote (2/9).
PERU	capó (9/10), capote (4/10).
ECUADOR	capó (10/10).
COLOMBIA	capó (13/15), capote (2/15).
VENEZUELA	capó (8/9), capota (4/9).
CUBA	capó (9/9).
DOMIN. REP.	bonete (6/6).
PUERTO RICO	bonete (9/9).
PANAMA	tapa (7/7).
COSTA RICA	tapa (6/6).
NICARAGUA	tapa (6/6).
HONDURAS	tonó (5/5).
EL SALVADOR	caperuza (6/6).
GUATEMALA	capó (6/6).
MEXICO	cofre (11/11).

5.2 Observations

Capó is the most commonly used term throughout South America and also in Spain. In South America, it is often spelled *capot*, and in a number of countries the variants *capote* and *capota* are also used.

Bonete is used in the Dominican Republic and Puerto Rico.

Tapa is used in Panama, Costa Rica and Nicaragua.

Tonó, used in Honduras, probably comes from English *tune-up*.

6. HORN

6.1 Data by Country

SPAIN	bocina (10/10), claxon (9/10), pito (7/10).
CHILE	bocina (8/8).
ARGENTINA	bocina (12/12).
URUGUAY	bocina (6/6).
PARAGUAY	bocina (5/5).
BOLIVIA	bocina (9/9).
PERU	claxon (15/16), bocina (11/16).
ECUADOR	pito (11/11).
COLOMBIA	pito (13/13).
VENEZUELA	corneta (9/9).
CUBA	claxon (6/9), pito (5/9), bocina (3/9), clauso (3/9), fotuto (2/9).
DOMIN. REP.	bocina (6/6).

PUERTO RICO	bocina (9/9).
PANAMA	pito (8/8), bocina (3/8).
COSTA RICA	pito (6/6).
NICARAGUA	pito (5/5).
HONDURAS	pito (7/7).
EL SALVADOR	pito (6/6).
GUATEMALA	bocina (7/7).
MEXICO	claxon (17/18), pito (10/18), bocina (3/18).

6.2 Observations

Bocina is used in South America from Bolivia to the south and in the Hispanic Antilles.

Pito is used in Colombia and Ecuador and in Central America with the exception of Guatemala.

In Spain there is evidence that *claxon* is the prestige term and *bocina* the more popular term in many regions. *Pito* appears to be more often used in southern Spain than in northern and central Spain.

Cuba is the only country in which more than three terms compete and Cuba and Venezuela are the only countries that have terms not used elsewhere.

The three Mexicans that gave *bocina* were all from southern Mexico (note that Guatemala is a *bocina* country).

The term *claxon* is used in Spain and also in three countries that historically have had close ties to Spain, namely, Peru and Mexico, where the seats of the colonial vice-royalties were located, and Cuba, which achieved its independence from Spain much later than most Spanish American countries.

7. HUBCAP

7.1 Data by Country

SPAIN	tapacubos (6/6).
CHILE	tapa (de rueda) (8/8).
ARGENTINA	taza (9/9).
URUGUAY	taza (4/4).
PARAGUAY	taza (3/3).
BOLIVIA	plato (6/9), taza (4/9), tapa de rueda (4/9), tapa (de) aro (2/9).
PERU	vaso (11/13), plato (7/13).
ECUADOR	tapacubos (9/9).
COLOMBIA	copa (9/11), plato (2/11).
VENEZUELA	taza (8/11), copa (5/11), tapa (2/11).
CUBA	bocina (4/6), tambora (3/6).
DOMIN. REP.	tapabocina (5/5).
PUERTO RICO	tapabocina (6/6).
PANAMA	copa (7/7).
COSTA RICA	copa (4/4).
NICARAGUA	copa (5/6), corona (2/6), plato (2/6).
HONDURAS	copa (4/4).
EL SALVADOR	copa (5/6), tapa (2/6).
GUATEMALA	plato (5/5).
MEXICO	tapón (7/11), copa (5/11), polvera (4/11), cubrepolvos (2/11).

7.2 Observations

Taza is used in Argentina, Uruguay and Paraguay.

Tapabocina is used in the Dominican Republic and Puerto Rico.

Copa is used in most of Central America and in Colombia.

8. INNER TUBE

8.1 Data by Country

SPAIN	cámara (4/4).
CHILE	cámara (4/4).
ARGENTINA	cámara (5/5).
URUGUAY	cámara (4/4).
PARAGUAY	cámara (4/4).
BOLIVIA	neumático (5/5).
PERU	cámara (7/7).
ECUADOR	tubo (5/5).
COLOMBIA	neumático (7/7).
VENEZUELA	tripa (9/9).
CUBA	cámara (6/6).
DOMIN. REP.	tubo (6/6).
PUERTO RICO	tubo (7/7).
PANAMA	tubo (4/4).

COSTA RICA	neumático (4/4).
NICARAGUA	neumático (4/4).
HONDURAS	neumático (7/7).
EL SALVADOR	tubo (5/6), neumático (3/6).
GUATEMALA	tubo (6/6).
MEXICO	cámara (8/8).

8.2 Observations

Cámara is used in Spain and the Southern Cone.

Tubo is used in Ecuador, the Dominican Republic, Puerto Rico, Panama, El Salvador and Guatemala.

Neumático is used in Bolivia, Colombia, Costa Rica, Nicaragua and Honduras.

Tripa is used only in Venezuela.

9. JACK

9.1 Data by Country

SPAIN	gato (6/6).
CHILE	gata (4/4).
ARGENTINA	gato (10/10), *cricket* (6/10).
URUGUAY	gato (4/4).
PARAGUAY	gato (5/5).
BOLIVIA	gata (5/6), gato (4/6).
PERU	gata (7/7).
ECUADOR	gata (5/5).
COLOMBIA	gato (7/7).
VENEZUELA	gato (10/10).
CUBA	gato (3/3).
DOMIN. REP.	gato (5/5).
PUERTO RICO	gato (8/8).
PANAMA	gato (5/5).
COSTA RICA	gata (5/5).
NICARAGUA	gata (6/6).
HONDURAS	gata (4/6), *jack* (3/6).
EL SALVADOR	mica (6/6), gato (4/6).
GUATEMALA	gato (4/5), *tricket* (4/5).
MEXICO	gato (9/9).

9.2 Observations

Usage generally centers around a choice between *gato* and *gata*, with the masculine term winning out in thirteen of the twenty countries.

There are two *gata* blocs, namely, the South American countries that have Pacific coasts and no Atlantic or Caribbean coast–Chile, Peru and Ecuador–and the Central American countries of Costa Rica, Nicaragua and Honduras.

Only in Argentina, Honduras, El Salvador and Guatemala have terms other than *gato/a* been registered.

For some Argentines *cricket* (CRI-que) refers to a smaller, manual jack, while *gato* is thought of as a more heavy-duty jack, whether manual or hydraulic.

In Honduras *jack* is pronounced (YA[C]) and in Guatemala, *tricket* is pronounced (TRI-que).

In Bolivia *gato* and *gata* compete. (Note that Bolivia is neither an Atlantic nor a Pacific South American Country, but rather lies between the two regions.)

10. LICENSE PLATE or AUTO TAG

10.1 Data by Country

SPAIN	matrícula (10/11), placa (3/11).
CHILE	patente. (7/7).
ARGENTINA	patente (12/14), chapa (9/14).
URUGUAY	chapa (6/6).
PARAGUAY	chapa (6/6).
BOLIVIA	placa (6/6).
PERU	placa (11/11).
ECUADOR	placa (11/11).
COLOMBIA	placa (11/11).
VENEZUELA	placa (7/7).
CUBA	chapa (9/9).
DOMIN. REP.	placa (7/7).
PUERTO RICO	tablilla (10/10).
PANAMA	placa (6/6).
COSTA RICA	placa (4/4).
NICARAGUA	placa (5/5).
HONDURAS	placa (6/6).
EL SALVADOR	placa (6/6).
GUATEMALA	placa (5/5).
MEXICO	placa (9/9).

10.2 Observations

Placa is used in Mexico, Central America, and also in South America as far south as Bolivia.

Chapa is used in Argentina, Uruguay and Paraguay, and also in Cuba.

Matrícula, the term that is predominant in Spain, appears to be a shortened version of *placa matrícula*. In several Spanish American countries *matrícula*, in the context of cars, refers to the "(motor vehicle) registration", rather than to the "license plate".

Tablilla is used only in Puerto Rico.

11. RIM (OF TIRE)

11.1 Data by Country

SPAIN	llanta (6/6).
CHILE	aro (6/8), llanta (3/8).
ARGENTINA	llanta (13/13).
URUGUAY	llanta (4/5), aro (1/5).
PARAGUAY	llanta (3/5), aro (3/5).
BOLIVIA	aro (7/7).
PERU	aro (12/12).
ECUADOR	aro (12/12).
COLOMBIA	rin (11/11).
VENEZUELA	rin (8/8).
CUBA	llanta (5/6), aro (2/6).
DOMIN. REP.	aro (6/6).
PUERTO RICO	aro (9/9).
PANAMA	rin (8/8).
COSTA RICA	aro (6/6).
NICARAGUA	rin (7/7).
HONDURAS	rin (4/4).
EL SALVADOR	rin (6/6), aro (2/6).
GUATEMALA	aro (7/7).
MEXICO	rin (11/11).

11.2 Observations

Aro is used in much of South America, the Hispanic Antilles, and parts of Central America.

Rin is used in Mexico and parts of Central America, as well as in Colombia and Venezuela; it is not commonly used in any country south of Colombia.

Llanta is used in Spain, Argentina, Uruguay, Cuba and Paraguay. In most of Latin America *llanta* means "tire" (see section fourteen below).

12. ROOF RACK

12.1 Data by Country

SPAIN	baca (9/9).
CHILE	parrilla (6/6).
ARGENTINA	portaequipajes (14/14).
URUGUAY	baca (6/6).
PARAGUAY	portabultos (3/4), parrilla (1/4), baca (1/4).
BOLIVIA	parrilla (8/8).
PERU	parrilla (11/11).
ECUADOR	parrilla (12/12).
COLOMBIA	parrilla (10/10).
VENEZUELA	parrilla (8/8).
CUBA	parrilla (6/6).
DOMIN. REP.	parrilla (6/6), tarima (3/6).
PUERTO RICO	*rack* (6/6).
PANAMA	parrilla (7/7).
COSTA RICA	canasta (5/5).
NICARAGUA	parrilla (4/4).
HONDURAS	parrilla (5/5).
EL SALVADOR	parrilla (4/4).
GUATEMALA	parrilla (7/7).
MEXICO	parrilla (14/18), canastilla (9-/18).

12.2 Observations

Parrilla is not used in Spain and the River Plate region, but is used almost everywhere else.

13. STEERING WHEEL

13.1 Data by Country

SPAIN	volante (6/6).
CHILE	manubrio (5/5), volante (3/5).
ARGENTINA	volante (6/6).
URUGUAY	volante (4/4), manija (3/4).
PARAGUAY	volante (4/4).
BOLIVIA	volante (7/7), manubrio (3/7).
PERU	timón (7/7).

ECUADOR	timón (3/5), volante (3/5).
COLOMBIA	timón (8/8).
VENEZUELA	volante (6/6).
CUBA	timón (2/2).
DOMIN. REP.	el guía (4/4).
PUERTO RICO	el guía (10/10).
PANAMA	timón (5/5).
COSTA RICA	manivela (4/5), rueda (3/5), volante (3/5).
NICARAGUA	timón (7/7).
HONDURAS	timón (6/6).
EL SALVADOR	timón (5/5), volante (2/5).
GUATEMALA	timón (4/4).
MEXICO	volante (8/8).

13.2 Observations

Volante is understood by educated speakers everywhere. However, it has competition from other terms in at least fifteen out of twenty countries.

Timón is common in all of Central America except Costa Rica and also in Peru, Ecuador and Colombia.

Puerto Rico and the Dominican Republic form a block with *el guía.*

Manivela is used only in Costa Rica.

In Colombia two informants from the Atlantic coast also gave the term *cabrilla* in the sense of "steering wheel".

14. TIRE

14.1 Data by Country

SPAIN	neumático (6/9), cubierta (4/9).
CHILE	neumático (9/9).
ARGENTINA	cubierta (11/12), goma (6/12), neumático (5/12).
URUGUAY	cubierta (5/5), goma (3/5).
PARAGUAY	cubierta (2/3), rueda (2/3).
BOLIVIA	llanta (7/7).
PERU	llanta (11/11).
ECUADOR	llanta (11/11).
COLOMBIA	llanta (13/13).
VENEZUELA	caucho (8/8).
CUBA	goma (8/9), rueda (6/9).
DOMIN. REP.	goma (6/7), rueda (3/7).
PUERTO RICO	goma (10/10), llanta (5/10), rueda (2/10).
PANAMA	llanta (6/6).
COSTA RICA	llanta (5/5).
NICARAGUA	llanta (5/5).
HONDURAS	llanta (6/6).
EL SALVADOR	llanta (4/4).
GUATEMALA	llanta (7/7).
MEXICO	llanta (10/10).

14.2 Observations

Llanta is used in all of Mexico and Central America and in South America from Colombia to Bolivia.

For many Argentines, *cubierta* is a more technical term referring specifically to the "tire" whereas *goma* and *neumático* refer to the "wheel" without distinguishing between its main components, the "tire", "inner tube" and "rim".

In Spain *neumático* can refer to the entire wheel and *cubierta* specifically to the tire, but many Spaniards do not make the distinction.

Caucho is used only in Venezuela.

15. TRUNK

15.1 Data by Country

SPAIN	maletero (10/12), maletera (2/-12), baúl (2/12).
CHILE	maleta (7/9), maletero (2/9).
ARGENTINA	baúl (11/11).
URUGUAY	baúl (6/6), valija (5/6).
PARAGUAY	valijera (6/6), baúl (3/6).
BOLIVIA	maletera (9/10), maletero (4/10).
PERU	maletera (12/12).
ECUADOR	maletero/a (10/16), cajuela (9-/16).
COLOMBIA	baúl (14/15), maletero (3/15).
VENEZUELA	maleta (8/9), maletera (4/9).
CUBA	maletero (9/9), baúl (3/9).
DOMIN. REP.	baúl (7/7).

PUERTO RICO	baúl (9/9).
PANAMA	baúl (6/7), maletero (6/7).
COSTA RICA	joroba (6/6), cajuela (2/6), maletero (2/6).
NICARAGUA	valijera (8/8).
HONDURAS	baúl (7/7).
EL SALVADOR	baúl (5/5).
GUATEMALA	baúl (5/5).
MEXICO	cajuela (13/13).

15.2 Observations

Baúl and *maletero/a* are the most common terms.

Almost all of the terms used are derived from one of four types of luggage containers: *baúl, caja, maleta* and *valija*.

In Ecuador *maletero/a* appears to be more common in the Costa and *cajuela* more common in the Sierra; one serrano also gave *cajón*.

In Costa Rica, *joroba*, according to some, comes from the shape of the Volkswagen Bug, which used to be the most popular car in that country.

16. SPEED BUMP

16.1 Data by Country

SPAIN	*NOT COMMON* (4/5), tope (1/5).
CHILE	*NOT COMMON* (4/6), lomo (2/6).
ARGENTINA	lomo de burro (13/13).
URUGUAY	(lomo de) burro (3/4), despertador (3/4), *NOT COMMON* (2/4).
PARAGUAY	mataburros (4/5), lomada (3/5), lomo de burro (1/5).
BOLIVIA	rompemuelles (6/6).
PERU	rompemuelles (11/11).
ECUADOR	policía acostado (11/11).
COLOMBIA	policía acostado (14/16), policía muerto (5/16).
VENEZUELA	policía acostado (13/15), burro (9/15).
CUBA	policía acostado (4/4).
DOMIN. REP.	policía acostado (8/8).
PUERTO RICO	muerto (8/8).
PANAMA	policía muerto (7/7).
COSTA RICA	muerto (5/6), policía dormido (5/6).
NICARAGUA	policía acostado (8/8).
HONDURAS	túmulo (7/7).
EL SALVADOR	túmulo (7/7).
GUATEMALA	túmulo (7/7).
MEXICO	tope (11/11).

16.2 Observations

(Lomo de) burro is used in Argentina and Uruguay.

Rompemuelles is used in Bolivia and Peru .

Policía acostado is used in Ecuador, Colombia and Venezuela

Túmulo is used in Honduras, El Salvador and Guatemala.

In Spain and Chile the object appears to be less common than in the rest of the Spanish-speaking countries, and two Uruguayans also stated that the object was not that common in their country.

In the countries where *policía* terms are used, other slang words for police officer are also used with the corresponding adjective. For example, in Ecuador *chapa* can mean "police officer" and *chapa acostado* can mean "speed bump".

One Paraguayan stated that *lomada* refers to a cement speed bump, whereas *mataburros* refers to a series of metal bars or wood beams that are placed on roads to serve the same purpose. A Uruguayan gave a similar description of *mataburros*.

17. STOP SIGN

17.1 Data by Country

SPAIN	stop (5/5).
CHILE	pare (6/6).
ARGENTINA	pare (7/7).
URUGUAY	pare (4/4).
PARAGUAY	pare (4/4).
BOLIVIA	pare (5/5), alto (3/5).
PERU	pare (8/10), stop (3/10).

ECUADOR	pare (10/10).
COLOMBIA	pare (6/6).
VENEZUELA	pare (8/8).
CUBA	pare (5/5).
DOMIN. REP.	pare (6/6).
PUERTO RICO	pare (9/9).
PANAMA	alto (6/6), stop (4/6).
COSTA RICA	alto (5/5).
NICARAGUA	alto (5/5).
HONDURAS	alto (7/7), stop (2/7).
EL SALVADOR	alto (5/5).
GUATEMALA	alto (5/5).
MEXICO	alto (8/8).

17.2 Observations

Pare is used in all of South America and the Hispanic Antilles.

Alto is used in all of Mexico and Central America.

In Panama stop signs generally say *alto*, but in the Canal Zone they generally say *stop* or occasionally are bilingual with both *stop* and *alto* appearing on the sign.

In Puerto Rico, the signs say *pare*, but on military bases they say *stop*.

18. DUMP TRUCK

18.1 Data by Country

SPAIN	volquete (5/6), volqueta (3/6).
CHILE	camión tolva (6/6).
ARGENTINA	(camión) volcador (9/11), volquete (4/11).
URUGUAY	(camión con) volcadora (4/6), (camión) volcador (3/5).
PARAGUAY	camión tumba (6/6).
BOLIVIA	volqueta (6/6).
PERU	volquete (13/15), volqueta (2-/15).
ECUADOR	volqueta (12/12).
COLOMBIA	volqueta (17/17), volteo (2/17).
VENEZUELA	(camión de) volteo (10/10).
CUBA	(camión de) volteo (7/8), volqueta (4/8).
DOMIN. REP.	(camión de) volteo (7/7).
PUERTO RICO	camión (de) tumba (3/3).
PANAMA	volquete (6/6).
COSTA RICA	vagoneta (6/6).
NICARAGUA	volqueta (5/5).
HONDURAS	volqueta (8/8), valastrero (5/8).
EL SALVADOR	camión de volteo (6/7), volqueta (2/7).
GUATEMALA	camión de volteo (6/7), camión de palangana (5/7).
MEXICO	(camión de) volteo (14/14).

18.2 Observations

Volqueta/volquete and *camión de volteo* are the most frequently used terms.

The terms *camión de volteo, camión volcador* and *camión con volcadora* are often shortened to *volteo*, *volcador* and *volcadora*, respectively.

The two Colombians that gave *volteo* were both from the Atlantic coast and their usage is consistent with that of much of the Hispanic Caribbean.

Chile, Paraguay, Costa Rica and Honduras appear to have terms not found anywhere else.

HOW CAN USAGE LABELS BETTER LABEL USAGE?

Jean Quirion
Université de Montréal

Keywords: Lexicography, Usage Labels, Dictionaries, Style, Register, Québec (Province of)

Abstract: As intensive users of uni- and bilingual dictionaries, translators acknowledge the importance of usage labels, but also recognize their ambiguous validity. From the dictionary-maker's perspective, the assigning of usage labels is certainly one of the most demanding task. The recent publication of two French general-language dictionaries in Québec has rekindled the discussion about labeling. It also shed new light on the dictionary's cultural role and suitability. This paper focuses on French stylistic labels (*littéraire, familier, populaire,* etc., the approximate equivalents of which could be *literary, informal, substandard*) as described and used in French dictionaries. The lack of clear label-assigning criteria -- in any language -- makes dictionaries easy targets for criticism. Because they do not always reflect social reality, stylistic usage labels deserve to be thoroughly reconsidered.

In an attempt to tackle these questions, we first state the role, the importance, and the typology of labels. The close links binding a dictionary and the society whose words it describes is then addressed. A study of the stylistic subcategories shows the *populaire* and *argot* labels to be suitable for depicting French society -- for the time being at least. But it also highlights their inability to properly represent the lesser-stratified Québec society. A look at labeling criteria supports the view that a typology of stylistic labels could be predicated exclusively on situations of communication.

Finally, we try to determine whether such typology can be used in various types of dictionaries (unilingual dictionary, bilingual dictionary, juvenile dictionary, etc.).

1. INTRODUCTION

When consulting a dictionary, have you ever questioned the usage label qualifying a word? For example, you may find it inaccurate that the word *about,* in the expression *he has a gun hidden about him,* be left unlabeled, because you may feel it is a formal use. Perhaps you would rather consider the synonymous expression *he has a gun hidden on him* as a neutral one. In another case, you might or might not agree that *admiration* is an archaic synonym of *wonder.* One can find thousands of such divergent interpretations of usage.

This paper deals with usage labels, a subject of much debate. What are they and why do they exist? Are they sufficiently reliable? What criteria determine their use? How can usage labeling be improved to ensure better reliability? These questions will be examined from the Québécois perspective, using stylistic labels as the focus. The relation of American English and Québécois French with, respectively, British English and French French will be discussed. The bitter dispute over the treatment of usage labels following the publication of the *Webster's Third New International Dictionary of the English Language* in the United States and the *Dictionnaire québécois d'aujourd'hui* in Québec will be compared.

2. USAGE LABELS

2.1 Typology of Labels

Generally, usage labels are divided into four categories: temporal (*archaic, obsolete*, etc.), stylistic (*informal, slang*, etc.), regional (*US, CD, GB*, etc.), and field labels (*chemistry, sports, building trade*, etc.).

2.2 What Are Usage Labels?

Usage labels are applied by lexicographers to indicate the association a word or expression carries. Most words are neutral and therefore bear no usage labels. These neutral words can be used in almost every context. Other words may carry informal or formal connotations, which restrict their use to certain situations. Labels indicate to the dictionary user the situations in which a word should (or should not) be used. For example, the word *car* is neutral, and can be used in any context. On the other hand, *horseless carriage* and *gas guzzler*, used to designate a car, will be labeled *humorous* and *(very) informal*, respectively. Unless they are used to produce a particular effect, these latter expressions are not employed in a formal or neutral context.

2.3 Why Are Labels Important?

Labels are important because they help translators make the right choice when writing or translating. They are the key to the association carried by a word or expression in the source language. Since these associations give clues as to the intent of the writer, examination of usage labels for key source text words is important during the decoding stage of translation. Experienced translators, therefore, often rely on labels to increase their understanding of the source text. In the same fashion, at the encoding stage, when the translator is in the process of rewriting the source text in another language, he needs to ensure that the words he has selected carry appropriate associations, both in terms of the source text and the target audience. Usage labels in dictionaries can help him make the right choices.

2.4 What Is Standard Language?

Usage labels are attached to words and expressions that deviate from standard language. Thus, any discussion of these labels requires a definition of standard language. But this is in itself a Gordian knot. While many definitions of language standards exist, the only characteristic they share is vagueness. Standard language is sometimes called good usage, at other times usage of reputable authors or of polite society, etc. But surprisingly enough, despite the fact that it is difficult to define, the concept of standard is nevertheless an operational one. A general idea of the standard guiding a dictionary writer can still be deciphered and debated!

3. DICTIONARY AND SOCIETY

3.1 Dictionaries As a Mirror of Society

Dictionaries are cultural objects. The choice of entries, spellings, definitions, examples, even gender in languages like French, and, of course, usage labels takes local considerations into account. For example, entries describing the educational and political systems, plant and animal life, etc., differ between Québec and France. The spellings are well known to vary between American and British English. The word *dépanneur* means «repairman» in France and Québec, but carries the additional meaning of «corner store» in French Canada. Examples citing geographical locations, historical facts, social activities are much more significative when adapted to the culture of the dictionary user. French say *le job* (masculine noun) while Québécois say *la job* (feminine noun). In similar fashion, usage labels need to be adapted, as we will see later.

The United States owe to Noah Webster the idea of writing American reference books distinct from the British models. His dictionary was the equivalent of a declaration of cultural independence from the mother tongue and country. Almost 200 years after Webster's work, Québec is now in the process of moving towards its lexicographic independence from the European French standards: two general dictionaries of Canadian French have been published in the last decade.

The making of Québécois French dictionaries has shed new light on usage labels. Of course, labels were to describe the current usage of French in Canada. For example, many words that are archaic in France are still common in Québec, and had to be relabeled accordingly. But Canadian lexicographers soon felt ill at ease with the stylistic label subcategories used in dictionaries from France. Their uneasiness gradually led them to question the system of labels. The temporal, regional, and field labels did not pose any particular problem; they were easy to transpose. The stylistic labels, however, were another story, as we will now see.

3.2 Stylistic Labels in France

Stylistic labels, like most of the information contained in a dictionary, reflect the culture they issue from. Stylistic labels in France can be broken down in the following subcategories:
a) **formality:** *littéraire* {formal or literary}, *familier* {informal}, and *très familier* {very informal}
b) **social class:** *populaire* {≈ substandard}, *argot* {slang}
c) **evaluative:** *vulgaire* {offensive or vulgar} and *péjoratif* {derogatory or pejorative}.

Any correspondence between French labels and corresponding American ones can only be approximate: the definitions of the French labels vary from one dictionary to another, as do those of the American labels.

3.3 Stylistic Labels in Québec

The French stylistic labels covering the formality and evaluative labels apply unchanged to Québec society. But both social class labels, *populaire* and *argot*, pose problems.

The *populaire* label

Populaire is defined in the following words in the introduction of *Le Nouveau Petit Robert*: "word or meaning prevalent in the spoken language of the working classes (often former slang that has acquired wider use), which would not be used at a higher social level. (to be distinguished from *familier*, which deals with a situation of communication.)[2]" As an entry in the same dictionary, *populaire* bears the following definition: "In linguistics, what is created and used by the working classes and is rarely used by bourgeoisie and educated people[3]." Interestingly, the label *populaire* has no logical counterpart *bourgeois*, most probably because *bourgeois* was considered unmarked, correct, i.e. standard French. No choice being candid in a dictionary, *bourgeois* was obviously the language of the upper class who wrote dictionaries.

The *argot* label

Argot is the other tag that causes difficulties, especially in the Québec context. Historically, the word designated the community of beggars, and became used to denote the language they spoke. Nowadays, it means either a language peculiar to a specific group of people (students, specialized workers, etc.) or the language of the poor. The use of *argot* in the sense of a language of a restricted group is questionable in a general dictionary, which must, by its very nature, concentrate on the most commonly used words. It is difficult to see the interest raised in a general dictionary by terms used and understood by very few. Moreover, once *argot* words become widely known, they fall into the (very) informal category. If the restricted group using the *argot* word or meaning belongs to a particular occupational field, the word could bear a field label instead of a social class mark. If necessary, a label indicating informality could be added. *Argot*, in the sense of the language of the underprivileged, is much like the *populaire* label in that it refers to a French conception of language use which is not very familiar to North Americans. French society, much like that of Great Britain, is still tinged with a long tradition of class divisions and aristocracy. This cultural trait comes to light in the levels of usage retained in French dictionaries. It is striking that labels have been left almost untouched since the first French dictionary was published in 1680. As Bernard Quémada states in the Universalis encyclopedia: "Right up to the present time, lexicographic methods have remained profoundly traditional and mostly empirical.[4]"

However, the conception of division by social class is less readily accepted in North America than in Europe. For instance, Thomas J. Creswell and Virginia McDavid (Random-House dictionary) discuss vocabulary variation in the following way:

> For the most part, vocabulary items are not individually correlated with socio-educational status. What chiefly distinguishes the vocabulary of the better educated or socially established speakers and writers is breadth and diversity. Educated speakers by no means always use polysyllabic, learned, or even "polite" words. Such speakers are instead capable of adapting vocabulary to topic, situation, and audience. Most of the words used by the educated are the same as those used by persons of lower educational attainment. But the

educated can, when appropriate, more easily vary their word choice. They are more likely to have command of a large range of synonymous words and grammatical constructions as well as a better perception of which synonyms are appropriate in a specific context[5].

Also, social mobility, although growing in Europe, is greater in America. A famous French lexicographer, Josette Rey-Debove wrote in the early seventies: "[...] in a democracy, compulsory public education and free access to any trade or profession mingles people from various origins and facilitates social mobility. It thus becomes extremely difficult, with merging lexicons, to characterize a word by the social class of those who use it[6]".

The above considerations clearly demonstrate that classifying words by associating them to distinct social classes becomes hazardous. Hence the labels *argot* and *populaire* are unjustified in Québécois French reference books.

4. CRITERIA FOR LABELING

4.1 Present Label Assignment Criteria

How do lexicographers, or those that make dictionaries, decide which label best suits a given word? Up until the present time, their linguistic intuition and profound knowledge of the language have been the main guiding force. More recently, however, lexicographers have been assisted in their task of labeling by electronic corpora. By studying key words in context, lexicographers can gain insight into the situation of communication involved in the use of a given word or expression. This somewhat empirical approach can suggest to the lexicographer a particular usage label. Two conclusions can be drawn from the present label-assigning process.

4.2 Lack of Objective Criteria

The first conclusion is that intuition and knowledge are rather variable from one person to another. They involve a great deal of subjectivity. This may explain why there is so much difference in labeling from one dictionary to the other, in the same language. One just has to take a few entries and compare their labeling in different dictionaries to see the diversities. The subjectivity can be attributed to the fact that the lexicographer has to leave his field of competence to enter the field of sociology, of interpersonal communication. Suppose that the average lexicographer is accurate 8 or 9 times out of 10 in assigning a label. Although this appears to be a respectable average, it still results in 9,000 wrongly labeled entries in a 60,000-entry dictionary.

4.3 Do Labels Depict Social Classes or Situations of Communication?

These considerations lead to a second conclusion, as to whether labels indicate social class division or situations of communication. Up to the present time, it appears to have been a mixture of the two. As a general rule, French society (encouraged by its dictionary writers) still tends to consider language in terms of social class. The Québécois society and lexicographers, on the other hand, have much more recent history and traditions, and tend to perceive language in terms of situations of communication. But as dictionary editorial staffs throughout the world rely more and more heavily (if not entirely) on corpora to decide on label, it is reasonable to believe that situations of communication, rather than social class divisions, will become the dominant basis for labeling, if it has not done so already. Labels depicting social classes could then fall into disuse, and labeling may thus become less subjective.

5. USERS' ATTITUDE

5.1 Dictionary Worship

If labeling is so difficult and often provides unsatisfactory results, then why do dictionary writers insist so much on it? Moreover, in a purely scientific approach, all words would be granted equal status. Would it not be better to describe language in a purely objective and scientific way? To ask these questions is to forget that dictionaries are primarily produced to be sold and that, commercially speaking, the buyer - the general public - is able to impose his demands. And that the customer has great expectations of the reference book. The general public uses, *grosso modo*, the same dictionaries as do language professionals. But it is more inclined to consider the dictionary as the ultimate reference tool. For non-linguists, it is generally a book that you buy once in a lifetime, twice if you are careless with it: it is never outdated. For the average reader, the dictionary cannot and must not fail. The following remarks made by Creswell and McDavid in the context of Anglo-American society are equally applicable to the French-speaking community:

> [...] the learned of [Samuel] Johnson's day [...] established the Anglo-American tradition of dictionary worship, which assumes that a dictionary or its editors are, in effect, the legislators of language. According to this view, a dictionary 'sanctions' or 'approves' usages and that approval is what makes the usages Standard English. [...] Laypersons are less likely to criticize a dictionary for unwarranted assumption of authority than for abdicating what they see as its responsibility to make decisions the public is bound to obey.[7]

5.2 What if the Dictionary "Fails"? Cases of *Webster's Third* and *Dictionnaire québécois d'aujourd'hui*

When dictionaries do not act as legislators of language, by providing usage labels for instance, they are considered "failures" by the general public. Both in the United States and in Québec, dictionaries have failed in the public eye in their labeling task. The disputable labels granted by lexicographers then confronted the opinions of the readers. Recognition failed and readers rejected both dictionaries. One of these was the *Webster's Third New International Dictionary of the English Language*. Published in the early 60's, it used a non-prescriptive approach. The editorial staff was accused of neglecting its responsibilities with respect to the establishment of a standard. The descriptive and normative approaches also clashed in Québec when the first edition of the *Dictionnaire québécois d'aujourd'hui* was published three years ago by the *Dictionnaires Le Robert*. As this dictionary marked a shift from a long-respected European French standard of labeling to a new Québec one, it stirred up strong reactions. Demands ranged from the ban of Québécois French words to requests for stricter labeling. The calling into question of usage labels led to a closer study of their definition and process of attribution.

6. CONCLUSION

In this paper, we stated the undeniable interest of usage labels for both the encoding and the decoding processes. We also discussed the recent publication of two general-language dictionaries made in Québec, which put into question the traditional classification system of usage labels. Two stylistic labels, *populaire* and *argot*, were found to be inadequate for marking Québécois words and expressions: they could be replaced or, more probably, discarded.

The discarding of the social class subcategory from stylistic labels leaves situations of communication as the only basis for labeling French Canadian words. This exclusion is coherent with the prevalence of electronic corpora as clues to usage in label assignment, because contexts inform more on situations of communication than on social classes of speakers. The mismatch of France's social stratification when applied to the Québécois society also coincides with exclusive resort to situations of communication.

The stylistic labels suitable in a Québécois French dictionary can be summarized as:
a) **formality:** *littéraire* {formal or literary} and *(très) familier* {(very) informal}
b) **evaluative:** *vulgaire* {offensive or vulgar} and *péjoratif* {derogatory or pejorative}.

The labels retained are very flexible and merely restrictive. Users need to be directed on the contexts in which words and expressions can be used. The general structure of labels and the process suggested for labeling is simple and thorough enough to suit any dictionary describing general language. Children dictionaries could benefit from this simplified version of stylistic labels. The description proposed could be used as a starting point for reference books specialized in a certain stylistic label subcategory (i.e. colloquial dictionaries). The writers of such dictionaries would probably tend to refine the part of the scale they are fully describing. The classification is simple and clear enough to be used in dictionaries for second-language learners and bilingual dictionaries.

6.1 Disparity or Consensus?

We underlined earlier the great disparity among dictionaries in the labeling of similar words or expressions. Paradoxically, these differences hide what we believe is a widespread consensus about the necessity of marking these same words or expressions. Although lexicographers often disagree on the label applied, they usually agree on what words to mark. Labeling obviously calls for a great deal of personal judgment, which leads to as many interpretations as there are dictionary writers. But how can agreement on the necessity of labeling be used to foster greater uniformity of labels? The reduction of the number of labels could lead to some consensus in labeling. Would the user lose precious information? Speakers have to be guided in the choice of words and expressions in an infinity of situations of communication. They basically need to know if a given word is neutral, high standard or low standard. Beyond this point, association or connotation is a matter of nuances and personal judgment. The assigning of labels could be based on a series of real or fictitious contexts, representative of the high and low standards. The adequation of the word or expression under study in some of the contexts could be sufficient for the lexicographer to assign a label. The contexts chosen could be explicited in the preface of the book, thus giving the user clearer indications of the criteria behind the labels choice.

6.2 Future Perspectives

Dictionary makers will rely more and more heavily on electronic corpora to assign labels. In the long run, it could standardize label-assigning criteria. But it still would be a far cry from uniformity of usage labels classifications between dictionaries in a given language or between different cultures.

Up to now, usage has been described by labels in order to save dictionary space. But the advent of electronic reference tools, for which shortage of space is a very relative concept, could very well set the stage for interesting new possibilities for the label-assigning process.

However, many labeling issues will remain to be resolved. Not only do the criteria for labeling need to be formalized, but the concept of standard language needs to be defined. It is only when all these issues have been addressed that usage labels can indeed better label usage.

NOTES

1. I wish to thank Dr Roda P. Roberts and Ronald Bissonnette for their valuable assistance in checking this paper for publication.

2. "Qualifie un mot ou un sens courant dans la langue parlée des milieux populaires (souvent argot ancien répandu), qui ne s'emploierait pas dans un milieu social élevé. (À distinguer de *fam.*, qui concerne une situation de communication)". Le Nouveau Petit Robert, p. xxiii.

3. "LING. qui est créé, employé par le peuple et n'est guère en usage dans la bourgeoisie et parmi les gens cultivés". Le Nouveau Petit Robert, p. 1 727.

4. Free translation. Encyclopédie Universalis, 1984, vol. 7, p. 390.

5. Thomas J. Creswell and Virginia McDavid. "Usage: Change and Variation", in The Random-House Dictionary of the English Language, p. xxii.

6. "[...] dans une démocratie, l'enseignement public obligatoire, l'accession libre à toutes les professions met en contact des personnes issues de milieux très divers et favorisent le passage d'une classe à l'autre. Il devient extrêmement difficile, dans ce brassage du lexique, de caractériser un mot par la classe de personnes qui l'emploient". Josette Rey-Debove, Étude linguistique et sémiotique des dictionnaires français contemporains, 1971, p. 92.

7. Thomas J. Creswell and Virginia McDavid. "Usage: Change and Variation", in The Random-House Dictionary of the English Language, p. xxiii.

REFERENCES

1. *The American Heritage Dictionary of the English Language*. 3rd ed., Boston, Houghton Mifflin, c1992. xliv, 2 140 p.

2. BARBER, Edwin. "The Treatment of Slang in 'Webster's Third New International Dictionary'". *American Speech*, vol. XXXVIII, no. 2 (May 1963), p. 103-11.

3. CARD, William, Raven I. MCDAVID and Virginia MCDAVID. "Dimensions of Usage and Dictionary Labeling". *Journal of English Linguistics*, vol. 17 (1984), p. 57-74.

4. CASSIDY, Frederic G. "Toward More Objective Labeling in Dictionaries". In *Studies in Honor of Albert H. Marckwardt*, James E. Alatis (ed.), Washington (D.C.), TEOSEL, 1972, p. 49-56.

5. CRESWELL, Thomas J. "Usage in Contemporary American Dictionaries". *Babel*, vol. XXIII (1977), p. 23-28.

6. *Dictionnaire du français Plus : À l'usage des francophones d'Amérique*. Claude Poirier, réd., Montréal, Centre éducatif et culturel, 1988. xxiv, 1 856 p.

7. *Dictionnaire québécois d'aujourd'hui : langue française, histoire, géographie, culture générale*. Sous la dir. de Jean-Claude Boulanger, Saint-Laurent (Québec), DicoRobert, c1992. 1 vol., pag. mult., cartes en coul.

8. DOBBINS, Austin C. "The Language of the Cultivated". In *Harbrace Guide to Dictionaries*, Kenneth G. Wilson, R.H. Hendrickson and Peter Alan Taylor (eds.), New York and Burlingame, Harcourt, Brace & World, 1963, p. 142-144.

9. EBLE, Connie C. "Slang: Variations in Dictionary Labeling Practices". *LACUS Forum*, no. 11 (1984), p. 294-302.

10. GOVE, Philip B. "Usage in the Dictionary". *College English*, vol. 27, no. 4 (January 1966), p. 285-292.

11. GUIRAUD, Pierre. «Français populaire ou français relâché». *Le français dans le monde*, n° 69 (décembre 1969), p. 23-27.

12. HARTMANN, R.R.K. "Style Values: Linguistic Approaches and Lexicographical Practice". *Applied Linguistics*, vol. II, no. 3 (Fall 1981), p. 263-273.

13. KENYON, John S. "Cultural Levels and Functional Varieties of English". *College English*, vol. 10 (October 1948-May 1949), p. 31-36. Reprinted in *Harbrace Guide to Dictionaries*, Kenneth G. Wilson, R.H. Hendrickson and Peter Alan Taylor (eds.), New York and Burlingame, Harcourt, Brace & World, 1963, p. 151-153.

14. MESSELAAR, P.A. «Les marques "familier" et "populaire" envisagées des points de vue lexicologique et lexicographique». *Cahiers de lexicologie*, vol. LIII, 1988-II, p. 91-106.

15. *Le Nouveau Petit Robert : Dictionnaire alphabétique et analogique de la langue française*, sous la direction de Josette Rey-Debove et Alain Rey, nouvelle éd. remaniée et amplifiée. Paris, Le Robert / Montréal, DicoRobert, 1994. xxxv, 2 467, [24] p.

16. QUÉMADA, Bernard. «Dictionnaire». *Encyclopaedia Universalis*. Paris, Encyclopaedia Universalis, 1984, corpus 7, p. 387-390.

17. QUIRION, Jean. «Les marques d'usage sociales dans les dictionnaires québécois : éléments de réflexion», communication présentée au 63e Congrès de l'Association canadienne-française pour l'avancement des sciences (Chicoutimi, 22-26 mai 1995) [inédit]

18. *The Random House Dictionary of the English Language*. 2nd ed., unabridged, editor in chief: Stuart Berg Flexner, managing editor: Leonore Crary Hauck, New York, Random House, c1987. xlii, 2 478, 32 p.

19. REY-DEBOVE, Josette. *Étude linguistique et sémiotique des dictionnaires français contemporains*, coll. Approaches to semiotics; 13, La Haye, Mouton, 1971. 329 p.

20. RYAN, William M. "Where Has All the Usage Gone?". *American Speech*, vol. XLIV, no. 2 (May 1969), p. 129-134.

21. *Webster's Third New International Dictionary of the English Language*. Unabridged, editor in chief: Philip Babcock Gove and the Merriam-Webster editorial staff, Springfield (Mass.), G. & C. Merriam, 1986. 72a, 2 662 p.

22. WOLK, Anthony. "Linguistic and Social Bias in the *American Heritage Dictionary*". *College English*, vol. 33, no. 8 (May 1972), p. 930-935.

Connections

Translation and Computers

PERFECT MT: LOGICAL CERTAINTY OR RECURRENT SELF-DELUSION?

(TEN FRAGMENTS AND THREE CONTENTIONS CONNECTED BY A SINGLE THEME)

Alex Gross
Cross-Cultural Research Projects, ATA

Keywords: Machine Translation, Human Translation, History of Computers, Knowledge Interfaces, Data Bases, Limits of Science

Abstract: The speaker will discuss Machine Translation as one of a number of language- and knowledge-organizing devices that have developed over time and in various cultures. His chief interest lies in the assertions of MT's pioneers--though still echoed by some specialists today--that perfect or near-perfect translation by computers can in fact be accomplished. Using overhead slides, he will move from early Chinese to classical Arabic to medieval Christian beliefs about the nature of knowledge and language along with various theories attempting to explain or control them. He will not neglect Swift's satirical Academy of Lagado as he moves closer to the computer age and attempts to show at least a few parallels between modern science and past systems of knowledge. Problems with various types of knowledge interfaces will be considered, including those for computers, and the speaker will conclude with some specific remarks about where MT is currently headed and how translators can best accommodate themselves to the kinds of work it does best.

I want to begin this paper with the all-important two-part proviso that has to accompany any treatment of MT at a translators' Conference. 1) No one opposes MT where it works, and 2) MT works quite well for those tasks where it is suitable. Similarly, there are two extremes which we must avoid at all costs:

a) MT is useless--it will never truly work; and
b) MT is inevitable, it will soon take over all
of translation, and we will all be out of jobs.

Now that I've made this abundantly clear, I'd like to discuss the question that I find most truly interesting. How is it possible that for nearly fifty years many perfectly intelligent and even ingenious engineers, linguists, and researchers could have ever supposed that something resembling perfect MT could ever come into being? And how is it possible, I might add, that quite a few of these authorities have still not totally climbed down from this position and continue to believe that it will sooner or later become possible to put most of the people in this room out of work? And, finally, how does it come about that a number of academic fields and even professional organizations are still aimed at precisely this goal? How has it become possible for any of this to take place and for any of these people to go on believing this?

To answer this multi-part question I want to take you on a private tour through a number of artifacts, images, or ideas from the past and/or from other cultures. I also want to describe a few encounters and episodes I've witnessed in our own age and culture. I'm calling all of these collectively "fragments," and I hope that by the end of this tour you will begin to share my view of the connections between them and see how they relate to MT today. What most of these fragments have in common is that they have something to do with a means of organizing either language or human knowledge or reality itself--or all three together. In the midst of our tour I will also introduce three outrageous contentions, which I mean to defend quite seriously.

The first such artifact [OVERHEAD SLIDE 1] is a Chinese geomantic compass, called a *fengshui luopan* or, literally, a wind and water compass. As you can see, it's a set of concentric circles and purports to demonstrate how individual human beings fit into the larger pattern of the universe. There is no way I can really explain *fengshui* without all of you assuming that it must be some form of astrology, even though I know for a certainty that it covers far more than what is properly considered as astrology. This wheel or compass provides a metaphysical, medical, and even a methodological guide to the classical Chinese universe.

But what does this have to do with computers, you may ask. If we look at the next slide [SLIDE 2], the answer should become clear. Once again we see a wheel or a set of wheels, but instead of Chinese characters we find English words, actually a translation from the Latin. This set of wheels, unlike the Chinese one, is credited by those in the field with providing the first crucial step towards the computer. The reason for this is that its various wheels move independently of one another, while the *fengshui luopan* was a single rigid piece. This is what patent lawyers call "the inventive step" and possibly defines that point where the East ends and the West begins, even though it may yet turn out that its inventor based it on an Arab original.

This inventor, whose life bridged the thirteenth and fourteenth centuries, was a truly great linguist and even the founder of language schools. He managed to be both a scholar and a popularizer, both a scientist in terms of his own age and a fanatical Christian apologist. And because he wrote in both Catalán and Arabic, for four centuries his works were known in both Christian and Muslim lands. His name in his native language is *El Beat Ramon Llull*, or the Blessed Raymond Llull, "Blessed" being the title one step below "Saint" in the Catholic hierarchy. He was also deeply interested in medical studies.

What do these Llullian wheels do? They supposedly illustrate the attributes of God according to various human and divine categories. To some extent they overlap on the territory of the Chinese Fengshui compass. But because the wheels move independently, what we are also looking at here is an early example of a relational data base, or at least of the hardware or software shell for such a data base.

This invention, as vapid and metaphysical as it may seem, pointed the way to the scientific age, which I think we've all heard about, and so I'll skip to the year 1726, when Jonathan Swift was busy sending up this entire movement in *Gulliver's Travels*. Here we see [SLIDE 3] the *Frame* or *Engine for Improving Speculative Knowledge* from the Academy of Lagado. Swift describes it as follows:

> The first Professor I saw was in a very large Room, with Forty Pupils about him. After Salutation, observing me to look earnestly upon a Frame, which took up the greatest Part of both the Length and Breadth of the Room; he said, perhaps I might wonder to see him employed in a Project for improving speculative Knowledge by practical and

mechanical Operations. But the World would soon be sensible of its Usefulness; and he flattered himself, that a more noble exalted Thought never sprang in any other Man's Head. Everyone knew how laborious the usual Method is of attaining to Arts and Sciences; whereas by his Contrivance, the most ignorant Person at a reasonable Charge, and with a little bodily Labour, may write Books in Philosophy, Poetry, Politicks, Law, Mathematicks and Theology, without the least Assistance from Genius or Study. He then led me to the Frame, about the Sides whereof all his Pupils stood in Ranks. It was Twenty Foot Square, placed in the Middle of the Room. The Superficies was composed of several Bits of Wood, about the Bigness of a Dye, but some larger than others. They were all linked together by slender Wires. These Bits of Wood were covered on every Square with Papers pasted on them; and on these Papers were written all the Words of their Language in their several Moods, Tenses. and Declensions, but without any Order. The Professor then desired me to observe, for he was going to set his Engine at work. The Pupils at his Command took each of them hold of an Iron Handle, whereof there were Forty fixed round the Edges of the Frame; and giving them a sudden Turn, the whole Disposition of the Words was entirely changed. He then commanded Six and Thirty of the Lads to read the several Lines softly as they appeared upon the Frame; and where they found three or four Words together that might make Part of a Sentence, they dictated to the four remaining Boys, who were Scribes. This Work was repeated three or four Times, and at every Turn the Engine was so contrived, that the Words shifted into new Places, as the square Bits of Wood moved upside down.

Six Hours a-Day the young Students were employed in this Labour; and the Professor shewed me several Volumes in large Folio already collected, of broken Sentences, which he intended to piece together; and out of those rich Materials to give the World a compleat Body of all Arts and Sciences; which however might be still improved and much expedited, if the Publick would raise a Fund for making and employing five Hundred such Frames in Lagado, and oblige the Managers to contribute in common their several Collections.

He assured me, that this Invention had employed all his Thoughts from his Youth; that he had emptyed the whole Vocabulary into his Frame, and made the strictest Computation of the general Proportion there is in Books between the Numbers of Particles, Nouns, and Verbs, and other Parts of Speech."[1]

In his description of the Academy of Lagado, Swift was of course exercising his satirist's right to exaggerate. Many of the experiments he singles out are totally ridiculous, though perhaps no more so than some experiments undertaken today. He was in fact satirizing all of Science, and this may be the reason why this episode has been cut from some abridged versions of Gulliver's Travels.

How mistaken Swift could be is proven by my next exhibit, a perfectly successful example of language-, knowledge-, and reality-management, dating from the year 1852. Once again there is a medical aspect, because its inventor was not only a doctor but served as Secretary and Vice-President of London's Medico-Chirurgical Society and later as a member and Secretary of the Royal Society. He also invented a slide rule, wrote a regular column on chess problems, and even created the first inexpensive chessboard. Perhaps most intriguingly--he worked long and hard during the 1840s on the invention of a calculating machine. But work on his most important contribution to humanity began only when he was in his seventy-first year. This invention was so successful that we still use it in one version or another today, 147 years after its creation. I used it in writing this paper.

It is of course [SLIDE 4] the "Thesaurus of English Words and Phrases" by Dr. Peter Mark Roget. In the preface Roget advanced both philosophical and practical claims for his work. Here once again we encounter the notion that an invention capable of organizing language can also affect human knowledge, perhaps everyday reality itself. Roget argued that anyone who used his Thesaurus would become more persuasive in argument and hence better able to influence events.

There's one important feature about his invention that I would ask you to note, as it has some relevance to MT. With all his interest in chess, slide-rules, and automatic calculators, Roget never suggested that his Thesaurus itself was automatic or that it could be consulted mechanically by a non-human user. The whole point of the Thesaurus is that would-be users must examine the various lists of words and make their own choices, *based on a specific task and context.* But even today not everyone gets the "point" of a thesaurus: there are those--mostly non-writers or others lacking verbal sophistication--who imagine it is a kind of failed dictionary. They want only the word that they imagine they want and may actually blame the Thesaurus for making them choose. I can't help wondering if some of their descendants have not now wandered into MT.

With Peter Mark Roget we are clearly only a few steps from his contemporary Charles Babbage, who was already at work--with aid from Lord Byron's daughter Duchess Ada--building the "Analytical Engine," which had it been completed would have qualified as a true prototype of the computer.

At this point, I think my next fragment can be none other than Alan Turing's famous statement in his paper *Intelligent Machinery*[2]. Here he foresaw the use of "television cameras, microphones, loudspeakers, wheels and handling servo-mechanisms" as well as some sort of "electric brain." It would be capable of [SLIDE 5]:

> "(i) Various games...
> (ii) The learning of languages
> (iii) <u>Translation of languages</u> (author's emphasis)
> (iv) Cryptography
> (v) Mathematics"

Now we are getting directly into computers, and my next example is going to be rather contentious or will appear as such to some people. It is in fact a set of three connected arguments I made two years ago at a New York Circle panel on MT. It goes as follows:

1) There will never, ever be a perfect computer interface that works satisfactorily for all purposes and for all people.

2) There will never, ever be a perfect hypertext system that permits ideal information retrieval for all people.

3) There will never, ever be a truly advanced system of machine translation that allows all texts to be adequately--not perfectly but no more than adequately--translated for all purposes.

These are the three contentions I mentioned in my sub-title. They are closely related, and most of my remaining fragments will be devoted to proving they are true. I will also have a few words for anyone who may be shocked by my use not only of the adverb never but of its colloquial cousin "never, ever," but I will save these for the Conference session itself.

My next fragment is an episode and has to do with CATNYP, the New York Public Library's computerized catalog. I was using this system recently when a young man sat down at the next computer. After spending about half an hour fiddling through the help screens, he was clearly close to despair. Finally, he turned to me and whispered "Look, how do you use this thing anyway?" I asked him what he was searching for, and he told me he needed descriptions about clothing worn by the middle classes in early nineteenth century England. "I've tried `clothing,'" he lamented, "I've tried `middle classes' and I've tried `England,' but none of it works."

For the next half hour we all did our best--soon two near-by researchers were also whispering solutions--to refine (or should we say pre-edit?) his question so that the computer could handle it. We ran through "garb," "apparel," "attire," and other generic hedges for "clothing," and finally our efforts were rewarded. CATNYP produced a screen listing an illustrated book that seemed to meet our friend's needs. We were elated by our collective success, but then he asked another question: "Hey, that's great, what a terrific system--now how do I bring the book up on the screen?" With some embarrassment, as though we ourselves were responsible for the system's shortcomings, we explained to him that he would have to fill out a slip, hand it in at the desk, wait ten minutes or so, and finally be handed a heavy, old-fashioned, page-ridden analogue book. He was clearly annoyed by this, and to some extent so were we.

But wait, I hear the cry ringing out, before you know it, the great computer miracle will soon have solved this. By tomorrow at the latest every single page of every single book ever written will soon be accessible, graphics, fold-outs, tables, and all--with just a few keystrokes. Those who suppose this will truly become possible, as with perfect machine translation, have not even begun to focus on the scope of the problem. In writing this paper, I made a few inquiries of librarians, and they confirmed my worst suspicions many times over.

The entire national library system long ago spent millions on "up-to date technologies" in the form of microfilm and microfiches. But even then, using a relatively inexpensive technology, they came nowhere near preserving all the world's books or periodicals but only a small fraction of the most valuable of these. Now both microfilm and microfiche are considered a passé technology, at least by computer advocates and salesmen. But by the time these new tools can possibly hope to record a comparable fraction of printed materials, what further technology will lie in wait with its own set of fabulous claims? Have we willy-nilly been placed in the position of the Red Queen in *Alice in Wonderland*, who had to run as fast as she possibly could simply to remain in one place?

A few decades ago some of us collected 78 RPM records, only to see them replaced by 33's and 45's, themselves later eclipsed by audio cassettes and 8-track stereo, all of them now replaced by CD's and/or their CD ROM cousins. I am once again happily collecting these last two products, but the word has been out for a few years that even their days are numbered. And at no point did any one of these remarkable media ever come close to exhausting the simply unbelievable wealth of classical music that exists in print or manuscript form. Is it any wonder that librarians are skeptical of some of the claims advanced by computer enthusiasts?

Now let's move in a bit closer to Machine Translation. I hope everyone understands that in computer terms there are distinct similarities in building a data base of any kind, whether it is to catalog books, build a thesaurus, set up an MT system, or create a model of the universe. Basically, in all these cases, what we are doing is constructing a data base, with just a few exotic (or perhaps not-so-exotic) differences thrown in. The computer doesn't care in the slightest which one you are doing. In fact, the computer never knows that it is word processing or accessing information or plotting a map or even telecommunicating or printing something on a page. In fact, the computer is so dumb that it can't even tell when it is displaying an erotic image.

When the time came to deepen the computer's relationship with printers, it had to be fed something called a page-description language--the most famous of these is "Postscript." When it was decided to store literary works in electronic form, a book-description language had to be invented--here the best known is "SGML." And with the advent of the World Wide Web, it even became necessary to produce a screen-description language--the now famous "HTML," which the experts are already discussing how to change. But in order for merely adequate machine translation to occur, it will certainly be necessary to create something far more ambitious, namely a "language-description language." Such a construct would have to take into account most phenomena that can occur in language, including semantic and contextual elements.

Here is the crux question: how is such a language-description language to be created? If we truly attempt to include a large number of linguistic aspects, then we will greatly increase the potential for error and also end up with something quite unwieldy. No matter how vast or fast our computers can become, it may still be unwieldy in human terms, and we're the ones who have to use such a program. If on the other hand, we attempt to include only a limited subset of language, then we will end up with something like the Controlled Languages evolved by Caterpillar and a few other firms and will have failed to reach our original goal. This entire conundrum brings to mind the problems encountered by cartographers in a one-minute parable by Borges, which I will now read in its entirety as my next fragment:

> "Of Exactitude in Science
> "...In that Empire, the craft of Cartography attained such Perfection that the Map of a Single province covered the space of an entire City, and the Map of the Empire itself an entire Province. In the course of Time, these Extensive maps were found somehow wanting, and so the College of Cartographers evolved a Map of the Empire that was of the same Scale as the Empire and that coincided with it point for point. Less attentive to the Study of Cartography, succeeding Generations came to judge a map of such Magnitude cumbersome, and not without some irreverence, they abandoned it to the Rigors of Sun and Rain. In the western Deserts, tattered Fragments of the Map are still to be found, Sheltering an occasional Beast or beggar; in the whole Nation, no other relic is left of the Discipline of Geography."[3]

Building a truly comprehensive language-description language is altogether likely to involve the same order of complexity and impracticality as this mission to build a same-scale, point-to-point map of the Empire.

But now let's turn to how Machine Translation really does work and see what lessons it holds for us. Essentially MT will work best--perhaps not perfectly even here--when you have what I call a Level Playing Field Translation setup. [SLIDE 6]

As you can see from the slide, what we have here is literally a "level playing field" with something like a tennis net in the middle. On one side, just to take one possible example, (you could have any scientific specialty at all, so let's be a trifle whimsical) you have Professional Japanese Hydraulic Biochemical Micro-Nuclear Space Scientists. On the other side, in the English-speaking world, we find Professional American Hydraulic Biochemical Micro-Nuclear Space Scientists. Here, except for the difference in language, you have an almost perfect match-up of knowledge and experience, so this example ought to be absolutely ideal for setting up an MT system, building terminology databases, constructing lists of new or unknown words, and pouring them into the original program.

But even here, in this nearly ideal MT setting, you can still encounter problems. What if lexical entries don't match up perfectly in the two languages? What if the scientists in the two countries aren't following the same procedures--in science or technology a near certainty? What if they don't even see themselves as performing the same steps for the same reasons?

Any or all of this could potentially happen even in our ideal level playing field example. But what if--let's just suppose--there were some factors present that tilted the playing field a bit to either side, or tilted it for some of our experts in one direction but for others in the opposite way? Some possible instances: supposing half of the people on the Japanese side turn out to be not Professional Hydraulic Biochemical *etceteras* but newly trained interns instead? Or if they are in fact professionals but come from related scientific fields with slightly different terminologies? What if they are students, or merely stockholders in the company, or investigative reporters, or members of the general public who have wandered in to find out what the company is up to? Much the same questions can be asked on the American side, and the answers to questions on one side can raise further questions on the other. What happens to the translation process in any combination of these conditions, even assuming *human beings* are providing the translation? But, most important, what will happen to an MT system under such circumstances?

Here we come to a crucial point which I have made elsewhere in other terms. Contrary to our facile belief that there can be such a thing as a "good translation" or a "correct translation" that will work in every case, no such thing as "generic translation" may exist at all. It may simply be a convenient fiction we have employed to shield us from the true complexity of the translation process and/or as a way of reassuring ourselves or our customers that we are in all cases capable of producing a "correct translation."

Let me say this another way: there is one other crucial factor involved in a translation besides the two languages involved and the nature of the subject matter--it concerns the audience and/or the occasion for such a translation. Wherever this audience or occasion changes even slightly, there may have to be a corresponding shift in the tone of the translation. Where either of these factors changes more than slightly, we enter the territory of rejected translations, possibly even charges of incompetence. But even the most conscientious translator or translation company may not always be prepared to meet every demand these circumstances are capable of hurling at us.

What we have run into here--or perhaps it has run into us with a big stick in its hand--is the true extent of the complexity of language. It is hard enough for humans to work under such circumstances--how can we expect machines to handle them? The real explanation here may well be that we all make some outrageously false assumptions about language and are totally unaware we are doing so. Once again, we assume that we are all walking around on a level playing field, where anyone can readily communicate with anyone else across a short and easily bridged distance.

But the truth is that we do not inhabit a level playing field at all where language is concerned. On the contrary, if we were to visualize ourselves and everyone around us as walking about on stilts of completely different heights, textures, and stability, so that even our very own two stilts are not necessarily of the same height or composition, we would have a better notion of how we actually move through linguistic space and communicate with others. You can easily persuade yourself that this is true by the way you react to others the next time you are in a social situation.

We each of us have our own store of linguistic tricks and devices, and we look out almost instinctively for those who have complementary tricks and devices. Whenever we meet such a person, we become flushed with enthusiasm, sometimes even love, and go on talking forever. But we just as quickly abandon those who do not respond to our conversational rhythms. True, we also carry on everyday conversations with persons who do not share our interests or language style, but we usually do not speak at length or in detail or about more than a few topics with them. What I am trying to suggest is that there is a whole universe of language habits we are simply unaware of. And if we are not aware of them, how can we suppose that a computer can gain such awareness?

Let me now penetrate to the core of practicality about MT: its place in the office environment. In every company over a certain size, there exists at least one individual whose sole duty is to make your existence as a translator extremely unpleasant. This person may be an office manager, an accountant, or perhaps even the boss's personal assistant. In all these cases, such a person will constantly be looking for ways of saving money. Almost invariably, their gaze will fall upon the translation department, whose employees are clearly being overpaid to do work that should be accomplished in a fraction of the time. Your doom may well be spelled if truly persuasive sales reps from an MT company pay a call, and this accountant (or whoever) falls under their spell. If you do get the sack or find yourself being retrained as an MT post-editor, you have only one consolation. As likely as not, three years later a completely new office manager (or accountant or boss's assistant) will be looking for ways to save again, and this time they will decide that the MT system costs too much and makes too many errors, and the time has come to retrain MT post-editors as humans. Your best bet in any such situation is to become as knowledgeable about MT as you possibly can be and learn how to play office politics, so as to influence the decision-making process before it happens.

Finally, I'd like to round out this paper in two ways. I started out by noting that several professional organizations and perhaps even entire academic fields are still dedicated to the goal of perfecting MT or still hold that such an option is viable. Here is a partial list of these groups and some of the conferences they have been sponsoring [SLIDE 7]:

REACHING FOR MIND:
FOUNDATIONS OF COGNITIVE SCIENCE
Call for Papers for the
Fourth International Conference on
The Cognitive Science of Natural Language Processing

LAST CALL FOR PAPERS
FORMAL GRAMMAR
in conjunction with the

European Summer School in Logic, Language and Information

GROUNDING REPRESENTATIONS:
Integration of sensory information in
Natural Language Processing,
Artificial Intelligence and Neural Networks
IEE COLLOQUIUM
IEE Computing and Control Division

The School of Applied Languages, Dublin City University

The Association for Computational Linguistics
Fourth International Workshop on Parsing Technologies

II International Conference on Mathematical Linguistics

Fifth International Workshop on Natural Language
Understanding and Logic Programming

Many of the members of these groups are highly respected and highly paid academics, and it might seem sacrilegious to some to suggest that they can possibly fail in their goal. I nonetheless believe that most of the people in this room--perhaps even most of the people at this Conference--possess deeper and more useful knowledge about language than most of the members of these groups, who could in fact profit greatly by listening to what many of you could tell them about language.

I hope that these fragments have now begun to fall into place to some extent. And now, to conclude, let's look briefly at our first two slides again. And in that context let's look at some of the diagrams produced by some MT advocates. Though only one is reproduced in the Proceedings, several will be shown at the Session itself. Someone from another planet might suppose that all these diagrams--MT, geomantic, theological, and satirical alike--shared certain characteristics. After all, they are all based on circles and other geometric figures, and they may also share in a certain circular reasoning in that they purport to represent clear and reproducible relationships between the abstract and practical realms, as do the magical drawings of many cultures and ages. But it is by no means certain that any of these diagrams fully succeeds in delineating--much less establishing--such a relationship. Each one may succeed up to a certain point only because of our self-serving desire that it should do so, even our need to believe in our own self-fulfilling prophecies. Yet each one also fails because it falls short of representing the complexities of both language and reality by several orders of magnitude and makes no allowance for the altogether variable identity of the human value at its center. Thus, it may still remain to be determined whether and/or to what extent our Doctors of Artificial Intelligence, Cognitive Science, and Mathematical Linguistics truly differ from the Professors at the Academy of Lagado.

[1]Swift, Jonathan. *Gulliver's Travels.* Part Three, Chapter 5.

[2]Hodges, *Andrew (*1983) *Alan Turing: The Enigma* Simon & Schuster, New York. p. 382.

[3]Ostensibly from *Travels of Praiseworthy Men* (1658)
by J. A. Suárez Miranda, actually a part of Jorge Luís Borges' *A Universal History of Infamy*, translated by Norman Thomas di Giovanni, London: Allen Lane, 1973.

Four Illustrations from this Session

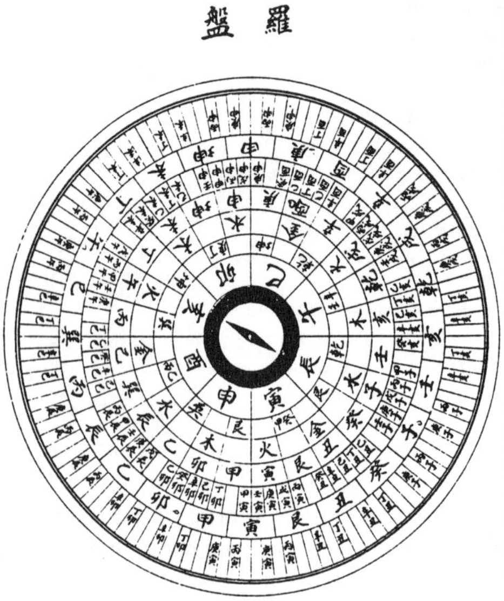

Chinese Geomancer's Compass

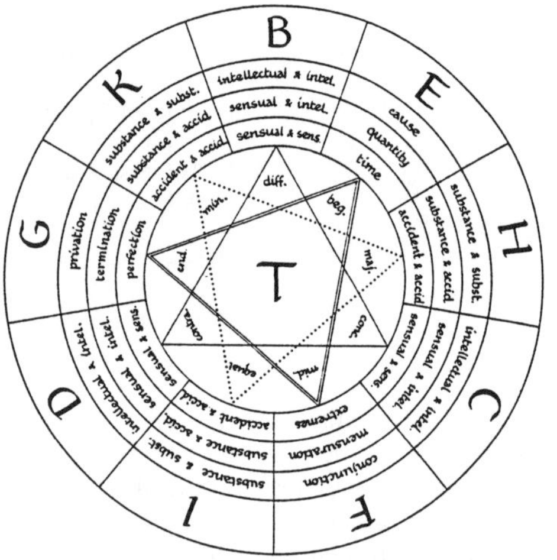

Llullian "Divine Engine"

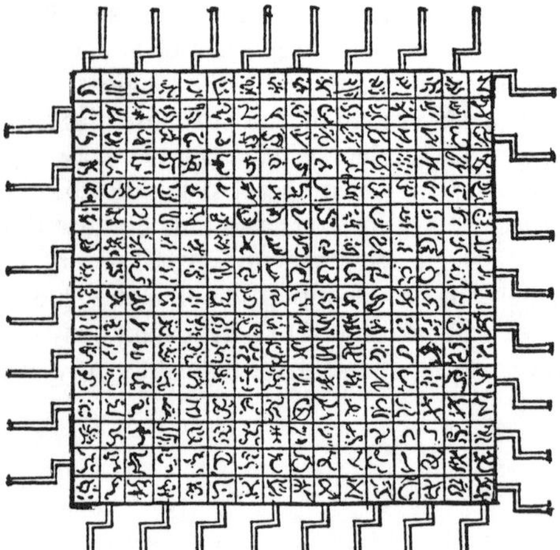

The Project for Improving
Speculative Knowledge,
Academy of Lagado

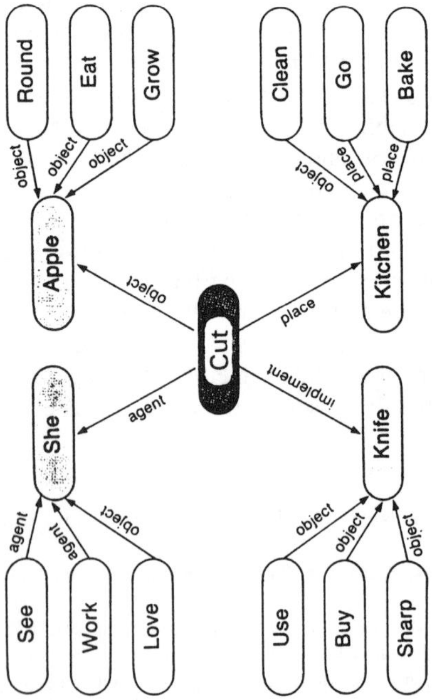

MT Dictionary in Schematic Form

MACHINE TRANSLATION: FRIEND OR FOE?

Alejandra Etel Koval
ATA Associate Member

Keywords: Machine Translation, Text Analysis, Dictionary Building, Customization, Post-editing, Final Review, Accuracy, Cohesion, Coherence, Readability

Abstract: Machine translation (MT) is occupying a more relevant place in our profession because it offers some advantages over manual translation. MT is especially useful when big volumes of technical text have to be translated consistently. The translator plays a vital role in building dictionaries prior to running the MT program, and in post-editing and making the final review of the translation. Although it is true that MT can do a good part of the translator's job, it can not be taken for a magic solution that will replace the translator and become a threat to his or her existence as a professional.

1. INTRODUCTION

I remember the day when I was called up and offered the job as a linguist at BTr. "If you like machines," the recruiter said, "this is the place to be." I did not say a word, but I was scared to death. "And after all, what is this thing about machine translation?" "How is it possible that a machine can perform a task that requires such creative thought?" I asked myself. Those and many other questions began to be answered after several weeks, when I was able to understand how the process works.

Machine translation has been a matter of controversial arguments and has brought about adverse reactions on the part of those who were —and those who still are—both unsure of the capabilities of MT and scared of being replaced by machines. In the same way that many people feared computers when those machines became popular, and stuck to time-consuming methods for doing jobs, today many translators do not trust the capabilities of a machine-assisted approach to do their work, and fear that they may be left jobless in the near future.

On the other hand, there are those who wrongly think that machine translation is a magical process, and that by just running a program for a few minutes, they will obtain a finished product that can be sent to the client without any work on their part.

Machine-assisted translation solves the problem of consistency when big volumes of material have to be translated, and as it is a mechanically triggered process, it can only work successfully with technical texts, like manuals, or reference guides, because of their (apparent) syntactic and stylistic simplicity. MT is an ideal tool for those translators who specialize in one particular field and are likely to continue working in the same field for some time.

In this paper, I define the role of the translator (I use "he" and "she" interchangeably) by describing the jobs associated with MT, and I emphasize the importance of his function by clarifying how machine-assisted translation can make his job simpler, rather than replace him. Dictionary building, post-editing and final review are three key activities if MT is to be used properly. The ideas in this paper and the work guidelines for post-editing and final review that I provide arise from my experience as a post-editor and final reviewer of machine-translated technical documentation in the field of telecommunications from English into Spanish.

2. THE NATURE OF MACHINE TRANSLATION

MT is a tool, and a very efficient one if used adequately. It aids the translator in her job of producing translated documents, and it performs tasks according to her instructions. It consists of running a text file through a program containing dictionaries.

Although MT software packages contain dictionaries, they are not comprehensive enough to allow for the output of good translations in specialized fields. Therefore, the first job for a translator who is using MT is to do the source text analysis before anything, and program the machine so that it will render an accurate translation. This entails populating the system with the translated terms that need to appear in the target language text. The machine can do very little by itself, and most of the outcome depends on the translator's prior programming.

MT basically performs direct translation from Language_1 to Language_2. That is, it transfers meaning by reproducing the syntactic pattern of the original, unless expressly instructed to do otherwise. If the sentence in the original is of the "Subject + Verb + Object" structure, for instance, the output will resemble that structure. However, there are many MT software systems that can perform Transposition[1] from the source language to the target language, which involves making syntactic transformations. For example, if there is a sequence of the "Adjective + Noun" type, the program has the capability of reversing the word order, and producing a sequence of the "Noun + Adjective" type. Some MT systems will transpose sequences with more than one noun modifying another noun, and a string of the type "Noun_1 + Noun_2+ Noun_3" will be rendered in Spanish "Noun_3 + Noun_2+ Noun_1," with the preposition "de" preceding both the second and the third nouns. "Transmission test function," for example, can be rendered by some MT programs "la función de prueba de la transmisión."

Some MT systems provide a word-for-word substitution, whereas others do the analysis of the parts of speech in the source text and generate the target language output. The translator can input not only single terms, but also full phrases, thus ensuring that the necessary changes will be performed during the translation process. This is why the system can be instructed to perform a "freer" translation. For example, a translation for phrases like "(do this) as directed" or "(do that) as instructed," so frequently used in technical manuals, can be input as "como se indica," so as to ensure that the phrases are translated as entire units.

However, it is important to remember that the purpose of using MT is not to perform a "free" translation. If the translator has to spend time instructing the machine to output a freer version (something that may or may not work out), he might as well opt for doing manual translation. This is why we can not think of using MT for literary translation, or even for translating legal or commercial documents, because the complexity of those types of texts requires the use of Modulation,[2] Equivalence,[3] Explicitation,[4] and other translation strategies that a machine cannot perform.

Technical manuals and reference guides that describe parts of machinery and working processes use mostly descriptive language, and they use the command form very frequently in order to instruct the user how to operate electronic or mechanical equipment, how to assemble parts, or install machinery. These are the most suitable text-types for using MT, because of their syntactic simplicity, and because the degree of syntactic transformation that is needed from the original to the target language is relatively small.

The quality of the source language text is very important —even more important than it is for manual translation— because machines cannot interpret text or make up for poor language. The source language has to be well structured, (that is, it has to be cohesive and coherent), and it should not be ambiguous. If it is, it will require a lot of work after MT has been run. The translator will have to "make sense" of an obscure source text for the translation to be intelligible.

3. REDEFINING THE ROLE OF THE TRANSLATOR

3.1 Text Analysis and Dictionary Building

It is true that machines can solve many of the problems that otherwise a human being would have to solve, and this is undoubtedly replacing human labor. However, for MT to be an effective tool, the translator has to do two important jobs first: *text analysis* and *dictionary building*.

The analysis of the text enables the translator to determine, by the specificity of the terms in the original, by the frequency in which some terms occur, and by the number of terms that are not contained in the existing dictionaries, which and how many are the terms she needs to introduce in the dictionaries.

Dictionary building or development is the process of creating appropriate terms in the target language for the source language terms. It consists of instructing the machine (through a menu-driven approach) how to do the text analysis; that is, to determine what parts of speech terms are, how they are inflected, and what their position is in the target text. The job includes inputting not only single terms, but phrases. Nouns are assigned gender, verbs are assigned inflection patterns, aspect, transitivity, the prepositions with which they collocate, and other attributes. By default, many systems output certain words with certain assigned specific categories. In Spanish, for example, in those systems, nouns are rendered with their corresponding articles, and adjectives are rendered in the masculine, and singular form. If the translator considers it necessary to modify the default occurrences, the words or phrases have to be run through the menu again and reassigned a new category.

If a certain term is not input in the dictionaries by the translator, the system will either render the translation from the built-in dictionary that the system software has or, if it doesn't find it there, it will render it in the source language; that is, it will not provide a translation at all. In other words, MT does not alter the text by providing incomprehensible output if it does not find the equivalent term or expression. It is the translator's responsibility to solve the problem later.

MT has the capability of allowing for customization, that is, adjusting the system to the user's needs. Customizing the system is the process of working with the dictionary utility of the system and expanding its capabilities, as long as the configuration of the system allows for an expansion.

With certain systems, the translator can build his dictionaries in the field in which he is doing the translations, and has the option of entering different definitions for one term in the source language. Thus, the output of the machine-translated (MT-ed) text will show one rendering for that term and, in parentheses, alternative translations for that term. Later, the post-editor needs to choose the most suitable translation. Other systems allow for only one translation of a source language term. Therefore, it is necessary for the translator to build different dictionaries and to establish sequences in which the system is going to look up the terms. The system is programmed to look up terms in Dictionary 1, then in Dictionary 2 —if it was unable to find the terms in Dictionary 1— and so on. Sequences are established from the more specific scope of application of dictionaries to the more general, and they are continually re-established according to the specific needs of the project in question. A user can build as many dictionaries as desired, but there is a limited number of dictionaries that may be used per translation job.

The translator can build dictionaries according to the degree of specificity of terms, and/or the frequency of their occurrence in the documentation. Thus, she can build a dictionary for general terms in a particular field (such as a dictionary of telecommunications terms), and other dictionaries that will be applicable for a specific project, a specific product, or a specific customer. For example, in a general Spanish dictionary, "drop" will be translated as "gota." For a project on cables, however, the term to be used is "cable de acometida," and it will be contained in a dictionary of more specific terms; "ring" will be translated "señal de llamada" in a general telecommunications Spanish dictionary, and "anillo" in a dictionary on hardware .

For Spanish documentation the translator may choose to or may be required to use different regional variants of the language: Peninsular or Castillian Spanish, Mexican or Central American Spanish, South American Spanish, or simply a "neutral" or "general" Spanish that may be intelligible to Spanish-speaking readers from any ethnic background.

Supposing a project is to be translated into the Mexican variant of Spanish for a project on outside plant cable installation, and the translator is using a system that allows for the establishment of sequences, he can establish the following sequence: Outside Plant Dictionary (where most specific terms are going to be found), Mexican Spanish (where the regional variants will be picked out), "Neutral" or "General" Spanish (where the rest of the terms can be found). Dictionary sequences affect the quality of the MT product; consequently, it is very important to establish the correct order in which terms are going to be looked up in the various dictionaries.

Dictionary building is a time-consuming yet rewarding job. It may take between two to three months to build a complete dictionary for a specific high-volume job. The more time a translator devotes to dictionary building, the more accurately the translation will be rendered, and the less she will have to work on the translation after it has been run through the MT program. The quality of the MT depends upon how extensive and precise the dictionaries are. This is why it is important to allocate as many resources (both human and technical) as possible to this aspect of the job. Very often, however, time constraints may force translators to run MT with a poorly built set of dictionaries, which jeopardizes the advantages of MT.

Once the dictionaries have been entered into the system, the MT program is run in order to test the system and analyze the output. But before running the program, the translator needs to prepare the text file for MT. Those terms and acronyms that do not need to be translated have to be protected with special signs, so that the machine does not produce any transformation with them. According to the MT outcome , additional work may have to be done in order to refine the job. Finally, the bulk of the work is run through the program.

As can be realized from the description so far, the role of the translator is critical for MT to be quick, accurate, and efficient as a tool, and a lot of dictionary updating and improving has to be done for better results. This is why the translator is in fact a *linguist*, because he does the text analysis first, builds the dictionaries, and programs the translation system according to the particular requirements of the job.

The translation program is run in a very short time, and it is also menu driven. After MT comes the second part of the job, which is to improve the machine output. This is accomplished through *post-editing* and *final review*.

3.2 The First Job After MT: Post-Editing

The main aspects that determine the quality and the linguistic and technical acceptability of a translation are its *accuracy*, the appropriate use of terminology, the correct use of syntactic, morphological, and semantic rules (its *cohesion* and *coherence*), and its *readability*. The translator who is post-editing and/or making the final review of an MT-ed text must carefully observe that all these aspects of the translation meet the standards for a good translation.

A translation is accurate when it conveys the exact meaning of the original. A translation that changes in any way the message contained in the source language text is not acceptable.

When a translator is working with MT-ed texts, she may encounter accuracy problems. These may occur due to the fact that, for certain terms, the machine renders a certain output by default, or it may also happen that the dictionary sequence is not extensive enough and that, therefore, the terminology is not the appropriate one in a given context. In Spanish, for example, the occurrence of the incorrect use of "ser" and "estar" may be due to the fact that the machine renders only one translation for the verb "to be" in a system that is not configured to render optional translations. To avoid inaccurate translations, the translator can input phrases with the verb "to be" in order for the system to render the appropriate "ser +

complement" or "estar + complement" phrases, depending on the meaning of the original phrases.

Other cases of inaccuracy may be the wrong rendering of the passive construction in Spanish, in which the machine outputs the preterit instead of the past participle of the main verb. Another example of inaccurate output is the rendering of one denotational meaning of a term for another. This may happen if the system has not gone through a complete sequence of dictionaries. For example, in a project on cables, "cover" may wrongly occur as "tapa" instead of "revestimiento."

During post-editing it is vital to make sure that the translation is *cohesive* and *coherent*. This means that the translator has to observe closely that syntactic, morphological, and semantic rules have been applied correctly, and that the ideas follow an orderly pattern. Nevertheless, a translator cannot be held responsible for a not-so-good translation if the original text is not well organized.

Syntactic rules are those that determine the organization of words into sentences, of sentences into paragraphs, and of paragraphs into an entire text. Each language has its own syntactic rules, and it is important to observe that the translation does not mirror the structure of the language in the original and, therefore, that it sounds foreign. In other words, it is essential to see to it that the target text is well structured, and that it follows the syntactic rules of the target language.

Morphological rules are those that establish how words inflect according to gender, number, tense, and mood. Morphological rules also determine spelling and accentuation patterns. Many times, a slight change in morphology can bring about a change of meaning. The goal of the post-editor should be to ensure that there is no change of meaning between the source language text and the translation; therefore, special attention must be paid to the morphological aspect of the translation. For example, an "s" in the wrong place may bring about a change of meaning if the MT system interprets an "s" for the plural form of a noun instead of the simple present form of a verb.

Semantic rules pertain to meaning. Many words have several different meanings. They are called polysemic words. On the other hand, one same idea (or meaning) can be conveyed in many different ways. It is up to the translator to determine what specific meaning is being ascribed to a word in the specific context in which it occurs, or from the intentionality (purpose) of the text. If the translator is post-editing a text on cables, for example, and the text contains the word "cover," he must be aware that in that particular context he needs to use "revestimiento" or "vaina" (as long as the cultural standards of the country for which the document is translated permit the use of this word), and not "cubierta" or "tapa," which would be appropriate renderings in other contexts.

During post-editing, the translator has to verify line by line, and word by word, that the translation conveys the exact meaning of the original. She has to change the word order in sentences, modify word endings, delete articles that are not necessary in Spanish, change, add, or delete prepositions and conjunctions, and apply a consistent approach to the capitalization of the expansion of acronyms. It is also useful for the translator to ask herself the following questions in order to check if she is post-editing the translation accurately:

* Do all sentences have a subject and a predicate, and do they have at least one conjugated verb?
* Are verbs well conjugated?
* Do nouns and adjectives agree in gender and number?
* Is the "s" in the original a plural form of a noun or the declination of the verb for the 3rd person singular in the Present Tense?
* Are modifiers in their right place?
* What is the scope of modification of attributes in the original, and consequently, are they correctly placed in the translation?
* Are articles, prepositions, and conjunctions used correctly?
* Are words accentuated according to morphological rules?
* Are acronyms capitalized and used in the singular form?
* Are expanded acronyms capitalized where necessary and according to the customer's request?
* Is there gender agreement between the article that precedes the acronym and the acronym?

The translator who is doing a post-editing job must also check that the format of the translation is in keeping with the original format, and that those terms and acronyms that were protected and did not have to be translated remain in the original language.

And last, but not least, he has to verify that no information is missing from the original and that no information has been added to the original during the processes of machine translation.

Certain translation problems cannot be solved with MT, such as the appropriate use of prepositions and conjunctions, or the correct insertion of articles, let alone the use of idiomatic expressions, collocations, explicitations, or expansions. This is a job that is definitively reserved for the post-editor. Besides, the post-editor must also verify that there is no redundancy in the target language text. It is acceptable to repeat a term in English. In Spanish, however, it is advisable to use synonyms and pronouns to avoid repeating words. On the other hand, it may be necessary to repeat a term or to explicitate it for the Spanish reader for purposes of clarity when there is no such repetition or explicitation in the source language text.

The phrases that will require more post-editing work are the ones that may have required a more complex transformation process from $Language_1$ to $Language_2$, and those are the cases that MT cannot solve with 100% accuracy.

Once the major aspects of accuracy, correct terminology, and text cohesion have been resolved, the translation has to be final reviewed.

3.3 The Second Job After MT: Final Review

What is the purpose of making a final review of a text? In the first place, many errors may have been overlooked during post-editing. Besides, some terminological problems may have been left unresolved; in addition, when the translator is post-editing a translated text, she is concerned about and focusing her attention on issues of accuracy and structure, rather than on the readability or the flow of the piece. This is why a second look at the text is necessary and why no translation can be handed over to a customer without a final review.

Final review is like doing "fine tuning" to the translated text. The translator looks at it as a piece of technical writing, and pays attention to detail. The units he takes for analysis and improvement are entire ideas (not smaller units like a sentence, a clause, or just a term or a preposition); he observes how the text reads in its entirety, and how ideas connect with each other.

During this stage, the original text is used solely for reference —in case the translated piece is ambiguous— and to make sure that the translation is technically accurate. Otherwise, it is even advisable to forget for a while that an original ever existed, and concentrate on the target language text as an independent piece. Since translation is re-creating the original, the quality of the translation will be determined by the translator's ability to make the translation read as naturally as the original.

When the translator is making a final review of a translation, she may choose to re-arrange paragraphs so as to make the target text more coherent and readable. Coherence in a text depends on the logical relation within parts of the text (a logical order of presentation of ideas and the way connectives are used). The readability of a text depends on the writer's choice of words (word usage), the register used, the use of emphasis, to mention just a few aspects. The translator has to ensure that the target text does not sound awkward or unnatural (that is, that there are no inappropriate words in that specific context), and that it does not sound "translated."

Other aspects of language the translator has to look at during final review are the use of repetition and ellipsis in the text. As stated before, repetition and ellipsis work differently in English and in Spanish. Final review is the right stage at which the translator can check these aspects again, and replace words by synonyms or equivalent expressions or pronouns, if necessary. By the same token, the translator may want to explicitate a term, to expand an acronym, or to explain a term for purposes of clarity. It is also necessary to verify if, during the process of post-editing, the translator inadvertently deleted or added information. And last but not least, it is also at this stage that spell-checking has to be done to ensure that there are no unintended typographical errors.

It is recommended that two different people do the post-editing and final review in order to ensure that, if the translator has overlooked any errors during the post-editing stage, they can be corrected during the final review stage. If this is not possible, the ideal thing to do is to let the translation "sit" at least one day, and to give it a fresher look for final review. By coming back to a translation after some time, the translator is better able to pay more attention to details and can be more objective when reading and assessing the document.

4. SOME FINAL CONSIDERATIONS

MT is a very useful and efficient tool, but not a panacea. Although the degree of accuracy that some systems render in the output is quite high, different problems may arise every time the program is run. This is why the translator who has adopted MT as a working tool needs to keep a record of the problems encountered, analyze those problems if he can, and prevent the further occurrence of problems of the same type. Supposing a certain word string keeps coming out in the wrong order, the translator should ask himself if it possible to obtain a better target text by modifying the input. The problem may be something the translator can prevent by inputting the information in the dictionaries in a different way, or it may be something that he himself will have to resolve anyway after the program is run. It is also advisable that the translator who uses MT consult an MT specialist who can instruct him how to use the system to its utmost advantage in order to obtain best results.

The process of machine-assisted translation can be summarized as follows: 1) text analysis, 2) dictionary building, 3) post-editing, and 4) final review. In all of these stages, the translator plays a fundamental role and is responsible for the outcome, therefore, there should be no reason to think that machines can or will replace these language experts.

What needs to be considered, instead, is how to allocate resources for the different stages of the translation process more wisely. As I mentioned above, text analysis and dictionary building require a good length of time and resources, especially if the translation projects are of a large magnitude. Therefore, most of the resources should be allocated at the front end, in order to ensure consistency in the terminology, good quality of MT output, and economy of time during the post-editing process. A well administered MT system can be a translator's friend—not her foe!

5. NOTES

1 Transposition is a translation strategy by which there is a shift in word class, category, or syntactic identity. For a detailed explanation of this strategy, see Vázquez Ayora, G. *Translation Studies I: Fundamentals of Translatology* , trans. Dr. Leland Wright, Jr., Copywright © 1980.

2 Modulation is a translation strategy by which meaning is expressed utilizing different symbols, and there is a modified point of view. See Vázquez Ayora, G.

3 Equivalence is a translation strategy that consists of using expressions that have, actually, a different lexical meaning, but refer to the same or similar cultural or figurative situation. See Vázquez Ayora, G.

4 Explicitation is a translation strategy that serves to make explicit in the target language what is implicit in the source language. See Vázquez Ayora, G.

MODEM TRANSFER: A DEMONSTRATION

Monique-Paule Tubb and Richard Tubb
Diplomatic Language Services, Inc.

Abstract: This presentation will be a practical, step-by-step demonstration of how to send and receive documents via modem. The goal of this presentation is to demystify the use of modems. Rather than a long discourse on the subject, we will perform a modem transfer live, showing on screen the actual steps so everyone can see that you do not have to be a computer whiz to operate a modem.

We will also circulate a handout with tips on choosing and installing a modem and the different communication software packages. Depending on the time available, this topic can also be discussed during the question and answer period at the end of the session.

My name is Monique-Paule Tubb, and this is my husband, Richard. I am the Director of the Translation Division at Diplomatic Language Services in Arlington, Virginia. We are here today to demonstrate how to transfer documents via modem. But first, I would like to go over a few basic facts.

Everyone is aware of the growing importance of having and using a modem to transfer written information, especially in a field like translation. It saves time and ultimately money, since time usually equals money, especially for independent translators. A modem has become as vital a piece of equipment for a translator as a typewriter or a phone was 20 years ago.

In spite of this, many translators have delayed buying a modem, or bought one and can't figure out how to set it up, or can't transfer a file to save their lives. This can be very frustrating, both for the independent translator trying to return a job or losing a job because s/he does not know how to use their modem, and for the client trying to receive the job. Translators, for the most part, do not care or need to understand all the intricacies of modem transfer. They simply want to know HOW TO MAKE IT WORK, and how to expediently and economically return that translation to their client. Richard and I will try to show you the steps you have to go through to effect your modem transfer.

Richard is a technical person who loves to figure out why something works and how it works. If you have "technical" questions, he will probably be able to answer them (he is not a computer engineer though!). I will play the role of the neophyte, a role which suits me well, since this is what I am. I have been using computers since PCs have been on the market, but I still don't like them. To me they are just a necessary evil. Therefore, I like to keep things simple. I do not try to get into technicalities. I just try to understand what I need to know to get the simplest and easiest use of these machines!

Let's now proceed with the live demonstration of file transfers.

Sending or receiving files via modem is fairly easy, but your computer can't do it without your help. You must give it certain information at specified times or nothing will happen. Assuming you bought the right hardware, installed it correctly and installed the communications software correctly, the computer can now send and receive files. But do you know the commands to give so it will do this?

Transferring a file by modem involves two simultaneous actions: sending and receiving. Sending is never going to be automatic, although receiving can sometimes be. To make it easier to understand, we have set up two computers side by side whose screens are projected on the big screen. One computer will send a file, and the other will receive it. I will give the steps necessary on both sides.

For this demonstration, we will use the communication software Procomm Plus. The specific keyboard strokes may vary with other programs; however, the process is the same.

First, you must agree with the sender/sendee on who will call whom. It does not matter which party initiates the call; once the connection is established, transfers can be done in either direction.

Before any call is made, you must agree on several "parameters" with the other party:

1) Line settings -- "N-8-1", or "no parity-8 data bits-1 stop bit" are the standard settings for most communications software;
2) Baud rate -- File transfer speed. 2400, 9600, and 14,400 are currently the most common choices, but 28,800 is not so uncommon anymore, and will probably become more and more popular as speed remains the end goal;
3) Protocol -- Method of checking the accuracy of data transfer. Kermit, X-modem, and Y-modem Batch are among the most common protocols. We use X-modem most often.

1. Calling

To access the Dialing Directory, press alt-D. If the contact's information has already been entered permanently, you can use the arrow keys and Enter to select it. If not, you must press alt-M and enter the information manually. After the phrase "Number to dial:" type in the modem number of the other party; include a "1" for long-distance numbers, and do not type in dashes. If you have call waiting, add "*" or "#70,,," before the telephone number. Press Enter. A box displaying the number you are calling will appear, and the computer will dial. When the connection is made, the message "Connect [baud rate]" will flash next to the phrase "Last call". You are now ready to begin the transfer.

2. Automatic answering

At the main screen, press alt-Y. You will get the message "OK". The computer will now automatically answer incoming calls. When the message "Connect [baud rate]" appears, you are ready to begin the transfer.

3. Uploading (sending) files

When you are ready to begin the transfer, press Page Up. The computer will present you with a list of protocol options. Select the one agreed upon in advance. The computer will then ask for the filename; you must type in the complete pathname of the file to be sent, including the drive and any subdirectory(ies), so make sure you know this in advance. A box will appear showing the status of the transfer, which you should monitor since the transfer might not go through at all or might be aborted due to errors. The message "COMPLETED" will flash several times when the transfer is finished, and the box will then disappear. To hang up the line, press alt-H. Check the file as soon as possible to be sure the transfer was successful.

4. Downloading (receiving) files

When you are ready to begin the transfer, press Page Down. A box with a list of protocol options will appear. Select the one agreed upon in advance. At this point, the procedure varies depending on the protocol used. For some, including Kermit and Y-Modem Batch, the computer uses the filename established by the sender and the receiver does not do anything further other than monitor the transfer. If you did not already get the filename from the sender, write it down now. For other protocols, the computer will prompt you for a filename, so you just need to give it a name. A box showing the status of the transfer will appear, which you should monitor, since the transfer might not go through at all or might be aborted due to errors. The message "COMPLETED" will flash several times when the transfer is finished, and the box will then disappear. To hang up the line, press alt-H. Check the file as soon as possible to be sure the transfer was successful.

5. Host mode

This is an answer mode that can operate without human input. You can activate it by pressing alt-Q at the main screen, and leaving the computer on. Callers can then dial in, log on and upload files to your computer. They can also download files, but they must know the filename and complete pathname. When setting up your communications software, it is wise to secure your directories so that they cannot be accessed by a caller.

6. BBS (Bulletin Board Service)

More and more, you will encounter companies using a BBS. This lets you upload a document at any time without having to first call and make sure your correspondent is available at that particular time. This is very useful in our profession, since when you live on the West Coast you might choose to return your translation as soon as you are finished with it. It might be 10 pm your time, i.e. 1 am in New York or Washington, not a good time to find your client waiting for a modem transfer. Bulletin Boards become very useful then.

Instructions for use of the Bulletin Board System (BBS) of Diplomatic Language Services, Inc.

Diplomatic Language Services, Inc. (DLS) has a 24-hour system for receiving documents. This enables freelance consultants to return assignments at any time without contacting DLS production personnel in advance.

The system is simple to use, but we have developed the following step-by-step instructions to eliminate any confusion. If you are a new user, please take the time to read all on-line information once you have logged on to the system. The instructions below assume you are a new user; if you have already established a user ID and password with our system, skip this description and proceed down to the FILE TRANSFER section.

Accessing the DLS Bulletin Board System

From your modem program, dial 703-243-7795. When you are connected, a message welcoming you to the system will appear.

Establishing a user ID and a password

After the Welcome text, you will be asked to type your user ID or the word "new" (without quotes). If you are an established user, you type in the user ID, the password when prompted for it, and proceed with the transfer.

If you are a new user, type new and press Return. Two paragraphs about the BBS will appear, followed by a question about whether a word is blinking. Type "y" or "n" then press Return. You will then answer a series of questions that will establish your user ID; you must answer these questions, as simply pressing Return will not allow to skip one. You will be asked for first and last names, address, machine (computer) type, your relationship to the company (translator, etc.), and the contact person (production coordinator for the specific job). You will be asked to select a user ID of 3-29 characters and then to select a password. Write down these two names and save them for use each time you enter the BBS. Press "y" when the system informs you there is new mail for you and asks if you want it. It will offer you a document labeled BBSUSER.DOC, which explains how to use the BBS. You can download this to your machine for later use if desired.

File transfer

The main menu will then be presented. To upload (to SEND DLS a file), press "L" for File Libraries. *Do not choose Quick Mail, as it is not the upload function.* In File Libraries, choose "U" for upload. Type the filename, **which should be as described in the Purchase Order for the project**, and press Return. Choose a protocol from the list of those available with the DLS system (choose one that *your* system has, i.e., Kermit, Xmodem, Zmodem, etc.)

You will get a message that you may begin the transfer. ***This does not mean the transfer is proceeding automatically. You now have to begin your own upload procedures (for example, with Procomm you press Page Up). When prompted, give the drive and specific filename of the document as it is in your system.***

After the transfer is complete, you will be asked if you wish to edit the file description. Answer accordingly. The system asks if you would like to upload (for instance, if you are delivering multiple files); proceed appropriately.

Exiting the BBS system

If you are finished with all transfers, press "x" to quit. You are still in File Libraries at this point. Press "x" again to reach the main menu. Press "x" to exit the main menu, and answer "y" when the system requests confirmation of termination of the connection.

7. Electronic Mail (E-MAIL) Services

Modems are used to transfer files directly from one computer to another, but they can also be used to send/receive files via networks and to access on-line services.

Use of E-mail services basically falls into two categories. One is for communication only, like the original Internet, in which individuals or organizations have "mailboxes" or "addresses." The other is for services, like those offered by CompuServe, Prodigy, or America Online. You will also find general information about some of these services in our handout. The descriptions are very concise, and only represent a sampling of these services; this is just intended to show the variety of resources available. Further information can be obtained from the individual services.

CONCLUSION

Modems won't improve your skills as a translator, but they can make receipt and dispatch of projects easier and cheaper. All that is required is knowing your options when buying one, installing it correctly and knowing how to send and receive files. I hope that our demonstration and explanations have given you the basic information you need. We will now answer questions for the remaining time.

Connections

Translation Theory

ANGOISSE, JOUISSANCE, AND VOLUPTÉ. LEVINAS AND TRANSLATION THEORY

Marilyn Gaddis Rose
Binghamton University

Keywords: Levinas, Lingis, Dire/Saying, Dit/Said

Abstract: Polylingual postmodern philosopher and Talmudic scholar Emmanuel Levinas (1906-) provides a conceptualization of human interaction that identifies the dislocation between saying (Dire) and said (Dit) and thus accounts for the dynamics in general language. His explicit affirmation of communication, it is argued, implicitly validates translation.

1. BACKGROUND.

It may be that the theory aficionados of the American Translators Association comprise a covert community of Hobbitts. Now I don't mean that we are necessarily short and furry. I mean that we seem to move about in a mid-region that can be detected only by especially attuned vision. I.e., we recognize one another, but, most of the time, we believe we are invisible to the others. Which others? Everyone except ourselves. In any event, it may be better that way.

Back on our home sites, typically a campus, philosophers do not expect us to understand their speculations, whether we extrapolate from continental or analytic traditions. Further, depending upon our own personnel berths, many of our colleagues in comparative literature or linguistics, prefer that we leave theory to them.[1] Since we are so taken up with practice anyway, we are considered adequate for teaching content courses. No matter how rapt our attention when our colleagues share their research with us, we are suspected alternatively of having the effrontery of measuring their theories against our practice or, worse, co-opting such theories to see whether any illuminate our own practice. (Speaking for myself, I do not expect any theory to help me translate.)

Yet when we come to the annual meetings, where the concerns of the membership at large must prevail, we find that we are not sufficiently engaged in practice for most of our fellow members to expect us to understand their priorities. So we carry on our discussions in a mid-world, off to the side, if not exactly on the margin.

2. INCITING ELEMENT.

That was why when I started reading Alan Melby's The Possibility of Language,[2] my elation soared with my enthusiasm. Here was a fellow ATA member not only willing to read and credit continental philosophy, but also willing to read it acquisitively, i.e., to see how or whether it illuminated translation. I had some prior inkling that Melby was a Francophile, but I should not have supposed that a computational linguist would suffer Emmanuel Levinas. Certainly it would not be generally imagined that a computational linguist would lead a comparatist like me finally to immerse myself in Levinas. But that is what happened: I had been postponing Levinas. It was Melby's endorsement that gave me the fortitude to plunge in.

3. LEVINAS AND TRANSLATION THEORY.

Why had I not plunged in earlier? Fear of drowning. From the excerpts I had been shown by colleagues and students with troublesome passages, I feared I might not have sufficient patience for the Levinas style. However, I was intrigued because I have been continually called upon to play the same role. I was always asked for help by readers having trouble with Alfonso Lingis's translations, and I was always expected to say there was a mistranslation, e.g., that while Lingis was ambiguous, Levinas was clear. I was never able to provide such reassurance. Nor can I do so now that I have put in time studying Levinas and comparing his texts with Lingis's translations. Levinas provides philosophical authority for the ambiguity and fluidity in general discourse--and by extension, an explanation for the failure of machine-translation procedures based on special-purpose discourse to accommodate general discourse. Levinas provides this authority in a style that illustrates, even imitates, his content, and Lingis reproduces the entire composite. [3]

Levinas reads as if he were an Old Testament prophet, mellowed by time. Perhaps Malachi who, at least mentions the Lord's love. (In a Talmudic scholar like Levinas, a biblical style is not surprising.[4]) He offers encouragement instead of threats. In the midst of the most amazing disasters, he persists in promoting love and ecstasy in language that has unmistakably strong sexual resonances. Mesmerizing in his eloquence, he nevertheless exasperates a hurried reader looking for topic sentences. Levinas writes from a firm infra outline with section headings, but his discourse is elaborately rhetorical as if he were writing Latin with French words. (I would not be surprised to be told that there is a Hebrew infra grammar.) His sentences seem to be struggling to find a suitable replacement for ablative absolutes. Lingis refers to this characteristic as "enchainements of appositions," where the "thought succeeds in formulating itself" (xxxviii). Lingis says he adapts these appositions to declarative sentences to keep from too much distortion of English grammar. But Levinas's struggle echoes through the English. Levinas's lexicon makes use of simple words stretched to their metaphorical limits, technical metaphysical terminology, and literary allusions which must be recognized but are not always identified. Lingis stays close, with cognates whenever feasible. As he himself says, "the present English version [of Otherwise than Being], more than most philosophical translations, is a transposition of the original text, and does not wish to sever its dependence on and subordination to it" (xxxviii). Lingis's translations of Totalité et infini (1961) in 1969 and L'Autrement qu'être (1974) in 1981 deliberately illustrated neo-literal translation, long before the term was coined.

My first reading of The Possibility of Language coincided with a search for current contexts for angoisse, jouissance, and volupté. All three nouns have to some extent come into English.

That is, in texts for college students, scholars both inside and out of the Academy, these words, words with body-sensation referents, are still italicized, but they are not explained. Anyone reading a text in which they occur is expected to know them. Sartre gave us angoisse, which is pronounced by inhaling and exhaling sharply. Irigaray[5] is identified with jouissance, pronounced while inhaling and simultaneously tightening the pelvic muscles.[6] Levinas certainly uses both as well as volupté,[7] which is pronounced by exhaling and relaxing those same muscles. (I guess I am claiming that to pronounce these nouns we weakly mimic the body response that they refer to.)

At the time of reading The Possibility of Language (spring 1995) I was teaching Baudelaire where I encountered these words frequently. Now, while I by no means reject applying current connotations to past usage, I do insist that we try to be aware that we are doing so. Jouissance in particular is used constantly in discussions where philosophy, literature, and Women's Studies intersect. Since Levinas's name often was mentioned at these intersections, I was flipping through Totalité et infini and L'Autrement qu'être both to see how he used these three words and how, when he did, Lingis, translated them. Melby's discussion indicated that Levinas's concept of language, arrived at by the traditional philosophical tools of reason, intuition, experience (of a more hermeneutic than empirical nature), and authority, validated the abandonment of extensive and extended structured experiments of machine translation for general literature.

Melby is on a track worth pursuing. There is a great deal about the Levinassian perspective that is not only attractive in itself but applicable and illuminating for translation. I have two provisos for what follows: first, Levinas himself uses the word translation only in broad, metaphoric terms[8] and, second, by next year I susect I may have moved on to some other source of illumination. In an excellent and clarifying application, Melby brings Levinas to bear on translation partly while demonstrating that Chomsky and Generative Grammar leave a lot of translating unaccounted for. I shall approach Levinas with the baggage of comparative literature, but I believe, the same kindly, confidence-inspiring face can be discerned along each of our paths. His truly is a countenance in the liturgical sense.

Indeed, we may wonder whether he is one of those singularly blessèd individuals whose very gaze illuminates its object. What is certain is that regarde is as ennabling for him as for his younger contemporary Sartre; only for Levinas the exchange of gazes between seeing faces (visages in face-à-face) is positively empowering, making possible love, friendship, social responsibility expressed in the "languages" of speech, gesture, and action. Thus, his is a world in which surfaces are interactively expressive.[9] This means that language--any language--conveys meaning, that love is requited, and through a network comprised of selves, others, and third parties, there is pervasive responsibility. All traditional virtues, traditionally defined, are real, for in this world, there is goodness, saintliness, holiness, and God, historically pre-existing or antedating formal languages.[10] In his preface to the German edition of Totality and Infinity in 1987, he is more explicit: "Human nakedness questions me insistently, questions the me I am, questions me from its position of weakness as nakedness, defenseless and unprotected, imperative and unarmed, the word of God and the word in the human face." He continues, emphasizing the power of the language of the face: "Visage, language already before words, original language of the human face stripped of the countenance it gives itself--or tolerates--beneath proper names, titles, lifestyles" (translation mine) (p.250).[11] Our signs are expressive human surfaces, encompassing, of course, any surface that a human can assume, create, witness, or use.

Above all, language, albeit inextricably linked to human dynamics and hence infinitely complex, is reliably linked and hence reliable. Here again, we must assume that Levinas is blessèd in his faculty for seeing both the good of communication and its reality.

His own multi-language world must be significant here. He was born in Kaunas, Lithuania in a Russian-speaking Jewish family which saw to his Hebrew instruction. There are three languages there: Hebrew, Lithuanian, Russian. (Could they have used Yiddish as well? Probably.) At the outbreak of World War I the family moved to the Krakov in the Ukraine and were there during the Russian Revolution as well. Add one more language: Ukrainian. Since he was receiving what his editors Catherine Chalier and Miguel Abensour term a classical education and was able to begin the University of Strasbourg in 1923, when he would have been 17, we should add four more languages: French, German, Greek, Latin. The University was German-speaking from 1872 until 1922 when French was reinstated as the medium of instruction. During this period he was studying Bergson. By 1928-29, he was in Freiburg and Davos to study Cassirer, Heidegger, and Husserl. He could not have kept up with them without Classical Greek and Latin. We have reached eight languages, not counting Yiddish. By 1930, with a move to Paris where he attended lectures by Brunschvicg and Kojève and collaborated with Gabriel Marcel,[12] he had chosen French as his primary language. He married, became a French citizen and was mobilized as an interpreter, spending the war in a prisoner-of-war camp in Germany. (His family members in Lithuania were exterminated.) He resumed serious Hebrew study in 1947. He did not receive a university appointment until publishing his dissertation (<u>Totalité et infini</u>) in 1961. This summary suggests that for Levinas, Babel, metaphorically, was no curse but an irrelevancy, if not a boon, to inevitably communicating creatures.

And now we reach, finally, the Levinassian concept which is the most relevant for translation. This is the symbiosis of the saying (the infinitive <u>Dire</u>) and the said (the past participle <u>Dit</u>). Levinas always capitalizes <u>Dire</u> and <u>Dit</u>. Lingis, as far as I could tell, never does. He usally translates the infinitive as "saying," occasionally as "to say," and the past participle uniformly as "said." Albeit verb forms, they are invariably nominalized. <u>Dire</u> is a state of communicative activity and a potentiality, both pre-existing and actual. <u>Dit</u> is an encapsulation, a provisionally static moment in communication. Either is bi-directional: "saying" becomes a "said," which, in turn, can lead to a "saying." Even though the dislocation between the situation of the saying (<u>Dire</u>) and its said (<u>Dit</u>) may be minimal, there is an incompressible nonsimultaneity as well as an inherent contradiction. Language is grafted upon the most invisible difference of time because of this dislocation.[13] This is, I extrapolate, the invisible semantic crevice that lets in change. Frankly, I was initially unimpressed by this concept, elaborated alternately rhpasodically and technically, especially in <u>L'Autrement qu'être</u>.[14] It seemed too self-evident to mention. Yet I have come to appreciate this Levinassian conceptualization as a thoughtful attempt to describe thoroughly and to qualify exhaustively with both individual affects and social implications the experience that writers, most emphatically including translators, have when they watch themselves bring their own thoughts into words. What we say, i.e., put into words, does not contain, cannot contain, all that we have to say and can never contain all that is to be said. On the other hand, the words (<u>Dire</u>) can say more or otherwise than what we wanted to say (<u>Dire</u>)

Yet--and this is heartening for translators--Levinas only rarely inserts a German word and even more rarely a Greek word into his French text. When he does, he appears to be doing so solely to help readers situate a concept, not to demonstrate that the concept is untranslatable.

To say, to be engaged in saying (Dire), to author a said (Dit) are responsible, and hence ethical activities. Ludwig Wenzler, the German translator of Le Temps et l'autre, in his postface, speaking about the responsibility of taking account of the other, implies that translation, mediating between author and reader, has doubled responsibilities, thinking in the time (or the tense) of both.[15] This insertion into another's time as an action and/or activity is an indispensable aspect of être, both ''to be'' and ''being'' and, as being, both ''existing'' and ''an existing entity.'' What has to be said, i.e., what comes into the domain of saying pre-exists (and will postdate) any natural language utterance. On the basis of Levinas' personal language biography, I extrapolate that once a saying (Dire) becomes a said (Dit) in one language, it can become a said (Dit) in any other language. Yet--and here I am proceeding by extrapolations of my own--a heightening of the dislocation of simultaneity, a lessening of correlation of the signifieds, and a greater possibility of saying otherwise in the said. Still, to return to Levinas, language is adequate for its tasks, and, by extension, I extrapolate, there is always the potential adequacy of translation. Rather than lament never being able to convey all that we mean or all that the text we are translating seems to mean in its originary language, we should rejoice in the plenitude and dynamism of the Dire our task makes manifest as we work towards the correlation of the saids (Dit) manifested. There is always more to say. Angoisse is inappropriate. On the contrary, translating situates us in the jouissance of profit and ecstasy and the volupté of total satisfaction.

Shouldn't we be grateful? I am reminded of Baudelaire's tribute to Liszt in ''Le Thyrse,'' the 32nd prose poem in Spleen de Paris. As a tribute, it fits Levinas amazingly well. Here is the conclusion in Edward Kaplan's Galantière prize Parisian Prowler:

Dear Liszt, through the mists, beyond the rivers, above the cities where pianos celebrate your fame, where printing presses translate your wisdom, in whatever place you may be, in the eternal city's splendors or the mists of the dreamy lands consoled by Gambrinus, improvising songs of delight or of ineffable sorrow, or confiding your abstruse meditations to writing, singer of Voluptuous Pleasure and Anguish eternal, [chantre de la Volupté et de l'Angoisse éternelles] philosopher, poet and artist, I honor you in immortality![16]

NOTES

1. E.g., so theoretical that some are now conjecturing an era when no translation was needed. George Johnson, ''Linguists Debating the Deepest Roots of Language'' (New York Times, June 27, 1995, pp. C1, 13) summarizes the current Nostratic debate in historical linguistics: Manamaster Ramer, Brian Joseph, Joe Salmons, Nostratic: Evidence and Status (John Benjamins, forthcoming); Allan Bomhard and John Kerns, The Nostratic Macrofamily: A Study in Distant Relationships (Mouton de Gruyter, 1994); Donald Ringe, '''Nostratic' and the Factor of Chance,''Journal of Diachronica (1995).

2. (John Benjamins, forthcoming).

3. In his introduction to Otherwise than Being (The Hague: Martinus Nijhoff, 1978), xxxviii,

Lingis states, "Levinas's composition in this book reflects the understanding of the work of language the book puts forth."

4. In Dieu, la Mort et le Temps (Paris: Grasset, 1993), he quotes Micah 1:iii-iv: "For behold, the Lord cometh forth out of his place, and will come down, and tread upon the high places of the earth.

And the mountains shall be molten under him, and the valleys shall be cleft, as wax before the fire, and as the waters that are poured down a steep place."

5. Sartre is a younger contemporary (1912-81). Irigaray's date of birth is never released. Her first publication was 1965 when she was already established as a psychoanalyst and psycholinguist. This would suggest that she was born in the late 1920s or early 1930s.

6. Margaret Whitford, ed. The Irigaray Reader (Oxford: Blackwell, 1991); Toril Moi, ed. French Feminist Thought (Oxford: Blackwell, 1987); Irigaray, Elemental Passions, trans. Joanne Collie and Judith Still; (London: Routledge, 1992).

7. In popular culture we have to give credit to Oscar de la Renta's perfume copywriters.

8. There is one instance I have come across:

"Le parler philosophique qui trahit dans son dit la proximité qu'il traduit devant nous, reste encore, en tant que dire, proximité et responsabilité;" L'Autrement qu'être (pp. 261-2).

But this is neutralized by Lingis: "The philosophical speaking that betrays in its said the proximity it conveys before us still remains, as a saying, a proximity and a responsibility "(p.168).

9. Totalité et infini. Communication is proof of language: "La mise en question de moi, coextensive de la manifestation de l'Autre dans le visage--nous l'appelons langage" (p. 185).

Totality and Infinity. "The calling in question of the I, coextensive with the manifestation of the Other is the face, we call language" (p. 171).

Grammar preserves the integrity of inter-relations: ". . . la structure formelle du langage annonce l'inviolabilité éthique d'Autre et, sans aucun relent de «numineux», sa «sainteté»" (p. 213).

"The formal structure of language thereby announces the ethical inviolability of the Other and, without any odor of the "numinous," his "holiness" (p.202)

Language has a veritably sacred origin: "Tout langage comme échange de signes verbaux, se refère déjà à cette parole d'honneur originelle. . . Il [le signe verbal] suppose donc déjà une authentification du signifiant" (p.221).

"Every language as an exchange of verbal signs refers already to this primordial word of honor" (p. 202).

10. This does not mean that there is no source or site of terror in Levinas's universe. In 1946,

he introduced the concept of il y a, a nominalized "there is" to refer to the something that is before and beyond nothing, which he compares to the "not to be" of Hamlet's soliloquy. De l'existence à l'existant (Paris: Fata Morgana), translated by Lingis as Existence and Existents (The Hague: Martinus Nijhoff, 1978).

11. January 18, 1987, reprinted in Entre nous. Essais sur le penser-à-l'autre (Paris: Grasset, 1991), p. 250: "la nudité humaine m'interpelle-elle interpelle le moi que je suis--elle m'interpelle de sa faiblesse, sans protection and sans défense, de nudité; mais elle m'interpelle aussi d'étrange autorité, impérative et désarmée, parole de Dieu et verbe dans le visage humain" . . . . Visage, déjà langage, avant les mots, langage originel du visage humain dépouillé de la contenance qu'il se donne--ou qu'il supporte--sous les noms propres, les titres et les genres du monde."

12. Henri Bergson (1859-1941), Léon Brunsvicg (1869-1944), Alexandre Kojève (1902-66), Gabriel Marcel (1889-1973). Although quite different, all could be classified as Idealists.

13. I am using the words of translator Simon Critchley for "Wholly Otherwise," (Tout Autrement, 1973) in Re-reading Levinas, eds. Critchley and Robert Bernasconi (Indiana University Press, 1991), pp. 3-10.

14. Esp. part 3 of Chapter I, parts 3 and 4 of Chapter II, parts 2 and 3 of Chapter V, amounting to slightly more than a third of the work.

15. Catherine Chalier et Miguel Abensour. Emmanuel Lévinas. Paris. Editions de l'Herne. 1991. pp.185-99. French translation by Guy Petitdemange.

16. (Athens: University of Georgia, 1989), p. 87.

REFERENCES

Baudelaire, Charles. *Oeuvres complètes*, I. Claude Pichois, ed. Paris. Gallimard. 1975. Trans. Edward K. Kaplan. *The Parisian Prowler*. Athens. University of Georgia. 1989.

Bernasconi, Robert and Simon Critchley, eds. *Re-reading Levinas*. 1991.

Chalier, Catherine et Miguel Abensour. *Emmanuel Lévinas*. Paris. Editions de l'Herne. 1991.

Hand, Sean, ed. *The Levinas Reader*. Oxford. Blackwell. 1989.

Irigaray, Luce. *Elemental Passions*. Trans. Joanne Collie and Judith Still; London: Routledge. 1992.

Johnson, George. "Linguists Debating Deepest Roots of Language," (*New York Times*. June 22, 1995. Pp. C1, 13.

Levinas, Emmanuel. *Entre nous. Essais sur le penser-à-l'autre*. Paris. Grasset. 1991.

--------. *De l'existence à l'existant*. Paris. Vrin. 1948. Alfonso Lingis, trans. *Existence and Existents*. The Hague. Martinus Nijhoff. 1978.

________. *L'Autrement qu'être ou au-delà de l'essence* (The Hague: Martinus Nijhoff, 1974). Kluwer Livre de Poche. 1990. Trans. Alfonso Lingis, *Otherwise than Being* The Hague. Martinus Nijhoff. 1978.

________. *Le Temps et l'Autre.* Paris. Fata Morgana. 1948. Trans. Ralph A. Cohen. *Time and the Other.* Pittsburgh. Duquesne University. 1987.

________. *Totalite et infini. Essai sur l'extériorité* (The Hague: Martinus Nijoff, 1961). Kluwer Academic Livre de Poche. 1994. Alfonso Lingis, trans. *Totality and Infinity. An Essay on Exteriority.* Pittsburgh. Duquesne University. 1969.

Melby, Alan. *The Possibility of Language.* Amsterdam. John Benjamins. Forthcoming.

Moi, Toril, ed. *French Feminist Thought.* Oxford. Blackwell. 1987.

Whitford, Margaret, ed. *The Irigaray Reader.* Oxford. Blackwell. 1991.

THE TRANSLATOR, HABIT, AND DUCTION (AB-, IN-, AND DE-): C.S. PEIRCE AND THE PROCESS OF TRANSLATION

Douglas Robinson
University of Mississippi

Keywords: Translator's Instinct, Experience, Habit; Subliminal and Conscious Problem-Solving in Professional Translation; Translator Training

Abstract: Much recent translation theory and pedagogy has assumed that translators operate with full conscious awareness of what they are doing; this paper is based on the assumption that, on the contrary, professional translators by and large operate unconsciously, letting their "fingers" do the translating, and making the leap into full conscious awareness only when a problem arises that their habitualized analytical processes cannot handle. This is how most professionals operate, in fact, growing in speed and competence both through the habitualization of skills and through short bursts of dehabitualized problem-solving. The model developed here has obvious implications for the training of translators: if it is true that professional translators operate more habitually than not, it will be important for us to help students internalize the analytical methods we teach them. The model is based loosely on the triadic thinking of the American philosopher Charles Sanders Peirce, especially his instinct-experience-habit and abduction-induction-deduction triads.

1. PEIRCEAN SEMIOTICS AND TRANSLATION

In 1994 Dinda L. Gorlée published her complex prolegomenon to a Peircean theory of translation, *Semiotics and the Problem of Translation: With Special Reference to the Semiotics of Charles S. Peirce*, roughing in a map to a most fascinating territory that bears a great deal of further study. Gorlée has done the translation studies community an enormous service by tracking down in Peirce's vast published and unpublished oeuvre not only hundreds of references to translation per se, but also numerous tantalizing parallels with other theorists of language and translation such as Ludwig Wittgenstein, Roman Jakobson, and Walter Benjamin.

Gorlée's book is a first approach to a Peircean theory of translation; rather than focusing on a single aspect of Peircean semiotics and fully exploring its applicability to translation, she attempts in her book to cover as much theoretical ground as possible, as suggestively as possible, leaving to others the detailed application of her findings to the study of translation. What I propose to undertake in this article, then, is to become one of those others--to cordon off a specific area in the territory Gorlée charts and develop it as thoroughly as I can.

The focus of my study is actually the convergence of two different lines of inquiry within Peirce's oeuvre: the instinct-experience-habit and abduction-induction-deduction triads. Like all of Peirce's triads, these two partly overlap: the very fact that they are Peircean triads, moving from Firstness through Secondness to Thirdness, guarantees a degree of theoretical congruence between them. But they also diverge in interesting and productive ways, and I want to map them onto a diagram depicting the process of translation so as to show forth (and to draw on for heuristic purposes) both their similarities and their differences.

First, however, a very brief introduction to Peircean semiotics may be in order. Specifically, the reader who is not conversant with Peirce's thought will need to know, at least in a very general way, how the triads work. The full complexity of Peirce's triadic thought is so complex as to test the patience of even his most devoted disciples; fortunately, it will suffice for my purposes here to sketch triadic thought with very broad strokes.

The first thing to note about Peircean triads is that they are dialectics, specifically post-Hegelian progressivist dialectics that eventually lead, Peirce believed, to truth--or to what he called the "fixation of belief." That is, the First and Second in any triad are polar opposites that are combined, synthesized, in a Third that is always perceived as an evolutionary advance over its predecessors. This is an ideological stance in which Peirce brooked no opposition; there is a strong messianic strain in his thinking that makes Gorlée's parallels between it and Walter Benjamin's "The Task of the Translator" extremely persuasive. Peirce's messianic "paradise" is more scientific than mystical, to be sure, a paradise in which reason and logic triumph over sloppy thinking; but the tenacity with which he assimilated his "scientistic" and "rationalist" thought to his progressivist eschatology suggests that some sort of "irrational" or mystical belief in a supranatural force governing human history remained essential to him.

The second and perhaps more informative statement to make about Peircean triads is that they always progress from a First or "immediate" element grounded in primary or primal responses, through a Second or "dynamic" element grounded in brute external objective reality, to a Third or "final" element grounded in a transcending of the First-Second opposition. Thus, for example, the most basic triad of Peirce's semiotics is that of sign (a First), object (a Second), and interpretant (a Third): a sign is anything capable of being interpreted, an object is anything capable of being represented by a sign, and an interpretant is anything capable of establishing a semiotic connection between a sign and an object.

But this is just the beginning. One complication is that each of these three elements is itself at least triadic: a sign, for example, can be an icon (a First based on artificial resemblance between sign and object), an index (a Second based on a natural connection or contiguity between sign and object), or a symbol (a Third based on logical precepts); each of these too in turn is triadic; each of these triadic subsubelements is again triadic, and so on, leading to a baroque proliferation of categories, hundreds upon hundreds of explicitly defined types of sign that themselves only stand in for the tens of thousands of categories that Peirce imagined (and dreamed of one day theorizing).[1]

Another complication is that each element in a triad can and often does function in terms of other elements as well. A sign, for example, to take the most important example, is a First in the sign-object-interpretant triad; but because a sign only becomes a sign by being interpreted as a representation of an object, the entire triad is itself a sign, which makes a sign a Third. Calling a sign a First, therefore, entails perceiving a Third in terms of its Firstness; or, conversely, recognizing the sign-object-interpretant triad as itself a sign might be thought of as perceiving a First in terms of its Thirdness. Calling a sign an index entails perceiving a First (or a Third) in terms of its Secondness--and so on. The repeated rapid level-shifts that this approach allows and encourages often make for heady and disconcerting reading: an element that is a First at one level is suddenly being discussed in terms of its Secondness on another level and its Thirdness on yet another level; each level-shift entails an act of semiotic interpretation that potentially (though usually only heuristically) recasts all the levels.

2. INSTINCT-EXPERIENCE-HABIT, ABDUCTION-INDUCTION-DEDUCTION

The two triads that I want to explore in some detail here, then, are the psychological triad of instinct-experience-habit and the logical triad of abduction-induction-deduction. In the former, instinct is a First because it operates in animals (including humans) prior to experience of the objective world, which is a Second; as instinct meets experience, behavior becomes more complex and is transformed into habit, a Third.[2] In the latter, which I find one of Peirce's most exciting and productive notions, the traditional reasoning processes of induction and deduction are limited not only by the either-or dualism in which they were conceived, always a strike against a concept for a triadic thinker like Peirce, but also by the fact that on their own neither induction nor deduction is capable of generating new ideas, with the result that both remain sterile. Induction, because it moves from a wealth of specific details toward the formation of general regularities, is a Second (the universe of objective reality); deduction, because it imposes general laws or precepts on specifics, is a Third (the universe of rational principle). But both of these reasoning processes must be fed raw material for them to have anything to operate on; and a dualistic logic that recognizes only these two ways of proceeding can never explain where that material comes from. Hence Peirce posits a third logical process which he calls abduction: the act of making an intuitive leap from unexplained data to a hypothesis. With little or nothing to go on, without even a very clear sense of the data about which s/he is hypothesizing, the thinker entertains a hypothesis that intuitively or instinctively (a First) *seems* right; it then remains to test that hypothesis inductively (a Second) and finally to generalize from it deductively (a Third).

Taking these two triads at their simplest, then, if only as a way of getting started, we might map them onto the process of translation as follows:

Instinct, experience, habit: the translator begins with a blind, intuitive, instinctive sense in a language (source or target) of what a word or phrase means, how a syntactic structure works (*instinct*); proceeds by translating those words and phrases, moving back and forth between the two languages, feeling the similarities and dissimilarities between words and phrases and structures (*experience*); and gradually, over time, sublimates specific solutions to specific experiential problems into more or less unconscious behavior patterns (*habit*), which help her or him to translate more rapidly and effectively, decreasing the need to stop and solve troubling problems. Because the problems and their solutions are built into the Thirdness of habit, and especially because every problem that intrudes upon the habitualized process is itself soon habitualized, the translator notices the problem-solving process less and less, feels more competent and at ease with a greater variety of source texts, and eventually comes to think of her/himself as a professional.

Abduction-induction-deduction: the translator facing an apparently insoluble problem leaps from the disjunctions between languages to a number of possible solutions and selects one (*abduction*); as s/he proceeds with the translation, or indeed with successive translation jobs, the translator tests the abductive solution in a variety of specific contexts, making sure that the word or phrase or syntactic structure chosen either has the desired meaning or impact or is flexible enough to be adaptable to local needs (*induction*); eventually the translator assimilates the tested solution into her/his methodological repertoire, habitualizing it as a kind of subliminal key or precept for the solving of entire classes of translational problems (*deduction*).

Obviously, I'm already tipping my hand here: my insistence on portraying deduction as a subliminal process reflects my assimilation of the latter triad to the former. There is no *logical* reason--i.e., apart from the fact that the former triad moved from instinct through experience to habit--to speak of habitualized deduction. (I think there are good empirical reasons for doing so, but let me hold off theorizing the empirical application and testing of these triads for a few more pages; I want to stay as close to Peirce as I can, here at the outset, the better to complicate and adapt him later.) Thus assimilatively mapped, both triads trace a movement from bafflement before a specific problem through a tentative solution to the gradual expansion of such solutions into a habitual pattern of response; both enable us to explore the translator's *Bildung* or "education," in a broad sense that includes both training (and training either in the classroom or on the job) and learning through personal discovery and insight. Both help us to flesh out a sense of the translator as at once (a) a *professional*, for whom many highly advanced problem-solving processes and techniques have become second nature, occurring rapidly enough to enhance especially the freelancer's income (the faster you translate the more money you make) and subliminally enough that s/he isn't necessarily able to articulate those processes and techniques to others, or even, perhaps, him/herself; and (b) a *learner*, who not only confronts and must solve new problems on a daily basis but actually thrives on such problems, since novelties ensure variety, growth, interest, and enjoyment.

So far, so good. But a number of things trouble me about this formulation. One of the most important is that it fails to explain how the translator becomes conscious, reverses the process of habitualization and strives to become as aware as possible of what s/he is doing and why. Most of the blame for this failure is mine, not Peirce's; I not only isolated these two triads for special attention, but gave the instinct-experience-habit triad the power to shape the abduction-induction-deduction triad more or less in its "subliminal" image. Peirce himself would unquestionably want us to see translators moving toward greater rational awareness of their actions, for example through conscious inductive and deductive processes.

But the instinct-experience-habit triad, especially when complicated with the abduction-induction-deduction triad, seems to me *empirically* to fit the semiotic process of translation far better than the traditional translator-ideal, which would have the translator working primarily through conscious deduction, knowing exactly what s/he is doing, holding in conscious reserve a whole arsenal of well-reasoned and -seasoned precepts and applying them systematically to each new situation as it arises. This ideal has historically both produced and reflected normative translation theories that attempt to legislate universal translation methods, in some cases even specific solutions to specific textual problems, for all translators to memorize and systematically apply. Traditional pedagogies of translation have similarly been grounded in the assumption that all translation proceeds rationally, logically, deductively, moving consciously and confidently from the precepts taught in school to the solution of specific translation problems. Peirce's instinct-experience-habit triad appealed to me initially because it seemed so much more realistic a portrait of how professionals in translation (and most other fields) actually operate than this rationalist ideal; and I initially assimilated the abduction-induction-deduction triad to the resulting concept of habitualization or sublimation in the same enthusiasm for its empirical descriptive power. In raising the issue of the translator's becoming-conscious, then, I am interested in working *within* the instinct-experience-habit model that I have begun to develop and not, say, simply throwing in the towel and following Peirce and the entire Western philosophical tradition in their idealization of conscious deductive reason as the end-all and be-all of human behavior. I want to stay close to my empirical observation that professional translators do characteristically tend insistently toward the habitualization of problem-solving--while also recognizing equally empirically that that tendency is repeatedly thwarted in the necessity of stopping to think, to reason consciously, to work things out. I am interested, in other words, in an empirical middle that might otherwise be excluded in a dualistic argument that idealized either the translator's perfectly conscious and systematic deductive reasoning or the translator's perfectly unconscious or subliminal habits.

And perhaps a word on empiricism is in order here. In Peircean terms, empiricism is a Second: a confrontation with real-world "objects" (including human actions, inclinations, beliefs, and words) that resist easy assimilation to large-scale systemic abstractions. Peirce's own powerful empiricism is clearly evident in his insistence upon leaving room in his systemic thought for the disturbing effects of empirical reality on abstract systems; rarely has a major philosopher given disruptive real-world phenomena such pride of place as did Peirce. Without the Secondness of external reality, all Thirdness is a sham, because it relies on human ingenuity to dismiss empirical complexity instead of dealing with it, making sense of it. And a large part of the incredible fecundity of Peirce's thought is his heavy reliance, throughout all his voluminous writings, on complex real-world examples, events, anecdotes, cases that, as often as not, undermine some systemic structure he is attempting to erect.

But Peirce is also, perhaps even primarily, a system-builder. He did believe in the primacy of deduction over induction, general principles over untheorized empirical details. The *primacy* of the one over the other, not the displacement of one by the other: general principles for Peirce were worthless if not grounded in a thorough study of the facts. But he was temperamentally far more inclined (a First) to impose the deductive patterns of his triads on the Seconds of empirical facts than he was to allow those facts to push him past, say, triadic thinking, or past the progressivist eschatology of his triadism. He never let his attention to Secondness deter him from his unswerving belief that all reality can and should be forced into a triadic logic, and that all reality must either be moving toward a messianic/scientistic future or dissipate into meaningless nonsense. The possibility that a given empirical phenomenon might make more sense if quadrated, say, seen as moving through four (or more) steps; that a given triad could be perceived as moving backwards, from Thirdness through Secondness to Firstness, without having to be charged with degeneracy; that in certain triads, in certain empirical situations, the First or the Second might be thought of as equally valuable with, or even more valuable than, the Third--all of these logical possibilities were systematically precluded by Peirce's rationalist (but not necessarily rational) attachment to progressive triadism.[3]

3. A NEW MODEL OF TRANSLATION

In what ways, then, will empirical experience of translation (a Second) force us to rethink both the consciously deductive translator-ideal and its tentatively empirical opposite as provisionally sketched out above, the model of the "subliminal" or "habitualized" translator? How can we account theoretically for the empirical observation--my own, based on my own experience as a translator, repeatedly confirmed by the translators to whom I have proposed the model in workshops and seminars around the world--that the process of translation is largely but not exclusively subliminal; that both subliminal competence and the eruption, from within that competence, of bursts of conscious awareness are integral parts of the translation process?

Let me suggest, therefore, a series of axioms arising out of the clash between the "habitualization" model tentatively outlined above as First and a more fully articulated experience of translation as Second. In a fuller exposition these would require considerable fine-tuning, and I hope to undertake such an exposition some time soon; for now, however, let me simply list them in their disruptive Secondness, the sense that I have of their power to complicate my "subliminal" or (from *The Translator's Turn*[4]) "somatic" model:

(1) Abduction, induction, and deduction are forms of experience. As such they might be considered not so much restatements of instinct, experience, habit as the First, Second, and Third of the experiential Second in that other triad.

(2) Abduction as a First in the translation process might entail the intuitive leaps

required to learn a foreign language, or to jump blindly from an expression in one language to an expression in another, in a language classroom, say, or in the bilingual child's burgeoning realization that two separate languages are spoken in the home and/or surroundings (what George Steiner calls "trust" [310]).

(3) The translator's inductive processes entail increasingly detailed experience of the real world in at least three areas: (a) experience of people and languages and cultures in general through reading and study and travel; (b) experience of the source and target languages and cultures in particular; (c) experience of various resource people in the source and target languages who might be able to help with specific problems in a given translation job.

(4) The translator's deductive processes always entail a theory or theorizing of translation, a generalizing about translation that may be partial-becoming-complete or unconscious-becoming-conscious, thus not a "theory" in the strict systematic sense but unmistakably part of the act of theorizing. Peirce's triadic thought reminds us that theoretical deduction is not merely deductive but steeped in the entire triadic process, including induction (close attention to what really happens when we translate) and abductive leaps born out of clashes between that close inductive attention and conventionalized deduction, traditional translation theories that restrict our awareness to certain narrowly circumscribed issues and techniques, like "fidelity" vs. "freedom," "foreignism" vs. "domestication." When inductive awareness overflows the bounds established by petrified deduction, both induction and deduction break down and must be brought into the shop for repairs by abduction, the creation of new theories of translation--which are then tested inductively and generalized deductively, until they too become fossilized and must be regenerated through new abductions (see Kuhn).

(5) The positioning of the abduction-induction-deduction triad within the Secondness of experience in the instinct-experience-habit triad suggests that all abduction, induction, and deduction in the translator's experience tends to become increasingly habitualized, subsumed into an expanded subliminal repertoire that becomes progressively more intelligent, competent, accurate, and fast, due to the experiential materials being incorporated into it. When translation works subliminally, then, it will work through abductive leaps, inductive scanning and testing of great volumes of real-world details, and deductive generalization--but unconsciously.

(6) This subliminal functioning is the homeostatic goal of translation practice; it enhances speed, accuracy, and enjoyment in the act of translation, and must be seen as the supreme good for the professional translator.

(7) Habitual homeostasis is disrupted, erupts into conscious thought, through the failure of subliminal problem-solving. This implies that the instinct-experience-habit triad sometimes operates in reverse.

(8) The need for abduction, for an intuitive leap from an apparently insoluble problem to a tentative solution, arises out of the failure of deductive and inductive reasoning processes. This implies that the abduction-induction-deduction triad too sometimes operates in reverse.

(9) These reversals of the triadic process, because they involve a desublimation or dehabitualization of the act of translation, slow the translator down and are typically experienced as hitches, snags, frustrations. As the translator's primary source of growth, learning, or expansion of repertoires, however, they are also extremely important and can even provide enjoyment--as when the abductive process yields the pleasure of new discoveries, new insights, creative solutions, and the like; or when the inductive process drives the translator out of her/his chair to fetch a reference work or walk to the fax machine, sending the blood circulating through stiff muscles; or when the deductive process produces an idea for an article.

(10) Formal translator training will be most effective if it works within *all* the parameters of this model: if it encourages students to engage in the creative intuitive leaps of

abduction, the inductive sifting through large quantities of potentially useful material, and the production of deductive generalizations, rather than simply presenting them with the predigested results of those processes as undergone by the instructor or other authorities; and if it helps students move from classroom experiences of all these types to the habitualization necessary for effective professional (especially freelance) translation.

The model may best be illustrated in a diagram:

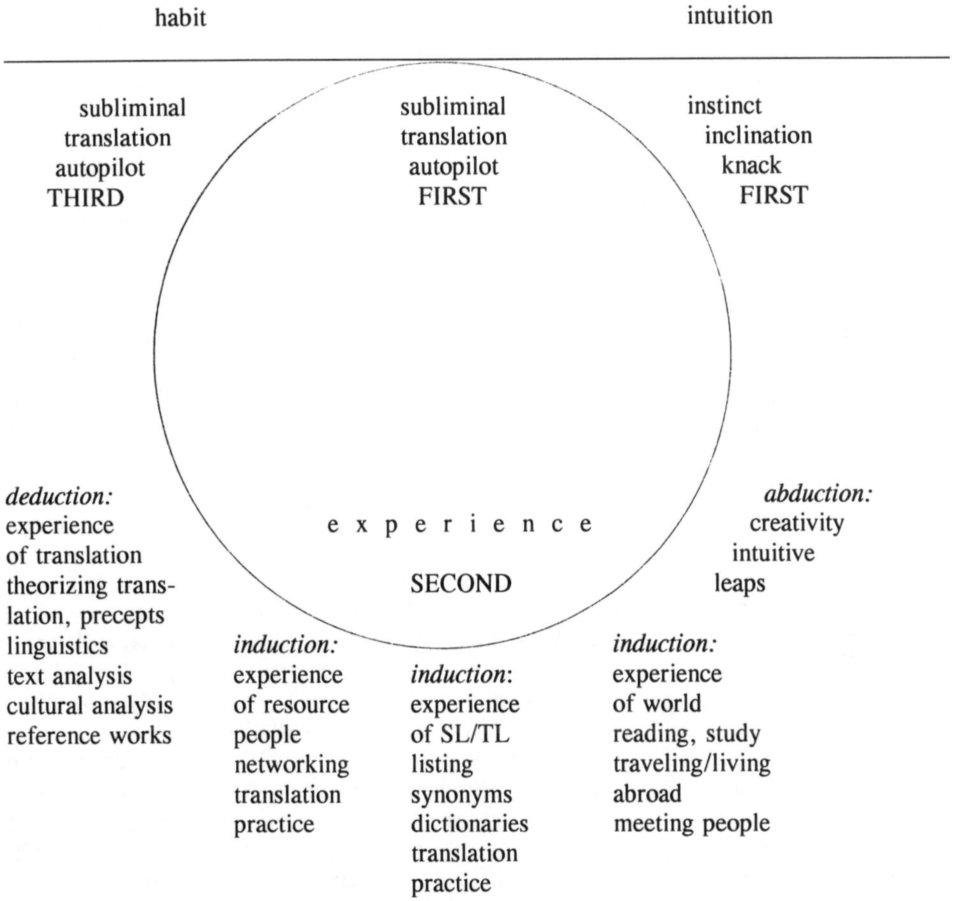

This diagram can be imagined as the wheel of a car, the line across at the top marking the direction of the car's movement, forward to the right, backward to the left. As long as the wheel is moving in a clockwise direction, the car moves forward, the translation process proceeds smoothly, and the translator/driver is only occasionally aware of the turning of the wheel(s). The line across the top is labeled "habit" and "intuition" because, once the experiential processes of abduction, induction, and deduction have been sublimated, they operate sub- or semiconsciously: the smooth movement of the top line from left to right may be taken to entail the smooth clockwise spinning of the triadic circle beneath it. This movement might be narrativized as follows:

The translator approaches new texts, new jobs, new situations with an intuitive or instinctive readiness, a sense of her/his own knack for languages and translation (a First) that is increasingly steeped in the Thirdness of habit.

On the cutting edge of contact with an actual text or job or situation, the translator likewise has an intuition or image of her/his ability to solve whatever problems come up, to leap *abductively* over obstacles to new solutions.

Induction begins with general knowledge of the world, experience of how various people talk and act, important for purposes of impersonation, the translator pretending to write or talk like a lawyer, a doctor, a technician, an advertiser, a novelist, etc., and experience of the differences among cultures, norms, values, assumptions, experience of the vast complexity of languages, etc. This knowledge or experience will often need to be actively sought, constructed, consolidated, especially but not exclusively at the beginning of the translator's career; with the passing of years the translator's subliminal repertoire of world experience will expand and operate without his/her conscious knowledge. Translation pedagogies that recognize the importance of this inductive process will encourage students to read, read, read, study and travel widely, live abroad, pay attention to how people talk, remember words, phrases, and registers, etc. Courses on the "social institutions" or "cultural norms" of two or more cultures will likewise fit in here.

Induction continues with direct experience of the source text, and generally of the source language and culture, and of the transfer parameters between that and the target language and culture. Here the translator will check dictionaries and other reference materials, or (better, because faster) rapidly list synonyms in her/his head, in the source and/or target language, looking for the best match. The learning principle in this area is: the more you translate, the better you translate. Translation pedagogies based on this principle will give students texts to translate and then offer critical but sympathetic feedback.

The final aspect of induction involves the building of a network of resource people to whom the translator can turn when knowledge of the world and of the source and target languages fails. Here one learns inductively not only who to call or fax or e-mail, but how to phrase one's queries so as to produce optimally useful answers. Translation listservers on the Internet, such as LANTRA-L and TRANSLAT, are extremely helpful here as well, as are translation conferences (especially between sessions, at the bar, or at dinner). This inductive aspect of translation is rarely taught in translation studies programs, but probably should be.

Induction will always be an integral part of any act of translation--there will always be novelties, surprises, unexpected words or phrases or contexts that will need to be dealt with on a bottom-up, trial-and-error basis. But gradually certain patterns will begin to emerge out of inductive work, and as the translator begins successfully to apply those patterns or regularities to novel situations, *deduction* begins, and with it the theorizing of translation. At the simplest level, this involves a repertoire of blanket solutions to a certain class of problems--one of the most primitive and yet, for many translators, desirable forms of translation theory. Because one's own deductive principles are typically built up through numerous circumnavigations of the circle (abductions and inductions gradually building to deductions, deductions becoming progressively habitualized), someone else's deductive solutions to specific problems--offered in a theory course or book, say, or at a conference--will probably be harder to remember, integrate, and implement in practice. At higher levels this deductive work will produce regularities concerning whole registers, text-types, and cultures; the theoretical modes emerging out of this sort of deduction include various linguistic forms of text analysis and various systematic analyses of culture, such as those developed by "translation studies" and "polysystems" scholars in Holland, Israel, the U.K., and elsewhere,[5] or more overtly politicized forms of cultural studies (feminist, materialist, postcolonial approaches) developed by people like Antoine Berman, Lawrence Venuti, Tejaswini Niranjana, and Eric Cheyfitz. Translation pedagogies based on this deductive aspect of the translation process have traditionally been authoritarian in nature, delivering precepts which students are expected to memorize and apply systematically in their work; but there is no reason why deductive translation pedagogies should not become more student- and process-oriented, encouraging students to develop their own deductive principles out of extensive experience with texts.

The last stage on the circle, Peirce's Third, is *habitualization*, sublimation, the gradual generation of a translation autopilot, which will enable the translator to work quickly and accurately, with minimal interruptions. It is important to remember that "translating on autopilot" *may* cause errors, as the common prejudice about this approach typically assumes, but that it is not inevitably doomed to erroneous translation. Precisely because it is in a loop with a whole series of abductive, inductive, and deductive reasoning and learning processes that engage the real world in a rich variety of ways, the subliminal translation autopilot that forms the Third of this triad is constantly growing, expanding, becoming more subtle and complexly accurate. Translation pedagogies that take this aspect of the process into consideration will need to develop classroom exercises for the effective internalization, sublimation, habitualization of the abductive, inductive, and deductive strategies that have been either presented predigested to the students or developed by the students themselves.

4. REVERSING THE PROCESS

This is the "perfected" model of the translation process, the process as we would all like it to operate all the time. Unfortunately, it doesn't. There are numerous hitches in the process, from bad memory and inadequate dictionaries all the way up through untranslatable words and phrases (*realia*, puns, etc.) to the virtually unsolvable problems of translating across enormous power differentials, between, say, English and various Native American or subcontinental Indian languages. My diagram allows us to imagine these "hitches" kinesthetically as stopping the car and throwing it into reverse, backing up to avoid an obstacle or to take another road. This might be traced as a counterclockwise movement back around the circle:

The subliminal autopilot fails; something comes up that the translator cannot solve with existing habitualized repertoires. In many cases the subliminal process will be stopped automatically by bafflement, an inability to proceed; in other cases the translator will grow gradually more and more uneasy about the direction the translation is taking, until finally being no longer able to stand the tension between apparent subliminal "success" and the gnawing vague sense of failure, and throwing on the brakes and backing up. It is also possible to build an alarm system, perhaps an automatic emergency brake system, into the "habit" or subliminal functioning, so that certain words, phrases, registers, cultural norms, or whatnot stop the process and force the translator to deal consciously, alertly, analytically with a problem. This sort of alarm or brake system is particularly important when translating in a politically difficult or sensitive context, as when the translator feels that her or his own experience is so alien from the source author's that unconscious error is extremely likely (as when translating across the power differentials generated by gender, race, or colonial experience); or when the translator finds him/herself in violent (or, perhaps worse, because less obvious, subtle) opposition to the source author's views. A translation pedagogy sensitive to the importance of stopping the subliminal process when problems arise will work to develop a sense not only of general cultural problem areas but of problem areas in each student's attitudes toward the world.

The translator's conscious regress back around the circle, through deduction and the various aspects of induction to abduction--the intuitive leap to some novel solution that may even fly in the face of everything the translator knows and believes but nevertheless *feels* right--should be clear. Every time one process fails, the translator moves to another: listing synonyms doesn't help, so you open the dictionary; the word or phrase isn't in the dictionary, or the options offered all look or feel wrong, so you call or fax or e-mail a friend or acquaintance who might be able to help, or send out a query over an Internet listserver; they are no help, so you plow through encyclopedias and other reference materials; if you have no luck there, you call the agency or client; and finally, if nobody knows, you go with your intuitive sense, generate a translation abductively, perhaps marking the spot with a question mark for the agency or client to follow up later. Translating a poem, you may want to jump to abduction almost immediately.

And note that the next step after abduction, moving back around the circle counterclockwise, is the subliminal translation autopilot: the solution to this particular problem, whether generated deductively, inductively, or abductively (or through some combination of the three), is incorporated into the translator's habitual repertoire, where it may be used again in future translations, perhaps tested inductively, generalized into a deductive principle, even made the basis of a new theoretical approach to translation.

There is, obviously, a good deal more to be said about this model; my schematic approach has, it seems to me, only begun to scratch the surface of what is potentially an extremely useful way of thinking about translation. In particular the pedagogical implications of the model need to be developed; the pedagogy of translation is still very much in its infancy, just now beginning to be theorized, the theories implemented and tested, largely because, I believe, so little work has been done to explore what translators actually do when they translate, how they work, and what works best for them *when* they work. I plan to do considerably more work myself on the process of translation and its implications for teaching; but I hope others will join me in that project as well. Perhaps you.

5. NOTES

1. See also, for a discussion of Peirce's interpretant triad, my "Intentions, Signs, and Interpretations."

2. Habit fascinated Peirce; hundreds of passing treatments of it are scattered all through his writings. Generally, for Peirce habit constituted a "promptitude of action" (5.477), "a person's tendencies toward action" (5.476), a "readiness to act" (5.480)--to act, specifically, in a certain way under certain circumstances. For a useful discussion of habit in Peirce's thought, see Greenlee (120-31); see also Gorlée, ch. 05.

3. Gorlée's argumentation is frequently plagued with the same tendency to privilege Peircean triads, perceived almost purely deductively as rationalist "thought experiments," over the Secondness of actual translation practice. The most flagrant example of this neglect of the empirical in Gorlée's book comes in her chapter on Roman Jakobson, where she assimilates Jakobson's three types of translation--intralingual, interlingual, and intersemiotic--to Peirce's triad of degeneracy: intralingual translation is a symbol, interlingual translation an index, and intersemiotic translation an icon, all on extremely shaky and barely articulated grounds, with only the most tenuous and superficial basis in empirical reality; therefore intralingual translation is "genuine semiosis," a Third, and interlingual translation (translation proper) and intersemiotic translation are forms of semiotic degeneracy.

4. I find myself inclined to believe that these complications were already present in embryonic form in that earlier work of mine; but then I notice the reluctance I have felt to theorize the importance of conscious inductive and deductive reason, say, for a subliminal or somatic model, and wonder ruefully whether their "presence" in that book might not be just a retroactive projection of my current thinking into earlier formulations.

5. For a useful discussion of these theories, see Gentzler.

6. REFERENCES

Berman, Antoine. *L'Épreuve de l'étranger: Culture et traduction dans l'Allemagne romantique* (Paris: Gallimard, 1984). Translated by S. Heyvaert as *The Experience of the Foreign: Culture and Translation in Romantic Germany*. Albany: SUNY Press, 1992.

Cheyfitz, Eric. *The Poetics of Interpretation: Translation and Colonization from* The Tempest *to* Tarzan. New York: Oxford University Press, 1991.

Gentzler, Edwin. *Contemporary Translation Theories*. New York and London: Routledge, 1994.

Gorlée, Dinda L. *Semiotics and the Problem of Translation: With Special Reference to the Semiotics of Charles S. Peirce*. Amsterdam and Atlanta: Rodopi, 1994.

Greenlee, Douglas. *Peirce's Concept of Sign*. Vol. 5 in Thomas A. Sebeok, ed., *Approaches to Semiotics*. The Hague and Paris: Mouton, 1973.

Kuhn, Thomas S. *The Structure of Scientific Revolutions*. 1962; rpt. Chicago: University of Chicago Press, 1970.

Niranjana, Tejaswini. *Siting Translation: History, Post-Structuralism, and the Colonial Context*. Berkeley: University of California Press, 1992.

Peirce, Charles Sanders. *Collected Papers of Charles Sanders Peirce*. Ed. Charles Hartshorne, Paul Weiss, and Arthur W. Burks. 8 vols. Cambridge MA: Belknap Press of Harvard University Press, 1931-1966.

Robinson, Douglas. "Intentions, Signs, and Interpretations: C. S. Peirce and the Dialogic of Pragmatism." *Kodikas/Code/Ars Semeiotica* 8.3-4 (1985): 179-93.

_________. *The Translator's Turn*. Baltimore: Johns Hopkins University Press, 1991.

Steiner, George. *After Babel: Aspects of Language and Translation*. New York and London: Oxford University Press, 1975.

Venuti, Lawrence. *The Translator's Invisibility*. New York and London: Routledge, 1994.

Venuti, Lawrence, ed. *Rethinking Translation*. New York and London: Routledge, 1992.

Connections

Translator and Interpreter Education

PILOT TRAINING PROGRAM FOR COMMUNITY-BASED INTERPRETERS

Betty Becker-Theye
University of Nebraska at Kearney

Keywords: Translator and Interpreter Training, Community Interpreters, University of Nebraska at Kearney

Abstract: This study describes a pilot program conducted by the University of Nebraska at Kearney in the spring of 1995 for training community-based interpreters.

1. INTRODUCTION

The State of Nebraska's location in the country's heartland--far from any border--may make it appear as though the need for language specialists would be slight. However, in addition to the state's population of Native Peoples, and descendants of immigrants from Germany, Czechoslovakia, and Sweden, there has also always been a significant population of Mexican origin which is, in fact, the state's largest minority. Yet even against this historical setting, when the small community of Lexington found itself host to some 3,000 new workers, many from Mexico, when a meat processing plant opened there, it also found its social and legal agencies, medical facilities, and schools unprepared to meet the language needs of the new population.

The need for language specialists was met there, as it is everywhere under such circumstances, by residents of the community and the work force who are bilingual. In Lexington, however, a group of these bilingual residents who were meeting the community's language needs, began meeting to organize themselves for the exchange of ideas and terminology and to initiate the first steps towards professionalizing themselves. Early in the fall of 1994, the group contacted the Director of the Program in Translation/Interpretation at the University of Nebraska at Kearney--the campus of the University which serves that part of the state--to meet with them to discuss their professional and training needs. At that meeting, they also learned of the American Translators Association and the then upcoming meeting in Austin which three of the group attended.

Later in the fall, the group again approached the University's Translation Program Director, this time requesting a special training course. It was agreed that the course would be a pilot program for community interpreters that would meet once a week for ten weeks for a two-hour session in the evenings. Initially, University personnel were to travel to Lexington rather than have the group members drive to Kearney until it was discovered that there were no language laboratory facilities at all in Lexington at either the high school, the two year college facility, or the recently closed parochial school building. The course would be planned, directed, and taught

degree aspirations to management positions with current employers or certification in federal court interpreting.
It was clear to the University personnel that this was a highly-motivated and talented group of participants from the interviews and information gathered at this first meeting.

4. PROGRAM OF STUDY

4.1 First Five Weeks

To familiarize participants with the equipment in the language laboratory and the various types of interpretation (simultaneous, consecutive, and sight translation), they were first given tapes in Spanish and English which describe various careers (real estate, medical, technician, etc., Ref.1). While these tapes were not prepared specifically as interpretation exercises, but rather to inform about careers, they run at a speed reasonable for practice, they use various national and regional accents in both Spanish and English, and deal with familiar subjects while injecting specialized terminology. The instructors used shadowing, dual task, and paraphrasing as well as interpreting techniques in these and all later sessions.

Materials for community and medical exercises were those actually in use in the hospitals, clinics, and agencies in Dawson county where Lexington is located. Some of the information was pre-recorded as tapes for laboratory practice; some was arranged by the University personnel into situations to be performed as consecutive practice; some was converted to transparencies to be projected onto a screen for sight-translation practice. A medical doctor from Kearney who works with the Spanish-speaking community but who speaks no Spanish came to the class to interact with the participants about medical interpeting from the doctor-user's point of view.

4.2 Second Five Weeks

The second five-week session focused on legal and court interpreting. Again materials were those in use in the family, civil, and criminal systems and courts in the county where the partipants live and work.

5. RESULTS

Of the nine who began the course, eight completed the full ten weeks and obtained an Award of Completion presented by the Vice Chancellor for Academic Affairs of the University of Nebraska at Kearney in a short ceremony, held the evening of the final class meeting. The group is eager to have the University or the courts sponsor an intensive training course for court interpreter certification.

6. CONCLUSIONS

This group of bilingual participants, already experienced as self-taught community cultural interpreters, quickly mastered the various techniques of formal interpretation training and were quickly ready for medical, legal, or social services specialization. Also it was observed that while the general subject matter and introductory legal and medical exercises were of interest to all, as the course progressed, the specific interests of the participants--particularly between those by the University's Director of Translation/Interpretation with the in-class assistance of one of the Program's instructors. It was further agreed that no more than fifteen participants would be enrolled.

2. COURSE CONCEPT

In the conceptual plan for the course, it was not imagined that such a short course would prepare professional interpreters, in and of itself, but rather that it would serve as introductory to the processes and concerns of professional interpretation, provide information about materials and resources, and give those enrolled a sampling of work from various fields of specialization in interpretation to help them discern their own interests and potential.

The ten-week course was divided into two five-week segments: the first emphasizing techniques of simultaneous and consecutive interpreting and sight translation using general and medical situations and terminology; the second emphasizing interpretation in the courtroom and other legal situations. Meetings would take place in both the classroom and language laboratory setting. There was also initial concern from the University personnel that the weekly evening drive to Kearney from Lexington--a round trip of seventy miles after a day's work--might not prove feasible for the participants for the entire ten weeks.

The cost of the program was set a $30 per participant. This fee, which is consistent with charges for the University of Nebraska's non-credit courses offered through University Extension, was used to cover the cost of materials and copying. The Translation/Interpretation Program Director planned, directed, and taught the course without compensation; the instructor who assisted was paid by the University's College of Continuing Education.

3. THE PARTICIPANTS

Nine participants enrolled for the course. All were totally bilingual in Spanish and English; several had considerable experience as interpreters; two had actual training as interpreters. All nine were employed: one as a paralegal, one as a minister, five in social services, one in the probation office, one as a language specialist and cultural trainer for the employer whose arrival had so stressed the community's bilingual resources. While a concern of the University faculty had been that the disparity in language ability among the participants might be great, making it difficult to select materials that all would find both workable and challenging, this was not the case. There was a shared language ability level among the participants that was exceedingly high. This factor, the complete bilingualism of the participants, plus their dedication and seriousness about improving their abilities as interpreters, were the most significant factors in the success of the pilot program and could be the most difficult to replicate in another pilot or regular program.

The first evening, the participants were asked to complete an enrollment form which solicited information about their work, educational level, and course and career objectives. Three had college degrees, another listed some college course work; all had completed high school. Five listed Spanish as their native language; the others listed English. All listed improvement of interpreter skills as a course objective. Other course objectives were to gain more confidence as an interpreter, to improve English or Spanish grammar skills, and to expand terminology--particularly medical terminology. Career objectives varied from advanced or undergraduate interested in medical and those interested in court specializations--made some sessions more interesting than others to those participants with specific interests. It would have been better had participants been able, beyond the segments of a course that were introductory and provided an overview of the profession and the areas of interpretation, to choose an area of specialization for further refinement of skills and terminology.

Materials which would be of use in a community interpreter program are those prepared by ACEBO in medical interpreting and court and advance court interpreting (Ref.2). The Cultural Interpreter Language and Interpretation Skills Assessment Tool project completed by the Cultural Interpretation Services of Ottawa for accreditation and candidate selection may be of use (or adapted for use in the U.S. environment) in assessing community and cultural interpreters. An assessment tool would allow training to focus on the needs of each individual.

Ideally, one would separate community interpreter training from medical or legal interpreter training but when working with a small, geographically isolated population--a small town, a small group of bilinguals--teamed with urgent, multiple needs in the community for interpreters, reality requires that the initial training be general and that it provide an overview of the various specialities so that the community's existing valuable human resource--its bilingual citizens--feel as confident and as prepared to meet these needs as possible.

7. REFERENCES

1 Starr, Charles. *Foreign Languages for Vocational Training.* E. Northport, L. I., NY: Audio Lingual Educational Press, Inc. 1983.

2. Mikkelson, Holly. *The Interpreter's RX.* Spreckels, CA: ACEBO 1994 and Mikkelson, Holly. *The Interpreter's Edge.* Spreckels, CA: ACEBO 1992.

ARE INTERPRETERS BORN OR MADE? REAL-LIFE SITUATIONS IN THE CLASSROOM

Krisztina Bohák Szabari
Eötvös Loránd University Budapest

Although interpreter training has been started in Hungary in the recent past, the question still arises from time to time even in professional circles whether good interpreters are born or made. I think you will agree with me in saying that quite a lot of good interpreters have come into the profession without any formal training, and they believe in learning by doing. These people are firmly convinced interpreting skills can be acquired only in real-life situations and they do not think that classroom practice can be of real use.

Being a professional interpreter myself I find this criticism partly justifiable: simulated interpreting situations are less convincing and motivating for the students than genuine ones.

As an interpreter trainer, however, I cannot fully share their scepticism concerning interpreter training: within a well-defined framework, by using a hypertext as Pöchhacker puts it and developing a well-targeted practice it is possible to improve the trainees› competence.[1]

In Hungary the interpretation market has developed a great deal in recent years and there is a growing demand for interpreters both in consecutive and in simultaneous. Although courses in interpreting have been offered for over 20 years now, there are many self-made interpreters on the market. They are convinced that learning by doing is the only way to become an interpreter.

Seeing the growing demand for international communication at the end of the 80s and the early 90s most universities and colleges were keen on introducing courses for translators and interpreters. The faculties of humanities and the business colleges established new courses for translation and interpreting and in most of the cases the teachers of foreign languages were asked to organise these courses. So interpreting and translation remained reduced to a linguistic or a language matter and the trainees had to gain their own experience in real life.

Some of these training schools provide post-graduate courses saying that candidates here can be expected to have achieved a requisite level of language skills, general and cultural knowledge as well as the maturity and the ability to work under stress.[2]

Some other schools offer courses for translators and interpreters at the same time saying a good interpreter has to be a translator as well. There are schools that believe that without being trained in consecutive interpretation students will not cope with simultaneous while others are firmly convinced that simultaneous and consecutive should be taught separately as both require different skills.

At the Training Centre for Translators and Interpreters of the ELTE University in Budapest where I come from we offer post-graduate courses in translation and consecutive interpreting at the same time with some practice in simultaneous.

Applicants to our courses hold a university degree and are required to pass an aptitude test. Interpreter training does not include language learning, trainees are expected to be sufficiently proficient in their B language. Candidates are also required to demonstrate a broad knowledge of cultural and current affairs.

We lay great stress on engaging practising interpreters as trainers because we agree with Danica Seleskovitch who wrote:

"... conference interpretation should be taught by conference interpreters. Strangely enough, while nobody would claim that medicine should be taught by somebody who has never seen an ill person, or music by somebody who has never been to a concert, there are, still now, quite a number of schools pretending to train interpreters where there is not one conference interpreter among the faculty, and where most if not all teachers have never been in an international conference let alone seen interpreters at work."[3]

It is not easy to find the right persons for teaching skills because a lot of interpreters are reluctant to reflect upon their work, they do not want to process their experiences for teaching or research purposes and very often good interpreters do not find the time for teaching either as autumn and spring terms coincide with the peak seasons of conferences.

I am one of those who are desperately trying to combine interpreting, interpretation studies and teaching because I firmly believe that it is not possible for any trainer to impart skills that he or she does not possess, nor can a trainer who requires students to produce a standard of performance that he/she cannot achieve, have very much credibility in the classroom. In a lot of professions it is perfectly natural that students learn from the expertise of accomplished performers. That should be so with interpreting as well. Trainer demonstration may serve as a very good complement to other teaching methods. Students of translation have at their disposal a standard of comparison, in that innumerable professional translations are easily available for analysis.[4]

Student interpreters have much less opportunity to compare their own performance with that of a professional because interpreters work with the spoken word and recording, if made, is not easily obtainable for them. Interpreters do not like to be recorded because their performance is not for general consumption but for a special audience on a special occasion at a special time and space.

It is not easy to find opportunities to take students to conferences in order to listen to professionals in action. And even if there is an opportunity to listen in, they cannot analyse and assess the performance afterwards.

That is why trainer demonstration is one of the best ways to motivate students, they get encouraged by listening to their teachers in operation noticing their weaknesses and realising that no-one is perfect, not even their teacher or trainer of interpreting. It is very important for their future life to make them realise that each interpreter can only aim at perfection and each performance could still be improved.

When teaching interpreting the teacher has to confront his/her students with life-like or even real-life situations to get away from the classroom atmosphere. And there are great differences between the classroom and real-life situation. In real-life situation we can find the initiator, the client, the speaker and the source-text listener as well as the interpreter and the target-text listener or the consumer or user. Pöchhacker suggests the term hypertext as a framework for the interpreting event.[1] Interpreting is a social act involving production of functional target texts in a specific situation of transcultural interaction. The target text produced by the interpreter at a certain time and place for a certain group of target-language speakers is mostly controlled by the skopos, i.e. the purpose or function to be fulfilled by interpreting.[5]

In the classroom the skopos will be automatically reduced to making a comparison between the source and target language text. In real-life situations it is not the case. The teacher-student relationship psychologically differs from real communication acts, the teacher assesses the performance of the students comparing the source- and target-language texts with one another. The students subconsciously know that they need not really convey the message because it is perfectly well known for the teacher and in most of the cases they shift their attention towards language structures and details which are not really relevant in real-life situations. Overcoming the traditional tutor/teacher - student/trainee relationship is a great challenge for both parties but its success determines to a great extent that of the interpreting course. The mistakes students make in the classroom situation are not the mistakes they would commit in a real-life situation. In the classroom the main goal of students is to meet the teacher›s expectations and these are often very different from those of the target-language listener in a given situation. In the classroom students tend to concentrate their attention on irrelevant details if the situation is not perfectly clear and well-defined, so memory and background knowledge do not play the same role as in a real-life situation. A lack of background knowledge due to the vagueness of the situation has to be compensated for by retention in short term memory and it results in a lower level of concentration on the main message or sense of the source-language text. In real life the success of interpreting does not only depend on the degree of so-called "equivalence" which dominates in the classroom but on the extent to which the target text functions as intended in its situation and sociocultural context.

In the classroom students are aware of the classroom-situation and it requires lots of preparation on the teacher›s part to change it by successfully simulating an interpreting event. Working with authentic material chosen by the teacher who him/herself worked as interpreter in the given situation and perhaps recorded him/herself enables to make the students aware of the framework. Students get a handout containing the former invitation to the conference which contains the title and the preliminary agenda, the list of participants and some information about the organisers. They can start preparations in order to become familiar with the topic. It is very important to make it clear that interpreting needs preparation in advance just as translation. The only difference between the two is that interpreters do not have the final text at the time of their preparations.

After having got familiar with the topic we have to set the scene in the classroom, this entails establishing by whom the speech was delivered, when, where, why and to whom. The next stage is for students themselves to do the interpreting which is recorded and subsequently compared with the tutor›s performance previously recorded at the conference. In spite of this preparatory work it often happens at the beginning of the course that students commit mistakes because they are not fully aware of the interpreting situation. They act like students in a classroom instead.

So in teaching interpreting it is most important from the very first moment that students realise the importance of being aware of the situation the interpreting event takes place in. Awareness is one of the main prerequisites for interpreters. It can be trained but you cannot produce it from one moment to the other, it has to be developed systematically in the class. The main task of the teacher is to build up this awareness in the students. It should become a natural demand for them to get acquainted with the surroundings of the interpreting event. The teacher possibly can do it more convincingly when he/she works with authentic material and real-life conference settings. As he/she is after the interpreting event and knows about every detail, he/she can set the scene for the students. He/she can explain them who says to whom what, how, with what effect, in which spatio-temporal setting. So the students will not just interpret away. The more they are aware of the situation, of the goals of the communicating partners (e.g. are they conflicting or shared goals), of the participants› involvement or non-involvement in the text, of the relation of power between them, of their social status etc. the better they can concentrate on the message to be conveyed and the easier they find the from the listener›s point of view new pieces of information in the source text.

This scene-setting is very important for the assessment of the performance of the students. It enables them to consider each other›s performance from the client›s or target-language listener›s point of view.

So I think better less interpreting prepared than more unprepared in the classroom. The holistic approach is needed and students have to go around the whole scene. Trainer presentation can be very motivating. Students feel as if they were part of the interpreting event themselves. Job satisfaction is an important question, it is not good to be frustrated. The main message teachers should convey to their students is that the interpreter is a communicator, an important person who knows the situation he/she acts is and not just a nuisance who is controlled all the time by his/her language teacher. And the audience is not just counting his/her mistakes as compared to the "original" but wants to get the message. Interpreters should be self-confident but remain humble to the situation and the message that should be conveyed.

References:

1 Pöchhacker, Franz. *Simultandolmetschen als komplexes Handeln.* Gunter Narr Verlag Tübingen, 1994, p. 47

2 Mackintosh, Jennifer. Postgraduate Training in *AIIC Bulletin* Vol. XIV, No. 1, Edition Speciale- Symposium de Strasbourg, pp. 10-11, AIIC Geneva

3 Seleskovitch, Danica. Teaching Conference Interpreting in *American Translators Association Scholarly Monograph Series* III. Binghampton: State University of New York, 1989

4 Altman, Janet. The _Role of Tutor Demonstration in Teaching Interpretation in L. Gran and J. Dodds ed. *The Theoretical and Practical Aspects of Teaching Conference Interpretation.* Campanotto Editore Udine 1989 pp. 237-241

5 Reiss, Katharina & Vermeer, Hans J. *Grundlegung einer allgemeinen Translationstheorie.* Tübingen: Max Niemeyer 1984

THE CHALLENGE OF TEACHING TRANSLATION IN DEVELOPING COUNTRIES

Cruz A. Cancel
School of Translation

Keywords: Sign Language, Total Communication, Oralism, Pragmatic Approach, Flexibility, Interdisciplinary, Terminology, Interpreting, Travel Abroad.

ABSTRACT

The challenge of teaching translation in a developing country is the focus of this paper. The discussion makes reference to the author's experience as a Fulbright Fellow at the University of Managua Nicaragua, School of Translation, and to other published papers on the topic in other countries. A study of students needs and adaptation of the course and teaching styles to meet those needs is discussed. A Pragmatic Approach is recommended to meet the needs of the target population and culture. Travel abroad is recommended for students and faculty.

1. INTRODUCTION

As I prepared to write this paper addressed to the challenge of teaching translation in developing countries, three statements made by Prof. Snell Hornby at the First Language Conference, Elsinore, Denmark, 1991, came to my mind (Ref.17).

Translation is "a sociocultural activity which presupposes not only language competence but also extensive factual and encyclopedic knowledge as well as the target culture " (Ref.17).

The Professional Translator of Tomorrow is one "who is not only a bilingual specialist, but also a bicultural (if not multicultural) specialist who works within the framework of related subjects" (Ref.17).

Our language is part of the **world** around us and our knowledge of that **world** is a **prerequisite** for translation." Prof. Snell Hornby 's last statement reminded me of my Translation thesis and my need for learning more about the world and the language around me, **Sign Language**. This is the means of communication of persons who are hard of hearing and deaf, with whom I have worked, and who I care for very much. The **target culture:** their **Sign Language**, was something I had rejected to use, to learn, and to study since the beginning of my education on deafness.

During my earlier years in the field of Audiology and Education for the deaf, I was, what educators called, an **oralist:** one who expects those who are deaf to use verbal communication only by means of lip-reading (also called visual hearing) and the use of a hearing aid for amplification of others' speech.

2. A SECOND THOUGHT

But the **world** changes! Now, I promote the use of **Total Communication** for those who are deaf. This is an inclusive term which consists of all avenues available for this purpose, and Sign Language is one of them. I wish I had learned Sign Language earlier in my professional career! This is what brought me one day to the School of Translation of the University of Puerto Rico.

I made an appointment with the Director to solicit enrollment in "**one or two courses**" which could help me translate into Spanish an English teaching guide book on Sign Language needed for the many Hispanic citizens who are deaf in Puerto Rico.

I still remember the director's smile when he told me very politely: "One must enroll as a regular student for the degree in order to be allowed to take any course in the Translation program".

Then, still smiling, he said : "You will need more than **one** or **two** courses to learn the skills of translation for that book".

The questioning continued without giving me a chance to answer one question, before he threw at me several more.

-"How much do you know about Sign Language?"
-"Do you speak in Sign language?"
-"How long have you studied the language?"
So, I had no choice but tell him the truth and face the consequences.
-"I did not have any interest in Sign Language, as a matter fact, I rejected it. It was not until **I was 'born' again ...**"
-" Stop right there! This is not a confession. You are not going to tell me that with the "**Grace of God**" you are going to become a translator".
I laughed! He calmed down and allowed me to tell him my story and the difference between the two approaches: **Oralism** Versus. **Total Communication.**

-"What made you be so stubborn before"?

-" Frankly, I blame my **oralism** stubbornness on my Hispanic culture and other influences in the various universities where I was studying Speech Pathology and Audiology in the United States. **Oralism** was the trend in most programs in South and North America".

-"What really happened and how long did it take you to change completely to a point that now you want to translate a Sign Language book?"

-" **A second thought**"!-
-" **A second thought ?**, Is that all?"-

-"The change was a slow process but I had no other choice. I was still an **oralist** when I received a Federal grant ($90.000) for three years at the Medical Sciences Campus of the University of Puerto Rico to organize the graduate Training Program for Teachers of the Deaf and a Model Program for Puertorican pre-schoolers who were deaf and hard-of-hearing. But there was a "catch"! I would have to promote Total **Communication.**"

-"It sounds like giving the "key to the thief!".
-"You may say so, but it worked!"
By the end of the first year of the grant I was already sold on the idea of having a Program in Total Communication. We recruited a Mexican-American professor in Deaf Education from New Mexico, Dr. Jacob Pino, who served as advisor in our developing program and teacher of short workshops for our public school teachers enrolled in the program.

After becoming a ***Total Communication*** promoter, my first step was to take an immersion course in Sign Language at the Bill's Rice Ranch for the Deaf, where I lived for a week among people who were mostly totally deaf. There I met with the author of the book which I wanted to translate, Mrs. Cathy Rice, and her staff. The few basic signs I learned were not enough. This experience helped me see how much I needed to learn about Sign Language and about Translation.

3: TRAINING AT THE SCHOOL OF TRANSLATION AT THE UNIVERSITY OF PUERTO RICO

My next step was to enroll at the School of Translation of the University of Puerto Rico, College of Humanities, to learn the translation skills necessary for my thesis requirement, the Sign Language book (Ref. 3). I took the ATA examination for English into Spanish and became a Certified Translator.

4. TEACHING TRANSLATION ABROAD

Four years later, I received an appointment as Linguist and Researcher, a Fulbright Fellowship award, at the School of Translation of the University of Managua, Nicaragua, C.A. This included teaching Translation and research on Sign Language in Central America (Ref.2). My experience in Nicaragua and my readings of some reports on programs in other countries made me raise the question: How can we apply what we learn from other programs to those institutions in developing countries?

4.1 Can we copy USA programs?

One may ask if it is possible and practical to apply what we learn from the USA programs to programs in some developing countries where the physical conditions, the technical and library facilities, and trained professionals are limited.

Several questions are raised when considering teaching translation in developing countries: Are there other alternatives to the student's needs: alternatives in teaching styles, in studying assignments, in revision , in class discussion, in selection of topics or class material for practice? Is it possible to provide a flexible interdisciplinary program?

Among some of the aspects that should be taken into consideration in the design of a program and in the preparation of the individual courses mentioned are: the student's cultural, pedagogical and economic environment, **the availability of resources** and their limitations in **time** to use those resources.

Time to study

In many of the developing countries, although education is free, the students do not have means to support themselves as is the case with many students in the United States. Most students need to keep a job while attending the university, and their time for laboratory and library work is limited.

Time was one of the major limitations at the Universidad Centro-Americana in Managua. The classes were offered in the mornings because most students support themselves and have to travel long distances to attend classes.

The next topics may give some alternatives for experimentation in developing programs.

5. FLEXIBILITY AND INTERDISCIPLINARY COLLABORATION

To make up for the limitations that one may encounter, **flexibility** and **interdisciplinary collaboration** may offer opportunities in other areas of expertise within the same university and location.

Professor Snell- Hornby considers advanced work in translation to be by nature interdisciplinary.(Ref.17). She insists that **flexibility**, as a key to future programs, specially at the advanced level, should have a high priority. Flexibility applies to various aspects of the interdisciplinary collaboration process. It applies specially to the creation of modular concepts essential for subject area expertise in the use of interdisciplinary collaborators.

Flexibility is also necessary in the procedure of negotiating the administrative arrangements at higher levels for the interdisciplinary service such as a course in another department and or the assignment of a professor to review and advise in the terminology of a thesis paper. Credit for the interdisciplinary course may require additional administrative consultation and arrangements necessary if it is to be awarded to a student who attends a course in another Department or School. This is usually the case if the course is not already listed in the required or elective courses of the student's program.

The simplest one of the various alternatives mentioned for **interdisciplinary collaboration**, one which requires less red-tape, is the availability of supervisors for specific topic areas for the student's thesis and special papers and interdisciplinary professionals to teach specific courses.

5.1 Interdisciplinary professionals as advisors or teachers

I experienced the benefit of an interdisciplinary team at the University of Puerto during the development of my translation thesis. In addition to the Director of the Program, Prof. Carmen Díaz, there were two members from other University Schools: Dr. Esther Strubbe, Professor in Speech Pathology and in Deaf Education at the Medical Sciences Branch, and Professor Matilde Alberti, from the Spanish Department, the author of the text book on *Redacción y Estilo* used as the text in the Translation program (Ref. 1). Also, I received assistance from the Director of the private English School for the Deaf, Reverend Rawlings, a Canadian professor of Total Communication, who reviewed my description of the signs. Mr. Alberto Negrón, a young graphic artist, a member of the Group of Adults who are Deaf, draw the hand signs which had to be adapted for the Spanish translation under the supervision of Professor Rawlings. In addition my course on Legal Terminology was offered by a lawyer.

An example of flexibility, a **pragmatic approach**, also offers means for practical opportunities for interdisciplinary collaboration in **terminology** and in **interpretation** from other areas of expertise.

6. THE PRAGMATIC APPROACH.

A pragmatic approach is considered to give more opportunities to our teaching program in a developing country where we must prepare the students with the proper skills to perform effectively as a translator according to the possible roles he or she will assume in his or her country.

Availability of library, and laboratory resources

This was one of the major limitations in Nicaragua since there were not many facilities, nor source books and dictionaries at the Main Library. All students received a pocket dictionary (Spanish-English) for their personal use which was their major source for terminology work. The Department had a small study room with several big dictionaries and a few other books, but it was only opened during working hours.

Practical work in the classrooms

The present challenge is to provide a program with opportunities to practice. This starts in the classroom and concludes with short take home assignments for the next session. Thus the student has two tasks as home assignments. 1) the revision of his previous assignment, translated in class, which he must revise and submit to the next class showing ways in which the translation was improved, and 2) his preparation for the translation of the next assignment, for which he has to do a study of terminology and initiate the first draft of the translation of the assignment.

Revision a means of Self-evaluation

There is great value in keeping the previous class assignment for revision purposes. In this respect, A. Alvarez, of Spain, suggests, that the student should get used to revise his or her own version (Ref. 2). A similar procedure is reported by Sainz, from the University of Uruguay, where there is a comparison of the student's own back-up translation with a published version (Ref.16). The improvement in the revisions may serve as a self evaluation measure of his or her own progress through the course.

I originally used this technique in Nicaragua with groups of five or four and was very effective. This solved the problem of the unavailability of photocopy materials for the individual study groups composed of four or five students each. I printed out from my portable computer four copies of short articles from magazines, newspapers or chronicle books, and airlines and other travel companies publications, one for each group to work on the blackboard. I brought several dictionaries to class and together they worked on the terminology and made their choices of the most appropriate terms.

Practicum experiences locally.
There were some opportunities for practicum locally, in Managua. A few of the most advanced students, who lived in town, searched for part-time work at goverment and comercial enterprises where they translated contracts and brochures, instructions, reports and announcements into Spanish. A few had the responsibility also of helping with the correspondence from Spanish into English. Several of the advance students in the Translation program in Managua, had temporary part-time jobs in governmental and commercial enterprises. Also, whenever there was an international meeting locally, which required an interpreter, those in charged solicited the assistance of students from the Translation program for this purpose.

Practicum experiences abroad.
To study abroad may be an important experience for the student who needs and wants to gain fluency in speaking and reading in his second language. The stay abroad may fulfill two purposes: it may serve to improve language competence (as in summer courses) and or may provide an opportunity for research or professional practice (as with exchange programs, research fellowships or scholarships, temporary posts as teaching assistantships or practical training in firms or other organizations. There are opportunities abroad for advance and graduate students, as well as for university professors from various programs in the United States and in other countries.

Two other areas of training: Terminology and Interpretation, which provide extensive opportunities for practical exercises and experience, are the next topic of discussion.

Terminology and Interpretation.
Terminology: This is an aspect of the training that relates to all practical exercises done in short articles as well as books. The practical form of the exercise occurs during the classroom translations and is a source of review in their take home assignments. It is a practical subject that does not require a great deal of theory. Sager describes the Translator as a decision making person in the task of choice of words and as a terminologist (Ref.15).

There are several tools available for terminology, in addition to the dictionaries, thesaurus, and the older ones of term banks and terminology software, CAT (computer assisted. terminology) and lately the CD-ROM. A recent source I found: *The Random House Menu*, organized in a series of concepts and words classification, could be very useful also as an additional source to locate terms (Ref.6).

However, to provide the students with the knowledge and skills required by the new computerized working environment is a challenge to present university translation programs. There is the possibility that some programs in developing countries are far from having the latest technology available, specially for the faculty, and even less for the student's training and use. Also, a well equipped laboratory is needed for Interpretation, the next topic of discussion.

Interpreting.
Guile, from the National Institute of Interpreters of Oriental Languages in France, made the comment that the general trend in Interpretation is that the majority of the students specialize in translation and later on take some courses in interpreting (Ref.8). Since a good laboratory for practice in all types of interpreting roles is necessary for Interpreting, it is possible that this has limited the establishment of programs in this area of Translation. In Nicaragua, although there was no interpreters program, as such, some of the most advanced students in translation, and the most fluent in the English language, were called on occasions to serve as interpreters in government and civic activities.

Success in spite of limitations.
I was able to predict success from most of my students because, in spite of the limitations they had, they were **highly motivated and determined to succeed** in their career. An asset to the program was that many of the students who enrolled in the School of Translation, originally were students of English. They had changed to the Translation program because they considered the School to be the best means for learning the English language in a practical way.

7. CONCLUDING REMARKS.

I would like to conclude that teaching in a developing country is a challenge. It is an educational experience where you gain more than what you give. I am proud to say that I enjoyed every minute I was working there in Nicaragua.

Those who made my teaching successful were the students whose main goal was to learn. They were very polite, and always had some way of expressing their appreciation for the efforts I made to meet their needs, and for the subjects I had chosen for translation assignments. Also, I am very grateful for the opportunity I received through the Fulbright program to teach and carry out research in other lands, with people from different cultures. I encourage all professionals, who like university teaching, to try teaching abroad and learn from "The University of Life".

REFERENCES

1 Albert, Matilde, *Redacción y Estilo*, Editorial Marle, P.R. 1984.

2 Alvarez, María Antonia,"Teaching Translation at a Distance: A Spanish Case Study, *Teaching Translation and Interpreting: Training, Talent and Experience. First Language International Conference*, Elsinore, Denmark, 1992,147-152.

3 Cancel, Cruz A. "Sign Language Dialects", National Conference of the American Translators Association, Wash. D.C.12/89, ***COMING OF AGE***, 97-103.

4 Cancel, Cruz A. *Lenguaje de Señas para Todos*, M.A, thesis, University of Puerto Rico, 1984.

5 *Chronicle of the 20th Century*, Clifton Daniel, Editor in Chief, Chronicle Publications, Mount Kisko, N.Y.

6 Glazer, Stephen, *Word Menu*, Random House, New York, 1992.

7 Gaddis Rose Marylin, "Translation Types and Conventions" in *Translation Spectrum*, State University of New York Press, Albany, 1981, p.31-40.

8 Guile, Daniel, "Basic theoretical components in Interpreters and Translator training". *Teaching Translation and Interpreting: Training, Talent and Exper-ience. First Language International Conference*, Elsinore, Denmark,1992,185-194.

9 Kongscott, Geoffrey. Bicentenary of Tyler's "Principles of Translation", *Teaching Translation and Interpreting: Training, Talent and Experience. First Language International Conference*, Elsinore, Denmark, Benjamins, 1992.

10 Lado, Robert. *Language Testing: The construction and Use of Foreign Language Tests*. McGraw Hill Book Co. N.Y. 1961.

11 Mauriello, Gabriella. "Teaching Tools in a Translation Class", *Teaching Translation and Interpreting: Training, Talent and Experience. First Language International Conference*, Elsinore, Denmark, Benjamins, 1992, p. 63-68.

12 Mossop, Brian. "Goals on a Revision Course". *Teaching Translation and Interpreting: Training, Talent and Experience. First Language International Conference*, Elsinore, Denmark, Benjamins, 1992, p. 81-90.

13 Ottinen, Rita, "Teaching Translation of Fiction- A Dialogic Point of View". *Teaching Translation and Interpreting: Training, Talent and Experience. First Language International Conference*, Elsinore, Denmark,Benjamins, 1992, p.75-80.

14 Rose, David Stephen, "Translation and Similarity" in *Translation Spectrum*, State University of New York Press, Albany, 1981, p. 8-30.

15 Sager, Juan C. "The Translator as a Terminologist", *Teaching Translation and Interpreting: Training, Talent and Experience. First Language International Conference*, Elsinore, Denmark, Benjamins, 1992, p.107-122.

16 Sainz, J.M. "Developing Translation Skills".*Teaching Translation and Interpreting: Training, Talent and Experience. First Language International Conference*, Elsinore, Denmark, Benjamins, 1992, p. 69-73.

17 Snell-Hornby, Mary. "The Professional Translator of Tomorrow: Language Specialist or All-round Expert". *Teaching Translation and Interpreting: Trai-ning, Talent and Experience. First Language International Conference*, Elsinore, Denmark, Benjamins, 1992, p. 9-22.

REFERENCES

DEVELOPMENT OF AN INTERPRETER TRAINING PROGRAM

John DiFiore
Union County College

Keywords: Interpreting, Interpreter Training, Curriculum Development, Court Interpreting

Abstract: Union County College is planning to implement a certificate program for interpreters in 1996. This program seeks to introduce bilingual individuals to the skills necessary for becoming an interpreter. The college is taking a rather unique approach by not restricting the program to any specific language combination. Courses will all be taught in English, so the most important requirement for admission is proficiency in one other language. This paper explains the background that motivated the creation of the program and describes the proposed curriculum. The author ends with a request for input from the readers.

1. MOTIVATIONS FOR IMPLEMENTING THE PROGRAM

1.1 Local Demographics

Union County has been experiencing a growth in the number of non-English speaking residents, especially in the urban area of Elizabeth. Since at least 1812, this city has been known for its non-English-speaking residents. At that time they were largely the French-speaking Huguenots (Ref. 20). Today there is a wide variety of foreign languages spoken in Elizabeth. This fact accounts for the growth of Union County College's Institute for Intensive English. The Institute's student population, composed almost exclusively of immigrants increased steadily throughout the eighties and into the nineties, reaching approximately one thousand students in each semester these days (Ref. 1).

The background of these students is quite varied. Ethnically they represent four continents and 68 countries. In addition to their ethnic variety, there is a wide range of educational backgrounds that the students bring to their studies. Some students come to the Institute with degrees or substantial training in higher education earned in their native countries. Occasionally, though, they may be unable or unwilling to work in their original professions once they begin a new life in the United States. Still their educational background remains with them. If students like those just described can reach a high enough proficiency in English (and many do), then they are excellent candidates for professional translating and interpreting. At least one author, Finlay, noticed that interpreters even more than translators are likely to be people who have acquired a bilingual ability either through emigration at an early age to education abroad. Immigrants and their children are very numerous in this profession (Ref. 8).

The students who attend the English Institute do so for many different purposes. One of the main purposes is to gain improved job qualifications through proficiency in English. Indeed, one of Union County College's main goals is to provide career training for its students (Ref. 13). So it will serve its students well by offering an interpreter training program.

1.2 Governmental Concerns

Earlier this year the State of New Jersey's Administrative Office of the Court (AOC) working under the leadership of Mr. Robert Joe Lee implemented its long-awaited Registry of Court Interpreters (Ref. 14,15). Under this registry anyone hired to be a court interpreter in the state must be approved and classified first by the AOC. Classification, following mandatory attendance at an AOC Seminar on Professional Ethics, depends strongly upon the interpreter's level of experience and training amongst other factors. The creation of this classification system comes as a result of the recommendations of the Supreme Court Task Force on Interpreting and Translation Services, which in 1981 began to foresee a tremendous problem in the state resulting from an increasing number of people who are not proficient in English and thus cannot be guaranteed equal access to the legal system (Ref. 7). This right to equal access was first established by the Federal Court Interpreters Act of 1978 and later reiterated by the New Jersey Supreme Court Task Force (Ref. 23).

In 1987, the State acting on the recommendations of the Supreme Court Task Force began another initiative aimed at improving the status and availability of professional, competent interpreters to serve in the court system and elsewhere. That state-wide project, begun by the former Department of Higher Education was called the Project on Legal Interpretation. Directed by Dr. Marilyn Tayler at Montclair State College it developed its *Curricular Guidelines for the Development of Legal Interpreter Education.* These guidelines make specific and comprehensive recommendations for institutions that intend to offer training in interpreting. One of the observations of the Guidelines was that "Little, if any, training is available to meet the need for legal interpreters, community interpreters, or bilingual/bicultural support personnel"(Ref. 5).

In fact only a small number of institutions of higher education in the state offer any organized programs leading to a recognized credential in interpreting. Many of these programs are conducted in sign language interpreting. For spoken languages, the only language pair taught is Spanish-English. The program to be held at Union will train interpreters in the theory and practice of interpreting without demanding that students in the class be bilingual in the same languages. A look at the proposed curriculum will help to clarify how this ambitious venture hopes to be carried out.

2. CURRICULUM

2.1 Bilingual Ability Required

One of the basic precepts of the interpreters training program is that the students must already be proficient in two languages at the beginning of their interpreting curriculum. This point is emphasized by many educators already involved in interpreter training (Ref. 6).

In order to insure the soundness of this policy, students who seek to be admitted to the program must be tested for proficiency in two languages. One of these two languages must be English since it will serve as the main language of instruction throughout the program. A test will be developed that will measure the applicants spoken and written abilities in both languages. Generally this procedure can be handled by a combination of short writing samples in the two languages and an interview designed to elicit spoken language samples in both languages. In the majority of cases the faculty member conducting the interview with the possible help of a language consultant (most likely a faculty member as well) can make reasonable judgments about the language proficiency of the applicant based on these samples. The process would be very much like the one that a translation manager or agency recruiter would use in predicting the future performance of job applicants. The person recruiting need not be completely proficient in a particular language in order to make educated judgements about a person's ability to work in that language. Another important indicator of the applicants suitability for the program will be his or her educational and employment background. If someone has a bachelor's degree or a significant amount of higher level education, it will indicate that the person is likely to have the necessary proficiency in that language of study.

2.2 Courses

Once study begins, students must complete thirty credits in order to receive the certificate. The thirty credits are divided into three areas, two of which contain required courses. The three areas are Interpreting, Communication, and General Education.

Interpreting courses

The Interpreting area requires the students to complete three courses for nine credits. All of these are new courses to be developed specifically for the creation of the interpreting program. Their titles are Interpreting 1, Interpreting 2, and the Role of the Interpreter. The first two are practical courses intended to improve the students' skills in interpreting. They are each built around one of the major two types of interpreting. Interpreting 1 around consecutive interpreting, and Interpreting 2 around simultaneous. The courses were similarly described in the *Curriculum Guidelines* (Ref. 5) and the skills taught have been identified by many authors (Ref. 3,4,17,18,22).

Interpreting 1 will develop consecutive interpreting skills. These skills include: listening skills, analytical ability, note-taking, short-term memory enhancement, paraphrasing, rapid language switching, and self-monitoring of accuracy as well as actual practice in consecutive interpreting. An introduction to sight translation will also be included. In both Interpreting 1 and 2, the students will be shown how to learn topic-specific vocabulary by compiling their own bilingual glossaries.

Interpreting 2 will continue to improve on the skills learned in Interpreting 1 by incorporating more difficult passages. As simultaneous interpreting is introduced, students will focus on the separate skills necessary for the task, such as determining lag time, improving comprehension by chunking, focusing on production by modulating delivery, and developing an intuitive ability in their target language through cloze-type exercises. An important aspect of this course will be to teach students techniques they can use to maintain and improve their skills long after they have finished the program.

The Role of the Interpreter will be a practical course on the many aspects of the profession that are not so obvious to the novice. The Interpreters Code of Ethics and Professional Conduct and its implications for impartiality will be stressed. Students will be encouraged to attend the required seminars conducted by the New Jersey AOC Court Interpreting, Legal Translating, and Bilingual Services Section. Cultural considerations for interpreting will be discussed, especially the notion of interpreting content and style rather than grammar and vocabulary. The benefits of belonging to professional translating and interpreting organizations will be an important part of the course. The various tools that a translator or interpreter needs to perform properly will also be covered. Valuable training will also result from students observing professional interpreters in a courtroom.

Although selections are not final, the following textbooks have already been suggested for adoption in the three interpreting courses: *The Bilingual Courtroom: Court Interpreters in the Judicial Process* by Susan Berk-Seligson, *Fundamentals of Court Interpretation: Theory, Policy, and Practice* by Roseann Duenas Gonzalez, and Danica Saleskovich's *Interpreting for International Conferences (Ref. 2,9,21).*

Other courses

The second section of the curriculum requires four courses totaling either twelve or fifteen credits in the communications area offered through the English Department. These are currently taught courses designed to enhance facility in the English language. Their utility for future interpreters was pointed out in the *Curriculum Guidelines (Ref. 5).* First of all, there is the requirement to complete the full sequence of basic writing courses, English Composition 1 and 2. In the case of students who speak English as a second language, English Composition 1 is actually completed by passing two courses. Then the student must complete courses two more courses in this are, entitled Dynamics of Communication and Public Speaking.

The final section of the curriculum allows students to select three elective courses, one in each of the remaining general education categories. They must complete one in math and science, one in social science, and one in humanities. The purpose of the general education courses is to stress the importance of having a breadth of knowledge in order for an interpreter to be prepared for the widest variety of interpreting assignments.

2.3 Faculty

It is quite rare to see an interpreter training program that is conducted for speakers of any language. Almost all programs that are running in New Jersey are geared for Spanish speaking students. William Paterson College offers a course for speakers of any other language beside Spanish, which demonstrates that it is possible to accomplish. The *Curricular Guidelines* state on p 18 "it is theoretically possible to have more than one language combination in a course or sequence of courses, though it is a difficult undertaking at best" (Ref. 5).

With the faculty resources at hand, the College feels that it can be accomplished at our institution. Although I have my educational background in Teaching English as a Second Language for fifteen years, I also have recent instructional and practical experience in translating and interpreting. My interest in translating and interpreting began during a Fulbright Teacher Exchange assignment in Hungary in 1991-92, when I met an ATA member who suggested the field and the organization to me and my wife. She soon joined the organization upon our return to the USA, and the two of us began a free lance business around interpreting and translating. The knowledge and skills I learned from developing this business with her, I eventually formulated into a course which has run for four semesters in the continuing education area.

That course, entitled Translating and Interpreting, teaches anyone who is interested how to enter the profession. The course has always been open to speakers of all languages since I teach it entirely in English. We discuss tools that interpreters use, ways to market oneself, the value of professional organizations, bookkeeping for free lancers, self-assessment, and techniques of translating and interpreting. When we do translations in class, students bring their own bilingual dictionaries and usually work in pairs so that everyone has a partner who is working in the same language combination. In their pairs the students can discuss the problems inherent in their particular language combination. After that, we discuss in English the difficulties that were encountered by each of them. We always have a practicing professional give a lecture in the course. And at the end students learn the value of networking with others to solve difficulties in translating and to find opportunities for professional work. It is based on this experience that I feel confident to move ahead with the creation of this interpreter training program.

As part of the Institute for Intensive English I and my colleagues bring specific skills that are necessary for the success this venture. We are all accustomed to working with speakers of many different languages simultaneously. Our classes always contain students from numerous nationalities. In addition, many of the faculty have had experiences similar to my Fulbright Exchange, which means they have lived abroad and are proficient in another language. Teachers such as these can serve as the language consultants or experts in a specific language combination when the need arises. They can evaluate the proficiency of incoming students in the foreign language, and even provide taped language models for classroom drills in the proposed interpreting courses depending on the composition of the students enrolled in any particular semester.

3. **INPUT REQUESTED**

The expected date to begin this program is September 1996. Necessary steps in establishing the program include circulating program announcements to other institutions of higher education and submitting the curriculum for review by the College's Curriculum Committee. During this time, the author wishes to receive criticism and comments from the readers of this article. Especially useful would be comments on teaching methodology, textbooks, and other materials to be used in the courses. Comments can be addressed to:

Professor John DiFiore
Union County College
Institute for Intensive English
12 West Jersey Street
Elizabeth, NJ 07202
E-mail comments are also welcomed at: difiore@hawk.ucc.edu

4. REFERENCES

1. *Annual Report 1994-1995.* Institute for Intensive English, Union County College, Dr. Helen Aron, dir. August 9, 1995.

2. Berk-Seligson, Susan. *The Bilingual Courtroom: Court Interpreters in the Judicial Process.* Chicago: University of Chicago Press, 1990.

3. Bowen, Margareta. "Building Up Speed (Simultaneous Interpreting of Read Speech)," in Edith Losa, ed. *Proceedings of the 34th Annual Conference of the American Translator Association.* Medford,NJ: Learned Information, Inc. 1992 p.57-62.

4. Bowen, David. "Building Up Speed (Simultaneous Interpreting of Extemporaneous Speech)," in Edith Losa, ed. *Proceedings of the 34th Annual Conference of the American Translator Association.* Medford, NJ: Learned Information, Inc. 1992, p.51-56.

5. *Curriculum Guidelines for the Development of Legal Interpreter Education,* New Jersey Consortium of Educators in Legal Interpretation and Translation. New Jersey Department of Higher Education, Project on Legal Interpretation. Montclair State College. Upper Montclair, NJ. 1988.

6. De Jongh, Elena M. "Foreign Language Interpreters in the Courtroom: The Case for Linguistic and Cultural Proficiency." *The Modern Language Journal*, 75, iii, 1991.

7. *Final Report: Equal Access to the Courts for Linguistic Minorities.* Supreme Court Task Force on Interpreter and Translation Services. Trenton, NJ: Court Interpreting, Legal Translating, and Bilingual Services Section of the Administrative Office of the Courts, May 22, 1985.

8. Finlay, Ian F. *Language Services in Industry.* London: Crosby Lockwood Ltd. 1973.

9. Gonzalez, Roseann Duenas et al. *Fundamentals of Court Interpretation: Theory, Policy, and Practice.* Durham, NC: Carolina Academic Press, 1992.

10. *Handout* from Seminar on the Code of Professional Conduct for Interpreters, Transliterators, and Translators. Trenton: Administrative Office of the Courts, Robert Joe Lee. June 7, 1995.

11. *Initiatives of the New Jersey Judiciary to Assure Equal Access to Courts for Linguistic Minorities.* Administrative Office of the Courts, Court Interpreting, Legal Translating, and Bilingual Services Section. Trenton, September 1, 1994.

12. *Interpreting and Translating as Professions.* Background Report #21: Supreme Court Task Force on Interpreter and Translation Services. Trenton, January 20, 1984.

13. *Master Plan 1990 to 1995.* Union County College. Cranford.

14. *Memorandum* to Persons Interested in Registering for the Registry of Free-lance Interpreters and Interpretation/Translation Agencies, Administrative Office of the Courts, Robert Joe Lee, Subject: Registry of Free-lance Interpreters. Trenton, February 22, 1995.

15. *Memorandum* to Members of the NCME Residing in or Near New Jersey. Administrative Office of the Courts, Robert Joe Lee, Subject: Search for Psychometric Consultant to Assist with the New Jersey Judiciary's Efforts to Assure Equal Access to Courts for Linguistic Minorities. Trenton, March 1, 1995.

16. *Memorandum* to All Court Interpreters and Agencies. Administrative Office of the Courts, Robert D. Lipscher, Subject: Amendment to R. 1:14 and Adoption of the Code of Professional Conduct for Interpreters, Transliterators, and Translators. Trenton, November 28, 1994.

17. Patrie, Carol J. "Learning Theory and Teaching Strategies for Interpreter Educators" in Edith Losa, ed. *Proceedings of the 34th Annual Conference of the American Translator Association.* Medford, NJ: Learned Information, Inc. 1992, p.205-214.

18. Patrie, Carol J. "Self-Assessment for Students and Teachers." in Edith F. Losa, ed. *Proceedings of the 33rd Annual Conference of the American Translator Association.* Medford, NJ: Learned Information, Inc. 1992, p.47-56.

19. *Release*: Administrative Office of the Courts, Robert D. Lipscher Trenton, August 16, 1993.

20. Rowland, Michelle Lindsay with Gwen E. Hall. *The Ship's Log: The 175-Year Voyage of the National State Bank.* Mountain Lion Books. 1987.

21. Saleskovich, Danica. *Interpreting for International Conferences.* Washington: Pen and Booth, 1978.

22. Schweda-Nicholson, Nancy. "An Introduction to Basic Note-Taking Skills," in Edith Losa, ed. *Proceedings of the 34th Annual Conference of the American Translator Association.* Medford,NJ:Learned Information, Inc. 1992, p.197-204.

23. *Schweda-Nicholson, Nancy.* "Language Planning and Policy Development for Court Interpretation Services in the United States" in *Language Problems Language Planning,* Summer 1986 V. 10,2, p.140-157.

24. *Statistical Tables on Court Interpreting Services in New Jersey: Calendar Year 1992.* Administrative Office of the Courts, Court Interpreting, Legal Translating, and Bilingual Services Section. Trenton, June 30, 1993.

25. Stromberg, Wayne H. and Gerald L. Head. "Court Interpreter Training in the Language Laboratory" in *Nalld Journal*, Winter 1984, p.6-20.

MULTIMEDIA TECHNOLOGY APPLIED FOR EDUCATION AND TRAINING PURPOSES IN JAPAN

Christian Galinski
International Information Centre for Terminology (Infoterm)

Keywords: Multimedia, Hypermedia, Hypertext, ISDN Communication Networks, Information Processing, Information Retrieval

Abstract: Multimedia technology offers hitherto unimagined possibilities for education and training applications—if properly applied. *Multimedia* as a concept have existed for a number of years—even several decades, if one takes its predecessor *hypertext* into account. From the point of view of information content, multimedia comprise traditional textual data, audio data, visual data (not only images, but also moving pictures and three-dimensional representations), and already or will soon also comprise the sense of touch and the sense of smell in a more or less integrated way. Multimedia thus appeal (or can appeal) to all human sense organs, and overcome the restrictions imposed on the symbolic representation of information (which developed over the millennia due to the inherent restrictions of *paper* as the major information storage medium). Multimedia are further developed into *hypermedia*, i.e., the full integration of multiple media into one system. Last, but not least, multimedia technology allows more interactive use of information technology.

In spite of all the potential that multimedia and hypermedia technology offer, there are still many problems (including some formidable technical, financial and legal, possibly also social ones) to be overcome in fields such as standardization, copyright and other intellectual property rights, nationwide and international ISDN communication networks, data safety and security, in addition to further miniaturization and the acceleration of data processing and transmission.

With respect to education and training, future generations will need a totally different approach to information and knowledge. It is to be hoped, therefore, that multimedia and hypermedia applications will spread in the educational system as rapidly as possible, as evidenced in many countries and regions (e.g., in the European Union). In view of the multitude of problems to be solved and masses of data to be made available to different kinds of users, exchange of information and increased cooperation are more important than ever. More emphasis needs to be put on methodology (e.g., with regard to the content-related ordering of information), not only on hardware- and software-related activities. National programs for research and development in the field of multimedia and hypermedia didactics as well as for the training of teachers and trainers are required in order to implement these objectives.

1. INTRODUCTION[1]

Thanks to a grant from the Japan Society for the Promotion of Science (JSPS) and to the cooperation of the Institute of Electronics and Information Science of Tsukuba University, the author had the opportunity to investigate the state-of-the-art of multimedia applications in Japan, in particular with regard to the state-of-the-art of multimedia technology as well as multimedia educational software development and its application in education and training. This report focusses rather on information science (i.e., content-related) aspects than on computer hardware and software aspects, although the latter have been duly taken into account in the course of the investigation. Although this report is not specifically oriented towards translation, the impact of these technologies on translation-related information media is considerable and will undoubtedly increase in the future.

What *exactly* do we mean by *multimedia*? This may sound like an unnecessary question. After all, the concept has been around for a number of years—decades, even, if you take its predecessor *hypertext* into account. A fairly straightforward definition of the term is possible from the point of view of information content insofar as multimedia encompass:

- traditional data
- audio data
- visual data (including movies and three-dimensional spaces as well as images)

Nevertheless, when we look at the *leading edge* of technology, we find that multimedia are also beginning to comprise sensory data, such as odors and touch. Multimedia have always had *eye-appeal*; we now find that they reach out to *all* senses. They let us throw off the shackles of symbolic representation, which has been unrivalled as a purveyor of information over the millennia, solely because of the restrictions imposed by paper as a storage medium. Indeed, so sophisticated is the development of multimedia that we now speak of *hypermedia*, i.e., the full integration of a multiple media into one system.

Contrary to conventional media and methodologies of information processing, such as browsing and other methods of retrieval, which only permitted a relatively passive reception of information, multimedia technology and methodology allow for increasingly *active* use of information and the conversion of any kind of data, e.g., symbolic data, numeric data, image data, speech data, music, etc., into any other kind of data. This became possible due to the following advances in computer hardware and software technology:

- digitalization of all kinds of information
- increased storage capacities thanks to the miniaturization of hardware components and the compression of information
- new computer languages and processing methods
- acceleration of data transmission, both inside a computer and among computers, even over long distances
- high resolution and high performance display units

With the emergence of broadband ISDN communication networks, even small businesses are beginning to wake up to the tremendous potential of multimedia and hypermedia technology. Outbreaks of euphoria are, however, still a little premature. Formidable technical, financial, and legal problems still need to be overcome. These include, for example:

- standardization of hardware, software, and increasingly important, methodology
- copyright and other intellectual property rights (IPRs)
- nationwide and international ISDN communication networks
- data protection (safety and security)
- further miniaturization and acceleration of transmission times, data processing, and data compression

These factors also pose implications for society in advancing the technological revolution.

At this point it is useful to remember that parallel to this development in computer and information science, a similar revolution has been taking place in brain research. Increasingly, the model of the computer trying to match the human brain is seen to have its limitations. Instead of trying to replicate the functions and performance of the brain, the better way forward is perceived to lie in developing computer-assisted methodology and the respective technology.

In summary, therefore, despite impressive technological advances, multimedia are still at best a combination of comparatively low-level computer and audio-visual technology for more or less simultaneous use by rather passive users. Methodologies for information ordering, information retrieval, and information representation are still not keeping pace with the ever-increasing amounts of information on the one hand, and the state-of-the-art computer hardware and software on the other hand, to such a degree that a truly efficient and user-friendly use by the average (i.e., unsophisticated and untrained) user is still beyond our reach. In addition, the nature of the social impact of multimedia technology in conjunction with the intended *information super-highways* remains a matter of speculation and will generate not only considerable research and planning activity, but probably new social mechanisms as well.

The potential of multimedia technology and methodology is evident, not least in education and training. Future generations are likely to have a totally different approach to information and knowledge. It can be argued that this will be essential. It is to be hoped, therefore, that multimedia and hypermedia application will spread in the educational system as rapidly as possible. In many countries and regions (e.g., in the European Union), this has been recognized and corresponding efforts are being undertaken. In view of the multitude of problems to be solved and the masses of data to be made available in the most appropriate way to different kinds of users, the exchange of information and increased collaboration is more important than ever. This report concentrates on the state-of-the-art of the use of multimedia and hypermedia technology and methodology for education and training purposes in Japan.

2. MULTIMEDIA FOR INSTITUTIONS AND ORGANIZATIONS

As can be gathered from the presentation of hardware, software and applications of Japanese companies at the multimedia exhibition *Multimedia'94* held in Tokyo, industry is the driving force behind *infotainment* and *edutainment*. About 60 enterprises (excluding associations and publishers of pertinent publications) offered their multimedia products and services. If one includes companies offering products and services that run under one of the systems provided by the giants of multimedia technology (such as Japan Apple or NEC), the number can be conservatively be estimated at around 100.

Given the commercial interests involved, it is no wonder that the Ministry of International Trade and Industry (MITI), the Ministry of Posts and Telecommunications, and the Ministry of Education Science and Culture (Monbusho), as well as some of the agencies and subsidized associations under their control, have a vested interest in multimedia and hypermedia developments.

Among industry associations and learned societies, the following organizations are particularly worth mentioning as having an interest in multimedia development:

- Information Processing Society of Japan (IPSJ)

- Multimedia Association of Japan (MMA)
- Japan Association for Promotion of Educational Technology (JAPET)
- Center for Educational Computing (CEC)
- Communication Industry Association of Japan (CIAJ)
- Electronic Industry Association of Japan (EIAJ)
- Japan Electronic Industry Development Association (JEIDA)
- Japan Computer Graphics Association (Nicograph)
- Japan Institute for Office Automation (JIOA)
- Association of Broadband Communication, Business Chance and Culture Creation (BBCC)
- Japan Personal Computer Software Association (JPSA)
- Japan Mobile Telecommunication Systems Association (JAMTA)
- Association for the Promotion of New Generation Network Services (PNES)

With the exception of the BBCC (located in Kyoto), all of them have their headquarters in Tokyo. The *Multimedia'94 Exhibitions Guide Book* lists about twenty more such organizations, most of which are certainly interested, but not necessarily deeply involved, in multimedia and hypermedia activities.

According to several experts who were consulted, the institutions and organizations particularly and explicitly involved in multimedia activities for education and training purposes are:

- Center for Educational Computing (CEC),
- Japan Association for Promotion of Educational Technology (JAPET),
- Institute of Information Science of Saga University

A few companies more or less exclusively focus on the topic of children and the computer, such as Nihon Futurekids KK™ and KK Datalink™.

In addition, there are some special advisory or study committees active in formulating future government policy with regard to educational software, which will certainly duly take into account multimedia and hypermedia software and applications. This does not seem like much in view of the size of Japan, its extensive educational system, and the development level of its electronic industry, but obviously the situation is very dynamic, and it is likely that time was insufficient to find out about all the developments pertinent to this report.

3. MULTIMEDIA AS A COMBINATION OF EXISTING DIGITAL TECHNOLOGIES

3.1 The Hardware Situation

With regard to the combinability of existing digital technologies within one integrated system or the linking of different more or less compatible devices together (whether by cable or wireless, with or without technological upgrading), more or less every digital device existing on the market could belong to, or fall under, the heading of multimedia technology. This makes every survey on multimedia hardware and software, as well as the application market, an extremely difficult endeavor.

From the point of view of a local PC-based multimedia workstation, such a system should comprise as a minimum the following components:

- CPU with about 16 MB memory or more, (high capacity) harddisk, audio input (microphone) and output (loudspeakers)
- a keyboard, mouse or other input support
- high-resolution color VDU
- high-resolution color printer
- high-performance scanner
- video and/or CD-ROM drive (separate or integrated into the CPU)

and—if required or considered useful:

- modem (for all kinds of external data communication)
- extended audio-visual input and output devices
- any other kind of electronic device or toy

The rest is accomplished with software. So actually, anything (including television) could fit into or be linked to this system environment—provided it is hardware and software compatible.

At present, the following development mainstreams can be observed world-wide:

- development of PCs into multimedia and hypermedia workstations
- concentration of hardware (CPU) development towards a limited number of standards
- CD development in the direction of more diversity and at the same time better integration into multimedia environments (including the integration of video into CD technology)
- development of additional input devices (e.g., writing pads) and methodologies (e.g., natural speech input)
- integration of television technology into PC VDUs
- integration of audio-visual technology into the PC environment
- development of all kinds of supplementary devices such as electronic books (especially lexicons, dictionaries and encyclopedia and other reference works), pocket notebooks, electronic games, and the like

This development offers business opportunities for enterprises that are optimizing existing technology, as well as for those developing new technology. Consequently, to the general user, the market will be as confusing, or even more confusing, for the next five years as it has been during the last five years!

3.2 The Software Situation

In spite of the confusing situation of hardware development, the real bottleneck is with software development. On the one hand, more software is being developed than ever before, but amazingly on the other hand, there is little regard for compatibility. The market for multimedia products and services as a whole, therefore, is not going to take off until the present more or less chaotic development of software is channeled into a few harmonized mainstreams.

This does not prevent some sectorial developments from becoming commercially viable, such as interactive cable television (CATV) and CD-ROM development, where there are many business opportunities for enterprises especially in the infotainment area. CD-ROM products and services are on the verge of taking off due to the demand for electronic games. Interactive CATV will possibly also get a boost due to the entertainment opportunities offered by the film and video industries. Nevertheless, it looks as if enhanced multimedia and hypermedia information and education applications still are hampered by several barriers so far as the general user is concerned, specifically the high price for such applications and the lack of intercombinability.

3.3 The Application Situation

Although much more high-level technology is available in the area of language education and training, it looks as if the audio-visual age has just reached a first peak of application in Japan. Many courses on cassette tape, video tape (e.g.,TOEIC) and CD-ROM (e.g., *Linguaphone*™) are offered to the language learner. Since last year, some simple CDI applications (such as *Linguabox*™ or *Quick English*™) have found a broader market.

On the one hand, language learning has always been extremely expensive in Japan, but on the other hand, more sophisticated tools such as CDI are limited to certain hardware and software environments (e.g., Macintosh), which makes their use as expensive as traditional ways of language learning with private teachers or at private schools (juku). Furthermore, learning with the machine does not offer any formal benefits, such as certificates, etc., and does not even guarantee success, since the learner is not under the "control" of a human counterpart.

On the other hand, there are many language laboratories and other facilities at schools and universities, and some places even boast of multimedia laboratories. Nevertheless, it appears as if technology is not as effectively applied as it could be (for many reasons, not the least of which is the lack of sufficiently trained teaching personnel). Software catalogues on CAI software and courses for use at schools and universities list thousands of titles, but again, for various reasons (in many cases a lack of budget for appropriate equipment), use at educational institutions still seems to be rather limited.

In the area of information activities, many efforts have been introduced, for instance:

- user-friendly information retrieval terminals (e.g., RUPO)
- the digital "library of the future" (such as the projects of NACSIS, Ariadne, RUKIT, etc.)
- information services operated via telephone lines and CATV

Although some of these applications have been quite successful (e.g., ticket reservation), they are generally not yet fully utilized due primarily to a lack of front-end user-friendliness.

As was reported by the author in 1992, there is some evidence (e.g., in the training of airline ground and flight personnel) that multimedia and hypermedia technology and methodology is impressively applied in the manufacturing and services industries very efficiently in Japan. Companies are, however, more often than not reluctant to publicize such activities, which give them a competitive edge over market competitors. Hence, the situation in private enterprises is very opaque, and investigation is difficult. Consequently, these efforts have been excluded from this report.

4. STANDARDIZATION NEEDS

Nothing is more annoying to the general (i.e., unsophisticated) user than to have to connect many devices by cables (worsened by the need to use adapters and converters, etc.) and to have to adapt software for any individual application. If the systems and components are not compatible with each other, it is even more annoying. Often systems and components claim to be compatible, but data and information structures are not, so that the information cannot be used—or used only after complicated conversion processes. The ideal solution would be multimedia and hypermedia workstations with integrated main components, which for maintenance and repair purposes should be easily replaceable with an automatic handshake, so that there is no need for "cable salad" and software adaptation by the user. In this area there is still a great and even growing need for official and industry standardization.

The situation in standardization with regard to multimedia is nearly as confusing as the system-development situation. At the international level, the International Electrotechnical Commission (IEC) and the International Organization for Standardization (ISO) finally joined forces during the 1980s and set up Joint Technical Committee ISO/IEC JTC 1: *Information Technology*, which has become by far the largest grouping within both ISO and IEC. Its sub-committee SC 29: *Coding of Audio, Picture, Multimedia and Hypermedia Information* is directly responsible for the standardization of multimedia aspects, but not exclusively. SC 18: *Document Processing and Related Communication* and SC 24: *Computer Graphics and Image Processing* also deal with certain multimedia-related activities. Under the umbrella of the International Telecommunication Union (ITU) there is another standardization framework, and at European level yet another one in the form of the European Telecommunication Standards Institute (ETSI). Since 1991, ISO, IEC and ITU have been attempting to coordinate their activities within the framework of the *Joint Technical Advisory Group 2 on Image Technology* (JTAG 2).

Within the ISO/IEC framework, several parts of the International Standards ISO/IEC 11172: *Information Technology — Coding of moving pictures and associated audio for digital storage media at up to about 1.5Mbit/s (MPEG-1 Systems)* and ISO/IEC 13818: *Information Technology — Generic coding of moving pictures and associated audio information (MPEG-2 Systems)* have been published or are under preparation. The latter provides for the multiplex and synchronization of multiple audio, video and data streams. It belongs to the series of International Standards on multimedia application being developed in ISO and IEC, in close cooperation with ITU, by the Moving Pictures Coding Experts Group (MPEG).

With new transmission technologies that allow for high bandwidth connections and switching technologies such as ATM (Asynchronous Transfer Mode), MPEG-2 Systems will form the basis for new audio and visual applications in coming years. Many industry groups have contributed to the development of the standard, which augurs well for its widespread implementation to the benefit of the suppliers of multimedia services, and to their customers. ITU has participated in development and is to adopt the standard as ITU-T Recommendation H.222.1, part of its broadband-ISDN audio-visual telecommunications terminal. Further within the ITU framework, Recommendation H.261 *Video codes for audiovisual services at px64kbit/s* was published in 1993.

Recently the working draft *DSM-CC (Digital Storage Media — Command and Control)* extension was developed and a call for proposals was issued for *MPEG-4 (AOE: Application and Operational Environment for very low bit rate coding)*. MPEG-4 will be an audio-visual coding standard covering new functions, such as support for manipulation of the content of audio-visual data. It will facilitate the integration of diverse multimedia and hypermedia developments, including entertainment, distance learning, remote monitoring, and home shopping.

The Multimedia and Hypermedia Coding Experts Group (MHEG) active within JTC 1/SC 29 is focussing its activities on multimedia application in PCs and especially on various interfaces among PC components and between PC components and external audio-visual components and systems, as well as on the user interface (in particular Graphic User Interfaces (GUI)).

Increasingly, electronic documents and publications can be combined with audio-visual data. This is the reason why the nature of *documents,* which traditionally consisted of *texts*, is changing rapidly. The International Standard *Information Processing — Text and Office Systems — Standard Generalized Markup Language* (SGML ISO 8879) published in 1986 has, therefore, been extended towards ISO 10744: *Information Technology — Hypermedia/ Time-based Structuring Language (HyTime)* in 1992. The same applies to the *Open Document Architecture (ODA),* which has to be extended in the direction of HyperODA to accommodate Computer Supported Cooperative Work (CSCW) by means of Audio Graphic Conferences (AGC) and Local Area Network (LAN) applications.

Even before standardization is accomplished, the INTERNET is providing a framework for using all the emerging technologies and methodologies, but not without certain risks. Inasmuch as standardization within the traditional frameworks of ITU, ISO and IEC just cannot cope with finding timely solutions to evolving problems due to the shear quantity of problems emerging in the course of multimedia and hypermedia development, new organizational frameworks such as the *Interactive Multimedia Association* (IMA), the *Multi-Media Communication Forum* (MMCF), the *Multimedia Communications Community of Interest* (MCCOI), the *Digital Audio-Visual Council* (DAVIC), etc. started activities in 1994.

ISO has addressed information presented in the form of documents and publications. Focussing on the methodology of handling information, ISO/TC 46: *Information and Documentation* is preparing quite a number of standards covering library and archive management. ISO/TC 37: *Terminology (principles and co-ordination),* whose secretary I am, is responsible for rules and methods for preparing, processing, exchanging, and harmonizing terminologies, as well as for the coordination of terminological activities. ISO/TC 37 International Standards not only treat subject-field communication, but also access to specialized information and the ordering of knowledge by means of terminology.

As a rule, the methodology-oriented standardization efforts of ISO/TCs 37 and 46 lag much more behind technological development than those of JTC 1 because they are primarily concerned with IT standardization. This situation can lead to serious problems with regard to user-friendliness in general and to the preparation and management of, as well as access to, information. Especially the need for new methodological approaches to multilingual information processing and access are still under-represented. This has been recognized within the European Standardization framework of the European Standards Committee, where CEN/TC 304: *Character set technology* was charged with the standardization of an amazing number of problems related to the processing of information written in different languages and writing systems.

Major components of the International Standards in the above-mentioned standardizing frameworks concern the identification of parts of (as well as kinds of parts of) information in multimedia and hypermedia environments. These efforts could and should become the basis for the formulation of a new copyright provisions framework, which the World Intellectual Property Right Organization (WIPO) is attempting to achieve, although at present WIPO seems to be taking a different approach, which might in the end make things more complicated than they are already.

5 MULTIMEDIA APPLICATIONS IN EDUCATION AND TRAINING

In theory, multimedia and hypermedia technology and methodology offer hitherto unexperienced and virtually unlimited possibilities for combining and extending conventional (rather static) representations of information and knowledge (on whatever medium) in the direction of highly dynamic—even interactive—forms of representation, combining levels of expertise in the process and overcoming all kinds of communication barriers. This trend does not, however, make teachers obsolete. Not all educational processes can be replaced by multimedia systems. In fact, they should be considered as complementary from the outset. Nevertheless, they are going to revolutionize more or less all aspects of our existing educational systems, probably less rapidly than expected, however, due to the inner inertia of our society.

In school education these technologies could provide solutions:

- at different educational levels (i.e., primary, secondary and tertiary school levels)
- in different types of schools (i.e., schools oriented toward the humanities, the natural sciences, etc.)
- for different types of students (from the point of view of ability, talent, etc.)
- in different subjects (e.g., musical education, mathematics, etc.)
- for special situations (e.g., students sick at home or hospitalized, etc.)

All kinds of remote education and training will—or could—become more efficient and effective beyond expectations. There are many indications in Japan and in other parts of the world that children are very receptive to computer-assisted learning and instruction methods and are quite eager to use computers. Some potential negative side-effects have already been recognized and could be counterbalanced by appropriate organizational and pedagogical measures.

Traditional school buildings and equipment could become obsolete in this environment. Teachers could handle more students and concentrate at the same time on a few who need special care (whether handicapped or especially talented, temporarily hospitalized or living in remote areas). Social relationships at school will also change. Advanced students— irrespective of age— could function as teachers, while teachers could study (e.g., in the form of further education or retraining or just for fun)—again irrespective of age—*under* someone who might still be a student in the present system.

Other changes may also be in the offing:

- Vocational training will also undergo many changes. Many subjects and learning processes will no longer need to be taught in the classroom.
- Tests could become more or less obsolete if interactive systems are designed to "recognize" student proficiency levels.
- All systems that convey *reference knowledge* in some way or other (e.g., museums, encyclopedia, dictionaries, scientific films, etc.) could be digitalized to a large extent.
- The preparation of teaching material in digital form for interactive use is already facilitated to quite an extent.
- Culture (in the form of theater, opera, etc.) can be put at the disposal of everybody.

The "dark sides" of these trends are still not well investigated and probably will surface only in the course of development. Furthermore, there are formidable barriers to rapid development towards a *multimedia society* (including education and training) such as:

- the current need for still unimaginable and incalculable amounts of investment for completely restructuring the educational system
- the need for thousands of teachers and trainers to retrain present teachers and to teach and train future generations of teachers in every major country
- the costs for equipment, software, and information necessary for each individual
- unsolved problems of assured energy supply, as well as network and data security
- the necessity to fundamentally rethink pedagogy and didactics
- the inappropriateness of the user-interface as it is today

Only some of these problems are tackled in the respective Japanese literature, which contains an immense volume of practical information, demonstration models, and investigations of existing situations (at home and abroad), but there is little material focussing on technology assessment aspects. Since multimedia development is going to revolutionize our society (including the educational system in all its aspects) on a world-wide scale, I would consider it highly instrumental to establish joint efforts to study the effect of this development carefully while at the same time increasing efforts to introduce the obviously positive aspects of multimedia technology and methodology into the existing educational frameworks (including vocational training). In this connection, research and development should—at least to a major extent—be concentrated on the development of multimedia methodology in education, i.e., on multimedia pedagogy and didactics, accompanied by empirical investigations of the respective effects on students. At the same time, a system for training teachers and trainers to use multimedia technology and methodology needs to be planned and implemented.

The literature consulted (research reports, government white papers, industry reports, public surveys, etc.) gives the impression that Japanese institutions and scholars alike are focussing their attention on technological development (and specifically on the positive aspects of it) rather than viewing the situation from the holistic and systemic point of view required in the light of the impact of multimedia technology development on our society as a whole.

6. CONCLUSIONS

There are many indications that the USA is still expanding its lead in the development of multimedia and hypermedia, together with related technology, software and applications. Japan, as well as Europe, is lagging behind also with regard to the most "intelligent" (i.e., sophisticated and at the same time user-friendly) and imaginative use of multimedia.

There is the real danger that the *entertainment* component in the concept of *edutainment* is taking precedence over the education and training aspects. It looks as if investment is poured into the infotainment development and activities rather than into the development of multimedia applications for education and training *per se*. Governments, therefore, are called upon to exert their regulatory role in order to prevent socially harmful developments.

Given the importance of multimedia- and hypermedia-based educational software and applications for the development of the human society, every country interested in multimedia application in education and training (every country should in fact be interested) should adopt a systematic approach to:

- standardization efforts
- evaluation efforts
- upgrading efforts.

It would be extremely useful to establish *Evaluation Criteria for Educational Multimedia*, if possible within the framework of international cooperation.

Some of the existing bottlenecks in and barriers to faster multimedia development include chaotic software development, lack of interchangeability and compatibility of hardware components, and excessively difficult handling (including the need for software adaptation) of individual devices. Standardization in the field of multimedia is therefore becoming more important than ever, but there is a definite need for increased coordination of standardization efforts, especially at the international level. There is a need for a better mechanism for (formally or informally) enforcing standards, and, last but not least, for coordinating multimedia and hypermedia standardization with intellectual property rights legislation.

More emphasis needs to be put on methodology, e.g., with regard to the content-related ordering of information, not only on hardware and software related activities. There is also a need for national research and development programs in the field of multimedia and hypermedia didactics as well as for the training of teachers and trainers for retraining present teachers and trainers as well as for training future generations of teachers and trainers in multimedia methodologies.

NOTES

1. The author wishes to acknowledge the work of Sue Ellen Wright and Alice Mitchell in editing and revising this article.

REPORT ON ARABIC-ENGLISH TRANSLATION WORKSHOP (2 RUNNINGS) GEORGETOWN UNIVERSITY ARABIC INSTITUTE, SUMMER 1994 & SUMMER 1995

Shuckran Kamal and Jacqueline Murgida
Georgetown University

Keywords: Translator Education, Arabic-English Translation, Georgetown University, Meaning-based Translation

Abstract: The authors planned and designed an intensive, workshop-style course in Arabic-English (A-E) Translation, which they taught in the 1994 and 1995 summer Arabic Institutes at Georgetown University, Washington, D.C. This paper discusses the need for such a course and the goals it hoped to achieve. It describes course content and methods, highlights the differences between the two runnings of the course, lists the changes that were made to improve the second running, and proposes options for future courses and workshops in Arabic translation.

1. THE NEED FOR EDUCATION/TRAINING IN ARABIC-ENGLISH TRANSLATION

1.1 Inconsistent Quality

The authors have worked for many years in the field of A-E translation as translators and as editors, revisers, and evaluators of translated texts. In performing these tasks over many years they have observed that while there are many excellent translators working from Arabic to English, a large volume of the translation work that is being offered by translators is poor and mediocre. An analysis of the quality of such translated texts indicates an inadequate command of English writing skills, Arabic reading comprehension skills, and/or sound translation principles.

The quality of A-E translations is inconsistent, in our experience. Most of the translations we were asked to evaluate required a certain amount of editing and re-writing. Some, even those completed by working, professional translators, were found to be completely unacceptable.

This appears to be due partly to the fact that most people working as A-E translators have had little or no formal training. We, ourselves, started out as translators with no training or preparation. Whether native speakers of Arabic or English, people are usually working in this field because they know both languages, and they are expected to learn how to translate on the job. Such people often had no desire to become translators; they became translators by default.

Such translators tend to receive on the job training from colleagues assigned to review their work or from clients or translation agencies who employ them or engage their services and offer them feedback on their work. Although there are talented individuals who have developed excellent translation skills through on the job training, such training is, by necessity, usually haphazard and not comprehensive. It is usually necessary or at least desirable to have some formal, focused training in various aspects of the translation profession to complement one's practical experience. In the absence of such training, working translators, including the authors of this report, have to rely on independent reading and study to improve and fine-tune translation skills. The few people we know of who have had some translation training before starting to work as translators took courses outside this country or took a general translation course in the United States in which they were the only students working with Arabic.

Many people assume that anyone who has a sufficient knowledge of two languages is automatically capable of translating between them. That is why they do not recognize that people who translate need formal training in translation principles. In our experience, many of our colleagues are working in a vacuum, with little or no knowledge of sound translation principles and an inadequate command of the source or target language (and sometimes both languages). In addition, a significant number of working translators are not aware of the advantages of membership in the major professional organizations for our field. Consequently, they deny themselves opportunities for professional development and enriching contacts with fellow translators.

Again, we, ourselves, at one time were in a similar state of ignorance about the art and the craft of translating, and we did not know about the existence of our professional associations. The major professional associations for translators which we have now in the United States either had not yet been formed or were in their infancy when we began work. Since then, outreach activities by the American Translators Association and other groups have been extensive. These associations have done a great deal to help translators take the initiative to enhance their professional qualifications and improve the quality of their translation work.

1.2 Lack of A-E Translator Education Programs

To our knowledge, there are no formal academic courses or certificate programs in Arabic-English (A-E) translation that are offered on a regular basis in the United States. Students of Arabic as a Foreign Language (AFL) are often exposed to translation exercises only as a class activity intended to demonstrate their understanding of an Arabic text. With the current emphasis in AFL classes on developing students' oral proficiency, the exposure that many students have to such translation exercises is minimal due to their instructors' attempts to use only Arabic in the classroom.

Even when AFL students are asked to do a translation exercise, the purpose of the exercise is usually to demonstrate that they understand the Arabic text. They are not expected to produce an accurate, idiomatic English rendering of the original Arabic. Furthermore, most of these translation exercises are oral, not written: an Arabic text is assigned, and students come to class prepared to read the text aloud and do "a sight translation" in class. This kind of in-class activity was the norm rather than the exception in a previous generation of AFL teaching.

Because the goal of AFL classes is certainly not to teach students how to translate into English or to write clearly and effectively in English, students generally emerge after several years of studying Arabic in the university at the undergraduate and graduate levels ill-equipped to translate. Although they may be able to carry on a conversation in Arabic and to read and comprehend Arabic texts, they may lack the extensive vocabulary and the firm grasp and command of the syntax of both source and target languages which they must have to analyze Arabic texts effectively and then translate them accurately into English.

Another group of translators to be considered is native speakers of Arabic. While some have acquired a high level of competence in writing in English, many have not. Some of those who lack such a high level of competence in writing in English are working as translators into English. Without adequate editing and revision, the work of such an A-E translator can be a recipe for disaster. The converse of this problem is that clients and employers who are unable to evaluate such "translation attempts" for accuracy and appropriate style seem to get accustomed to a poor quality of English in translated texts. Some go so far as to blame the Arabic original for causing the translation to read poorly.

1.3 Demand for A-E Translation Services

The need for high-quality A-E translation in business, government, and industry is urgent and recognized. There is also a need for competent translators who can translate Arabic literary works into eloquent, well-written English that conveys the intent of the original as truthfully and completely as possible. All types of A-E translation, if done well, can increase Americans' understanding and appreciation of Arab culture and civilization and can provide needed information for American industry and government. Few Americans study Arabic, and among those who do fewer still make the effort to achieve the skill levels in both Arabic and English that are required for translator competence. Therefore, the shortage of highly-skilled professionals who can provide needed, quality translation services in the United States is real.

1.4 Decision To Develop Course

We therefore decided to develop a program of study in Arabic-English translation. We started with a half-day workshop which was presented in the 1992 Middle East Studies Association (MESA) annual conference. Almost two years later, we were ready to offer the first intensive summer A-E translation workshop at Georgetown University in Washington, D.C.. We believe that a series of workshops, courses, and seminars offered at regular intervals could provide a remedy for the problem of poor and inconsistent quality in Arabic to English translations in the United States. Ideally, a certificate program offered by a university translation program in the United States would go a long way to alleviate the problems described above.

2. COURSE GOALS

The goals set for this course by the instructors were:

--To provide working translators and advanced AFL students with a structured learning atmosphere that will enable them to improve their own translation competence and work more effectively with employers, clients, publishers, and colleagues.

--To impart to students a methodological understanding of the nature of the translation task and a better appreciation of sound translation practices and theory developed by scholars and working translators. This would help them acquire a conscious awareness of what they are doing when they translate. Instead of translating "by the seat of the pants," they would know what steps they are taking in translating a text and why.

--To take the first step in a longer term program that would eventually give students the tools and the ability to pass professional accreditation exams or to receive certification on a par with that of Georgetown University's Division of Interpretation and Translation.

--To provide a first-level translation skills course that focuses exclusively on skills associated with the translation process to bridge the gap between on the job training and/or advanced AFL courses and translation courses for people who already possess the requisite linguistic skills in both languages.

3. COURSE DESIGN, CONTENT, AND METHODS

3.1 The Following Elements Were Shared By Both Runnings of the Course. Differences are addressed in section 3.2 below.

--The workshop format: Instructors emphasized in-class translation exercises and conducted discussions of translations done as homework assignments.

--Students were assigned translation homework on most class days.

--Students were asked to read at least the first 268 pages of Meaning Based Translation, by Mildred Larson, covering "Overview of the Translation Task," "The Lexicon," and "Propositional Structure."

--The instructors and invited guest speakers lectured briefly on specific topics relevant to translation, in general, and to A-E translation in particular.

--The instructors prepared handouts on specific topics (e.g., suggestions for identifying and translating problem words in Arabic; articles on how to use dictionaries and reference works effectively; and a basic bibliography, which included selected dictionaries and reference works helpful in A-E translation, as well as general works relevant to translation).

--The instructors placed considerable emphasis on Larson's maxim "translate meaning, not words." Consequently, they devoted considerable time in class to exercises on analyzing Arabic sentences and "unpacking" grammatical structures and lexical items to determine meaning. They also emphasized restatement exercises which gave the students opportunities to practice conveying the same (or similar) meaning in different ways in English. Significant time was also spent on discussions aimed at creating an awareness among students of the different meanings, nuances, or emphases achieved by different restatements.

--The instructors tried to create an awareness of effective use of dictionaries, reference works, and other resources, through discussions of student translations and focused presentations by the students themselves.

--The students were introduced to different genres and text types through assignments and in-class exercises. To do this, the instructors used news reports, editorials, political analysis articles, academic diplomas, a court brief, the text of a law, and literary passages.

--The instructors provided exercises in editing and revising translations and in working as a team. The aim was to have native-English/native-Arabic translators use each other's strengths to edit each others' work and produce a quality translation.

--The instructors assigned a class project which required students to select one of two texts to translate as a team; students were asked to divide the work among themselves and to decide on the translation, editing, and revision steps they would follow.

--The instructors and invited guest speakers made presentations and led discussions on the effective and appropriate use of computer technology for translation-related tasks. Topics covered included word processing, terminology software, Internet and online services, online dictionaries and references, and the expected advent of Arabic online dictionaries and machine translation.

--The instructors and invited guest speakers made presentations and led discussions about professional associations. Topics covered included membership benefits, certifications, and kinds of services and opportunities offered by such organizations as the American Translators Association, the American Literary Translators Association, the Translators Guild of the AFL-CIO, and the professional translators' groups in the United Kingdom.

--The two instructors exemplified the team approach they are trying to promote: one is a native speaker of Arabic, and the other is a native speaker of English. Both were present at all classes. The experiences and strengths of each complemented those of the other and enhanced the competence of both.

--In both runnings, the number of students was small (four in 1994, three in 1995).

--The instructors stressed the importance of the appropriate use of all types of dictionaries (including monolingual, bilingual, and specialized) and other reference works.

--The instructors placed on reserve at the library selected works from the bibliography.

--Students were encouraged to adopt a holistic approach to the text to be translated and to the task of translating it, even though it was necessary in a pedagogical context to consider the interlocking facets of the text and the translation process in isolation.

--The course carried 3 credits, in intensive summer session, workshop format.

--As anticipated, there was considerable variation in the level of students' language skills (in Arabic and English) and the level of their translation skills and experience per se. Also as expected, each class had at least one native speaker of Arabic.

--Some students were working translators, taking the course to enhance professional qualifications, and some were students adding to their academic preparation. One student was in both categories. Most of the students continued working in their regular jobs or in a summer internship while attending the course.

3.2 Differences Between the 1994 and 1995 Workshops

After receiving student feedback and conducting their own evaluation of the first running, the instructors made several changes to both the format and content of the 1995 running. Other factors also affected the changes that were made.

--The format of the 1994 workshop was very intensive. The class met for 3 hours per day (9:00 a.m.-12:00 noon), 5 days per week, for 3 weeks (July 11-29) during the second summer session at Georgetown University. Consequently, time in between classes was insufficient for students to complete both a translation and a reading assignment and absorb the material presented in class. The instructors found themselves with no choice but to reduce planned reading assignments. Accordingly, they covered more material in class presentations and discussions and decreased the number of texts assigned for translation outside class.

For the 1995 workshop, the format was changed as follows: class meetings for 3 hours per day (9:00 a.m.-12:00 noon), 3 days per week, for 5 weeks (July 10-August 11). Feedback from students and the instructors' observations indicate that this format was much more effective. Students were comfortable with the pace of the class and the work load of their assignments, even though two out of the three students did continue working at jobs during the course.

--There were four students in the 1994 workshop: two native speakers of Arabic and two native speakers of English. They represented a wide range of linguistic ability and translation experience/competence. Their level of competence in understanding written Arabic and in writing English was quite disparate. Such disparities necessitated detailed and prolonged discussions of ambiguous terms and complex grammatical structures.

In the 1995 workshop there were three students: one native speaker of Arabic, and two native speakers of English. The disparity in students' abilities in both languages and also in translation skills was not as great as it was the previous year. As a group, the 1995 students required less time to absorb the concepts that were introduced in discussion of ambiguous terms and complex grammatical structures.

--In 1994, individual texts chosen by the instructors for translation exercises and assignments ultimately drove the course content. Because the students could not handle reading and translation assignments, topics were covered as they came up in class discussions of the translation assignments and exercises. At times, these discussions were too diffuse and unfocused. Consequently, the students as a group were prone to go off on tangents, and discussions about ambiguous terms or complex sentence structures were sometimes tedious and prolonged.

In 1995 translation texts and invited guest speakers were presented (as much as possible) according to the sequence of assigned readings in Larson's text. Class discussions focused on a limited number of individual topics (e.g., lexical choice, idioms and metaphors). Other topics associated with the texts under discussion were deferred and addressed later at a more appropriate time. This approach to structuring the various topics of the course and the progression of these discussions worked well. The students and instructors were more comfortable with the course content, organization, and pace than they were in 1994.

--In 1994, the course description in the university's catalogue cautioned the students about the homework they would be expected to complete outside class.

In 1995, information on the amount of work students would be expected to complete outside class was reworded and given boldface emphasis to give prospective students a more realistic idea of how much work outside class the workshop would require. This was done to discourage students from making excessive commitments that would interfere with their performance in this workshop.

--In 1994, four speakers were invited to address the class and answer the students' questions on the following topics: translating and interpreting at the U.S. State Department; legal/commercial interpreting and terminology development; translator organizations; computer technology, Internet, and machine-aided translation.

In 1995, three speakers were invited to address the class and answer the students' questions on the following topics: translating and interpreting at the U.S. State Department; legal/commercial interpreting, terminology development, and translator organizations; literary translation (This was not a formal talk, but rather a discussion of literary translation assignments with an accomplished, prize-winning literary translator who contributed his unique insights to the discussion.)

4. FUTURE POSSIBILITIES

The instructors would like to develop follow-on courses and different formats for A-E translation training to reach a wider range of prospective students. They have identified the following possible formats and approaches to expanding the available training and education opportunities for people seeking to learn about A-E translation and for A-E translators seeking to improve their translation performance.

The same basic content could be offered in roughly the same amount of time (a semester or less), but at different times of the year. Students who are not able to take such a course in the summer could do so at other more suitable times.

More advanced courses could be developed. These could be designed to prepare students for taking certification exams. Such advanced courses would have no students who still need ESL or AFL study before they are capable of doing adequate translation work.

4.1 Modularization and Customization

The basic content of the 1995 course could be broken up into modules. Each module would have a specific focus and would target students at certain levels. The content of each specific module would be modified accordingly. More advanced or specialized topics can also be offered in modular formats. The following are some possibilities:

--Analyzing Arabic sentences -- identifying the propositions and the relations between them in preparation for translation, restatement, editing, and revision.

--Lexical items and terminology -- techniques for finding lexical equivalents, determining the correct translation of domain-specific terms, and developing terminology and glossaries.

--Writing/editing/revising/proofreading skills for A-E translation.

--Teamwork -- how native speakers of English and native speakers of Arabic can work together to translate texts and edit each others' translations for the purpose of improving quality.

--Computer-assisted translation, translation software, and using Internet; software demonstrations by vendors.

--Technical and semi-technical translations

--Translating in specialized fields, such as advertising copy

--Literary translation

--Getting the most out of dictionaries, reference aids, and other resources

4.2 Possible Formats

These modules could be given in a number of time formats which students in academic programs, working translators with full-time jobs, and free-lance translators would find convenient for their schedules. For example, a few half-day or all-day sessions could be offered in conjunction with academic and professional conferences, such as the annual conferences of the American Translators Association (ATA) and the Middle East Studies Association. Modules could be offered in clusters on consecutive days or over a series of weekends or weeknights to accommodate working people. Some modules (standard or customized) could also be offered during working hours on the premises of a business or organization or close to one.

The modules could also be offered in other geographic locations. They could also be sponsored by national or local translator associations and offered in conjunction with their regularly scheduled activities.

5. Conclusion, Acknowledgements

The instructors believe that they have made a small but encouraging beginning in what they hope will be a longer-term effort to provide opportunities for students and professional translators to receive the education and training that are necessary for providing clients with quality Arabic-English translations.

The instructors would like to thank Georgetown University for giving them the opportunity to teach these two workshops over two consecutive summers. They especially would like to thank Professor Karin Ryding, Professor Belkacem Baccouche, Mr. David Mehall, and Ms. Elizabeth Salernou of the Arabic Department. They would also like to thank Dr. Esther Ryder and Ms. Jessica Madrigal of Georgetown University's School for Summer and Continuing Education. The instructors wish to thank Dr. Marilyn Gaddis Rose, David and Margareta Bowen, and Dr. Peter Krawutschke, as well for their encouragement and suggestions during the early days of planning the workshops.

Last but not least, the instructors would like to thank Dr. Mildred Larson. Her book, "Meaning Based Translation," has been the mainstay of their translator self-education efforts, as well as the cornerstone of the two workshops. Without it, their task of teaching the workshops would have been immeasurably more difficult and tedious.

INTRODUCTORY TEXT SELECTION FOR INTERPRETER TRAINING

Nancy Schweda Nicholson
University of Delaware

1. INTRODUCTION

Choosing appropriate materials for use in the interpretation classroom is a task which faces every instructor. The level of difficulty must be graduated, progressing from straightforward texts with few terminological problems to increasingly more complex speeches/testimony/interviews. The trainer must strive, especially at the introductory stage, to create a delicate balance of (a) challenging her students and (b) building their confidence, both at the same time.

The current paper (1) offers general guidelines for introductory text selection and (2) illustrates the characteristics of suitable beginning materials via an excerpt of a videotaped speech which was delivered at the Delaware-Panama Partners of the Americas "Democracy Education Seminar" held at the University of Delaware, November 2-9, 1991. The goal of the seminar was to provide twelve Panamanians with an historical background as well as a current overview of democracy in the United States. Participants were primarily educators at all levels but also included a sales manager for a hospital medical plan as well as a Special Assistant to the President of the Republic of Panama. These individuals were selected because they were judged to be in a position to influence democracy-building in their communities when they returned to their home country.

Transcribed in Appendix 1, the excerpt chosen for analysis and comment was taken from a speech entitled "Democracy in Latin America" presented by Dr. John T. Deiner, Associate Professor of Political Science at the University of Delaware.

Although the author has used this text for both consecutive interpretation (CI) and simultaneous interpretation (SI) practice, the current paper describes its suitability for beginning long CI. Prior to the introduction of such a text, students have practiced note-taking with both long sentences and paragraphs of increasing length.

2. GUIDELINES FOR INTRODUCTORY TEXT SELECTION

2.1 The Text[1] Itself

When choosing material for initial practice in either SI or CI, the instructor should consider the following relevant factors which are specifically related to the internal composition of the text:

(1) topic: It is a good idea to select a topic that is general in nature and one which is familiar to the students. The subject matter of the current text is "democracy in Latin America". The concept of "democracy" is a basic one and does not present a particular challenge for trainees. "Democracy in Latin America" is also an issue which is frequently in the news, often in the context of the establishment or the overthrow of a military government. It is generally known to students that democracy is a goal in Latin America, and that a struggle in this regard has ensued there for many years.

(2) terminology: In keeping with the selection of a general topic, the words used to describe that topic should be relatively basic and not typified by specialized terminology. Of course, it is not possible to avoid all potentially problematic lexical items, but the primary goal is that they should not appear regularly. The current text includes very few such difficult words: "ousted" and "dictatorship" are really the only two that may pose a problem. It should be noted, however, that prior training in "cognitive flexibility" has given students practice in accessing many different ways to express a single idea (Ref. 1). In this connection, "ousted" can also be rendered as "overthrown", "thrown out", "toppled", and so on. Trainees are taught not to be tied to the words, but rather to focus on the meaning and take advantage of the great flexibility which languages provide to represent synonymous concepts. "Authoritarian" might be troublesome in some working language combinations but, in English, Spanish and French, the word is a cognate.

(3) argument development: Argument development should proceed logically and, when relevant, chronologically as well. It is helpful to choose a text that is easy for the students to follow, perhaps one in which the speaker/author delineates points clearly ("I will cover four major topics during my talk today. They are: ______."). Prof. Deiner's introduction offers an historical perspective on the status of democracy in Latin America. He begins by mentioning the long tradition of both democratic and authoritarian rule. He speaks about the 1950s, then about the 50s and 60s, proceeds to mention the 60s and 70s, refers to the past decade, and finally brings the listener into the present by briefly alluding to the situation in Haiti. Dr. Deiner then sets the stage for the discussion which follows in the second section, one which highlights the basic elements of democracy.

The author has written before on the concept of "contextualized" vs. "non-contextualized" information (Ref. 2).

It is worthwhile to briefly mention its relevance to topic, terminology, and argument development. Contextualized information (that which is often predictable and follows logically in the argument development) is the primary type of information contained in this text. The speaker mentions past and present situations, and the transitions from one to another are easy to grasp. In other words, the material is basically well-connected and flows along smoothly. Examples of non-contextualized information (statistics, dates, proper names) are included, but they are quite straightforward. As a result, the students are not challenged beyond their fledgling capabilities. Some of these non-contextualized items (which help to tell the story but which are largely unpredictable and, unless noted, will probably not be retained in one's working memory) include the dates ("50s", "60s", and "70s") and a couple of the proper names ("OAS" and "Prof. Boyer"). The other proper names are extremely common and would undoubtedly pose no threat ("United States", "Caribbean", and "Haiti"). All in all, from the contextualized vs. non-contextualized perspective, this text is, in general, highly contextualized and presents no serious problems to the trainees. What Prof. Deiner **could** have included which **would** have constituted difficulties are names of country presidents and military leaders, as well as names of revolutionary groups, involved statistics, and so on. None of this type of information is present, which helps this text to be an excellent one for a beginning consecutive note-taker and interpreter.

The aforementioned three factors play a role in the selection of any practice text, whether it be written (such as a newspaper article which is edited and prepared as a talk) or oral (such as the instructor speaking extemporaneously about a favorite pastime or a videotaped conference speaker addressing an audience on his life's work). Choosing a videotaped speech of someone other than the primary instructor or teaching assistant is an excellent idea. It exposes students to a variety of voices, accents, formats for content presentation, and speaking styles.

2.2 Delivery Format of the Text: The Speaker's Role/Input

The instructor's decision regarding text selection is only half of the picture. The other half (**how** that text is presented to the audience) includes a number of components:

(1) voice: When selecting a video for introductory interpretation practice, the speaker's voice should be loud and clear. There should be no incomprehensible concepts or sections because the speaker's voice is problematic. It is wise to select a native speaker of the source language (SL).

In this video, the speaker's voice is strong and he speaks loudly enough so that all can hear. Moreover, he articulates clearly, and no material is lost as swallowed words at the ends of sentences. The speaker is a native English speaker, so has no foreign accent.

(2) format for content presentation and delivery speed: There are essentially three possible formats for content presentation: (a) true extemporaneous speech (no notes); (b) primarily extemporaneous speech with the aid of notes or an outline; and (c) oral reading of a written text. It is also appropriate to add (d), which is any combination of (a), (b), and (c) above, inasmuch as speakers may alter their manner of presentation during the course of a speech. From the beginning interpreter's perspective, true extemporaneous or extemporaneous speech with the aid of notes is the best speaker format. This is not to say that **all** talks delivered in either of the above two ways will always be "do-able"; however, the novice has the best chance of success with these two because lecturers who present their material in this fashion are formulating their thoughts while speaking. Conversely, speakers who read prepared texts often deliver the material very quickly. Their often untenable speed is usually due to the fact that the speakers are not thinking about what they are saying when they read. As a result, the stage of ideation and message planning while concurrently speaking is virtually eliminated. (For detailed information on how to train students to prepare and work with texts, see Ref. 3. Moreover, for an interesting article on various kinds of speakers, see Ref. 4).

The speaker in the sample video is presenting his material extemporaneously with the aid of notes. Very importantly, he is **not** reading a prepared text word for word. His delivery speed is just about ideal, as it is neither too slow nor too fast. As illustrated in Appendix 1, Dr. Deiner's speed was 131 words per minute (wpm) during the first minute, 128 wpm during the second minute, and 64 wpm during the final 34 seconds.

(3) speaker style: Closely related to Point #2 above, the speaker's style includes elements such as eye contact, smooth or choppy delivery (pauses, hesitations, false starts, and so on), repetitions, rephrasings, gestures, and facial expressions. It goes without saying that eye contact is helpful, and a smooth delivery is more desirable than a choppy one (although pauses and hesitations can be time-gaining devices for interpreters - they buy more analysis time). It should also be mentioned that false starts can be confusing for interpreters and require them to do mental backtracking just as they begin to analyze an incoming idea. Repetitions (word-for-word reiteration) and rephrasings ("let me put that another way"; "in other words . . . ") can also be viewed as time-gaining devices. They help to reduce processing demands because the "new" information load is reduced. Gestures and facial expressions are viewed as extralinguistic components and may or may not be helpful as the interpreter strives to decode and understand. They could also be viewed as neutral, in that they neither add to nor detract from comprehension.

The speaker in this video maintains excellent eye contact with the audience which, the author believes, aids in decoding and understanding. His delivery is generally smooth with normal pauses and hesitations. There are none which interfere with the listener's ability to follow the meaning of the message. No false starts are present. Repetitions, however, are quite prevalent throughout the 2 minute 34 second excerpt. (They will be discussed in greater detail in the next section.) The speaker's style is characterized by a few gestures (i.e., when he talks about "conflict") and his facial expression is relatively constant. The presence or lack of gestures and facial expressions is generally related to the content of the material and the tenor of the delivery. The information imparted by the speaker is not emotional in nature but rather factual with a neutral tone. Consequently, the lecturer's message simply lends itself neither to wild gesturings nor marked changes in facial expressions.

2.3 The Role of Repetitions

If one examines the text in a more detailed fashion (see Appendix 2), one notes that there are a number of repetitions in this short 2 minute 34 second block. For example, the words "democracy" or "democratic" are repeated no fewer than 10 times. Moreover, "Latin America" or "Latin American" is also reiterated 10 times during the excerpt. The word "tradition" is spoken 6 times within the first 32 seconds.

These repetitions have several implications for novice interpreters. First of all, the reiterations reduce the processing load because students are not bombarded with a constant flow of new information. This permits more analysis time and also represents an excellent opportunity for trainees to focus on note-taking strategies. With the repetitions cited above (and others which will be treated shortly), students can put into practice what they have been taught about note-taking vis-à-vis the representation of repeated material. In other words, they already know that "Democracy in Latin America" is the topic of the talk. Before even beginning to listen to the speech, they may decide upon "D" to represent "democracy" or "democratic" and "LA" to stand for "Latin America" or "Latin American". Inasmuch as this is the primary subject matter of the speech, trainees can expect that these words will be mentioned frequently (and they are). With respect to "tradition", the note-taker may choose "trad" the first time it is spoken and then easily switch to "T" for subsequent occurrences. This strategy also works with "country/countries" (repeated 4 times) - "coun" could be selected initially, and could become "C" in succeeding utterances. In this connection, "Latin American country" then becomes "LAC" in the interpreter's notes.

The speaker also repeats "military governments" 4 times and "dictatorships"/"dictatorial" 3 times. For three of the four repetitions, Dr. Deiner uses these words together. The first time "military government" is heard by students, they may choose to abbreviate it as "mil gov". The second time, though, they can easily change to "MG". "Dictatorship" may be initially abbreviated as "dict". Realistically, there should be no concern about confusing this abbreviation with the word "dictionary" because knowledge of the relevant context should prevent the trainee from making this mistake. When it is heard for the second time, the natural choice would be to opt for the upper case "D"; **however**, "D" has already been selected for "democracy". The simple solution so as not to confuse these two completely opposite words is to decide upon "Di" or "Dic" for "dictatorship".

The text is also characterized by additional repetitions, although they are not as numerous as the ones mentioned above. Examples include: "There have been times in the (recent) past" (2 times); "many people are hopeful" (2 times); "an end to" (2 times); and "It's a very optimistic time" (2 times). The final example also constitutes an easy analysis/note-taking situation. The speaker says: "So, it's a very optimistic time for Latin America. It's a very optimistic time for democracy". Very little needs to be noted here, not only because of the repetition of the lead-in phrase, but also because of the abbreviations which have already been decided upon for "Latin America" and "democracy".

Another example which may be considered a type of repetition is Prof. Deiner's final statement: ". . . what kinds of lessons, especially, the United States can learn from the situation of democracy in Latin America, and what kinds of lessons Latin America can learn from the situation of democracy in the United States". Inasmuch as the second part of the utterance is a simple "vice versa" of the first part, this sentence is not only easy to analyze (as it follows very logically in a parallel format), but also requires very little note-taking.

There is also an example of rephrasing: "During the last ten years. . . during the last decade. . . ." Although not a word-for-word repetition as cited above, consecutive interpreters can select **either** expression for their rendition, because they mean exactly the same thing. Once again, the rephrasing aids novices by reducing the incoming processing load.

The repetitions in this speech excerpt allow beginning interpreters to focus a bit more of their attention on note-taking strategies without being under extreme pressure to analyze continuously new input. It is hoped that, with practice, these note-taking decisions will become almost automatic, thereby permitting most attentional resources to be devoted to listening, decoding, and understanding.

3. CONCLUSION

This short paper has examined strategies for introductory text selection for interpretation practice and has illustrated these strategies via a videotaped speech excerpt. To reiterate, it is critical that students not be given "undo-able" texts in the early stages of their training, as the instructor risks discouraging them. Trainees need to work on building confidence right from the start, and using a text like this one for practice helps to orient them toward achieving the important balance between listening/analyzing and note-taking. A "graduated difficulty" approach has proven to be most successful with students over the years. It is challenge enough for most novices to stand up in front of everyone else and give their rendition. Most trainees are neither schooled nor experienced in public speaking, and this aspect of consecutive interpretation is a source of anxiety for them. To add to their nervousness by requiring them to work with material which is too difficult is a counterproductive technique. It must also be remembered that, at these initial stages, students are still experimenting with note-taking as well.[2] Trainees are also working on further strengthening of their short-term memory (STM) as they explore its interplay with the notes they decide to take.

All in all, texts like the one selected provide novices with the opportunity to get on the right track when it comes to effectively dealing with the attentional demands inherent in the interaction of SL listening, analyzing, note-taking, and formulation of target language (TL) output.

NOTES

[1]"Text" is used to refer to either written or oral material.

[2]Building and refining one's personal note-taking system is, of course, an ongoing process which extends well beyond the formal training period and continues throughout one's career.

REFERENCES

1. Schweda Nicholson, Nancy. "A New Look at the Cognitive Flexibility Factor in Interpreter Training," in A. Leslie Willson, ed. *Proceedings of the 31st Annual Conference of the American Translators Association.* Medford, NJ: Learned Information, Inc. 1990, p. 107-117.

2. Schweda Nicholson, Nancy. "Consecutive Note-Taking for Community Interpretation," in David and Margareta Bowen, eds. *Interpreting: Yesterday, Today and Tomorrow.* Binghamton, NY: Center for Research in Translation, State University of New York. 1990, p. 136-145.

3. Schweda Nicholson, Nancy. "Documentation and Text Preparation for Simultaneous Interpretation," in Deanna L. Hammond, ed. *Proceedings of the 30th Annual Conference of the American Translators Association.* Medford, NJ: Learned Information, Inc. 1989, p. 163-182.

4. Taylor, Christopher. "Towards a Typology of Speakers," in Deanna L. Hammond, ed. *Proceedings of the 30th Annual Conference of the American Translators Association* . Medford, NJ: Learned Information, Inc. 1989, p. 183-188.

Appendix 1

TRANSCRIPT OF DEINER TEXT EXCERPT

When you think about democracy in Latin America, um, it's a very long tradition, and it's a tradition that in Latin America also competes with an authoritarian tradition, so there are really two traditions that have been in conflict for hundreds of years in Latin America. Right now, many people are hopeful that the democratic tradition will be the one that triumphs and, as Prof. Boyer said, many people are hopeful that the democratic tradition will become firmly established in the future in Latin America.

There have been times in the past, especially in the 1950s, when military governments were ousted from office. There have been times in the recent past when many people thought there would be an end to military governments in Latin America, an end to dictatorships, (**1 minute = 131 words**) and the establishment of democracy. But, democracy didn't succeed every place in the 1950s and 1960s and, in fact, Latin America suffered through a series of military governments and dictatorships during the 1960s, into the 1970s.

During the last ten years, though, during the past decade, we have seen military governments, dictatorial governments fall in just about every Latin American country. So that at the present time, civilians are in power in just about every Latin American country. And even in Haiti where the situation is confused, there is a very strong effort under way by the OAS and by the Caribbean countries to try and restore an elected civilian government in that country.

So, it's...it's a very optimistic time for Latin America, (**2 minutes = 128 words**) it's a very optimistic time for democracy, but at this time, I think it's important to think about what democracy means, what's needed for democracy to succeed and what kinds of lessons, especially, the United States can learn from the situation of democracy in Latin America and what kinds of lessons Latin America can learn from the situation of democracy in the United States. (**34 seconds = 64 words**)

Appendix 2

REPETITIONS/REPHRASING IN SPEECH EXCERPT FROM "DEMOCRACY IN LATIN AMERICA"

(Time of Excerpt: 2 minutes 34 seconds)

Repetitions:	**Number of Times**
"democracy"/"democratic"	10
"Latin America/"Latin American"	10
"tradition"	6
"country/countries"	4
"military governments"	4
"dictatorships"/"dictatorial"	3
"There have been times in the (recent) past. . ."	2
"in just about every Latin American country."	2
"It's a very optimistic time. . ."	2
"Many people are hopeful that the democratic trad."	2
"an end to . . ."	2
"what kinds of lessons. . ."	2
"can learn from the situation of democracy in. . ."	2
"United States"	2

Rephrasing:

"During the last 10 years . . . during the last decade"

SEQUENCING IN INTERPRETER EDUCATION

Carol Patrie
Gallaudet University

Keywords: Teaching interpreting, sequencing activities.

Abstract: Effective interpreter education relies on a carefully integrated sequence of instructional units in order to teach students how to gain control of their interpretations. As Schweda-Nicholson (1988) points out the whole of the interpreter training program experience does not always equal the sum of its parts. Thus, it is important to sequence instructional materials and experiences in a way that is supportive of cognitive development in adults. When educational experiences are organized in a sequential manner which meets the cognitive needs of adult learners, evaluation of student progress becomes less problematic and more useful. Materials designed to be integrated with this sequential approach may also have multiple uses throughout the curriculum. There are three tiers of variables which must be appropriately sequenced in effective interpreter education. These variables are; type of skill, student developmental level, and type of training material.

INTRODUCTION

The basic premise of this work is that effective interpreter education is componential rather than wholistic. It is analytical rather than gestalt. It relies on mastery of skills which are quantitative in nature rather than on successive approximations via imitation. Recently developed programs tend to follow the componential and analytical approach. These newer programs tend to be focused on structural patterns. Students are encouraged to internalize these hierarchical patterns rather than simply memorize and demonstrate learned behaviors.

These underlying premises have direct implications for the design of Interpreter Education Programs. Programs which are designed in accordance with these premises are more likely to be able to accomplish the extremely difficult and complicated task of training effective interpreters in a limited time frame. As Gile, (1995) points out, the goal is to maximize the training experience. One of the main ways to accomplish this is to separate out the discrete components of the interpreter education program and to sequence them in an appropriate manner. There is a growing body of literature to support the notion of sequencing specific skills within the interpreter training program (CIT Proceedings 1984; Lambert 1989; Roberts 1992).

This work suggests a framework which is based on a sequence of fourteen specific components within the interpreter education program. This proposed system may be modified to meet individual program needs. They are suggested here in a logical progression from simple to more complex. The eleven components that are addressed here are:

1. Language competence
2. Aptitude
3. Extralinguistic knowledge
4. Intralingual skills
5. Cognitive techniques
6. Consecutive interpretation
7. Simultaneous interpreting
8. Evaluative instruction
9. Backup interpreting
10. Professional teaming skills
11. Professional skills

These proposed components are sequentially independent by nature and thus, each has separate implications for time order and the development of related instructional tasks and materials. These components are embedded in a matrix which provides for variation in student skill level and difficulty of source material. A brief delineation of eleven components is presented. A fuller description of the student typology and materials will appear in *Interpretation: A Teacher's Manual* (Patrie, in preparation).

1. LANGUAGE COMPETENCE

Roberts (1993), "suggests that language competence includes the ability to manipulate, with ease and accuracy, the two languages involved in the interpreting process and is a prerequisite for successful interpreting . . . " The presence of language competence must be clearly evident for both Language A and Language B, where language A and language B represent the language pairs in the Interpreter Education Program. Language competence is ordinarily thought to include expressive proficiency. This includes familiarity of linguistic structure Gile, (1995) and adequate vocabulary to deal with various types of topics.

Language competence includes but is not limited to, comprehension of the source and target languages at fairly sophisticated levels. During the process of language acquisition, comprehension is ordinarily in place before expression of language. Also needed is a familiarity with a variety of linguistic structures such as topic-comment sentences, interrogatives, rhetorical questions and others. Language competence also assumes the ability to deal with information presented in a variety of sentence lengths, which in turn, range from simple to complex. One indicator of language competence is the ability to perform well on cloze tasks.

1.1 Cloze

Cloze exercises were originally designed as an approach to determining language competence. In "cloze" testing words are deleted in a systematic fashion. For example, every 10th word or every 5th word is deleted. This type of test is assumed to evaluate lexical, syntactic and semantic aspects of language processing." (Longley, p. 107. 1989) Students will predict the possible lexical items, phrases, and entire utterances in given places in a text based on grammatical constraints and context. This teaches students to predict incoming messages when parts of the message are missing. Cloze exercises are also appropriate for use in developing intralingual skills.

2. APTITUDE

After determining the presence of language competence, it is crucial to directly address aptitude for developing interpretation skills. It is a disservice to students to begin a training program without checking for aptitude in the areas related to interpretation.

As Longley (1989) suggests, the ideal candidate for interpretation training has a "natural or instinctive capacity or tendency" . . . to be able to demonstrate the characteristics which predict success in learning to interpret. Longley (1989) suggests the following seven aptitudes:

1) Excellent knowledge of A, B and C languages and cultures.
2) The ability to grasp rapidly and to convey the essential meaning of discourse, irrespective of the language spoken.
3) A memory which recalls the links between logical sequences of discourse.
4) The ability to convey information with confidence, complied with a pleasant delivery.
5) Broad general knowledge and interests, a curiosity and willingness to acquire new information.
6) Ability to work as a member of a team.
7) The ability to work under stress for long periods.

The presence of these aptitudes can be checked for via a series of tests. For additional information on these procedures, see Longley, 1989.

3. EXTRA LINGUISTIC KNOWLEDGE

According to Gile (1995; p.84) "The contribution of extralinguistic knowledge to comprehension does not level off at a certain point. . ." Gile suggests that the more one knows about a topic, the better able they will be to disambiguate the source text and "read between the lines" (p. 85).

Further, Gile notes that there are two types of extra linguistic knowledgy (ELK). The first is pre-existing ELK and the second is "knowledge acquired from the text itself " (p. 85). By being able to maintain high levels of attention, Gile suggests that the interpreter can gain and process new information during the interpreting event which can assist in the disambiguation process.

4. INTRALINGUAL SKILLS

Apart from facility in comprehension and expression, and next in the hierarchy of skills, is the area of intralingual skills. These skills include a variety of exercises in which the response is in the same language as the prompt or stimuli. Ideally, these exercises are given in both languages. Intralingual exercises include shadowing, paraphrasing, outlining, prediction and cloze, to name a few. As we have seen, cloze can also be used as an indicator of language proficiency. Three types of intralingual exercises are summarized below.

4.1 Paraphrasing

The goal of paraphrasing is to render the meaning of a message in words or signs other than those used in the original. It is also possible to provide exercises in syntactic paraphrasing, in which the order of clauses may be reversed or altered, provided that the meaning is not altered. Paraphrasing can be accomplished by lexical substitution or rephrasing.

4.2 Abstracting

The goal of abstracting is to summarize the main idea while ignoring extraneous detail. This prepares the student to synthesize and summarize information appropriately. The length of summary may be varied.

4.3 Prediction Tasks

The goal of prediction exercises is to encourage students to predict the possible content of texts based on limited contextual information. This is designed to provide students with the opportunity to prepare for interpreting assignments with a minimum amount of information and with maximal flexibility.

5. COGNITIVE TECHNIQUES

Cognitive techniques is the general term used to represent the various mental manipulations which an interpreter must be able to handle with ease. They are designed to enhance the ability to deal with divided attention tasks. Interpretation requires the ability to successfully manage division of attention across competing stimuli.

5.1 Auditory and Vision Screening

As a preliminary measure vision and audition should be evaluated in the student of interpretation, in order to determine that the source language can be perceived without distortion.

5.2 Shadowing

Shadowing is another term for repeating what is seen or heard. The ability to shadow provides an indicator of the ability to listen and speak simultaneously. In this instance the message is not altered, nor should it be processed for meaning extraction at beginner levels. At more advanced levels, memory for content of shadowed material may be evaluated.

5.3 Decalage

Decalage exercises are designed to provide practice in retaining information by artificially imposing a delay on the response time. These exercises are thought to increase management skills related to short term memory. Cokely (1992) suggests that interpreters with greater lag, or decalage, tend to make fewer errors. Exercises which require students to reproduce utterances of varying lengths while consciously maintaining a time or message interval behind the source language can increase control of processing time.

5.4 Dual Task

Dual task exercises require students to attend to more than one task at a time. These are done both intralingually and interlingually. When they are done intralingually, one or both tasks can be assessed for accuracy. For example if the two competing tasks are reciting a well-known poem and writing numbers (backwards or forward) then one or both of these may be checked for accuracy.

5.5 Digit Processing

Digit processing exercises require students to deal with texts which contain digits. Digits cannot be precisely predicted or inferred from context. They must be remembered exactly. For beginning level students approximations may be allowed, such as "less than one hundred for the digit 98."

6. CONSECUTIVE INTERPRETATION

Consecutive interpreting can be used as a training tool and as a professional skill. At first, glance it may seem inappropriate or undesirable as a professional tool to those accustomed to simultaneous interpreting services. This reaction is primarily due to a lack of preparation in consecutive interpreting at the training stages. In spoken language interpreting, consecutive interpreting enjoys a well-respected and unquestioned place. It's standing is not as secure in ASL - English interpreting, because its techniques and applications have not been explicitly taught. As a result, an entire generation of interpreters may have overlooked the value and appropriate use of consecutive interpreting in professional settings.

It may be that sign language interpreters who were working as professionals in the late 1960's and 70's were familiar with this techniques and used it as needed. This group of interpreters tended to be those people who had deaf parents and no formal training interpreting. Most for whom this description is accurate, were using consecutive interpreting even as children. This type of consecutive interpreting can commonly be seen in settings where children are raised bi-lingually.

6.1 Consecutive Interpretation of Monologues

Single speaker events provide the starting point for skills development in consecutive interpretation. There are several variables to consider when planning instruction in consecutive interpretation. One is the length of the utterance to be interpreted which can range from sentences to short speeches. A second variable to be considered is rate of delivery. Other variables include, but are not limited to, density of topic, presence of reported speech and accented speech.

6.2 Consecutive Interpretation of Dialogues

Training in consecutive interpretation of dialogues may begin with information regarding turn taking, which includes information which relates to overlap, interruption and non-attention by one or more parties within the dialogue. Another area of exploration which is directly relevant to consecutive interpretation is questions and answers.

Experiential instruction may be prepared for the following types of dyads. The first scenario is two speakers of Language A who are conversing with each other in the presence of a user of Language B. The second scenario is two users of Language B who are conversing with each other in the presence of a user of Language A. The third scenario is a speaker of Language A and a speaker of Language B wishing to converse with each other through the use of an interpreter. Each of these specific scenarios will require attention to the cultural and linguistic aspects of these types of dyads as well as the interpreting requirements for each of these scenarios.

6.3 Consecutive Interpretation for Groups

Within consecutive interpretation for groups, training must be provided in both Language A and Language B. A scenario where this may be used is often seen in large conferences where a speech may be delivered in its entirety or may be delivered by paragraphs and the interpreter is required to interpret it to an audience which may be large or small.

The variables which are listed under consecutive interpretation of monologues must also be included here. Participants of these groups should also be delineated and explained. Therefore, the groupings of audiences mentioned in section 6.2 may also be addressed when preparing instruction for consecutive interpretation for groups.

7. SIMULTANEOUS INTERPRETING

Within the components of simultaneous interpreting will be found interpreting for monologues, dialogues and in group settings. The factors delineated under consecutive interpreting for these three areas monologues, dialogues, and groups must also be addressed within the context of simultaneous interpreting. At this level, the issue of message preservation in the presence of simultaneity is most important. It is assumed that the information presented in relation to consecutive interpretation has been mastered at appropriate levels prior to moving into simultaneous interpreting.

8. EVALUATIVE INSTRUCTION

Evaluative Instruction is the component which deals with the study of models of interpretation. Although, models may be presented at the beginning of either consecutive or simultaneous interpretation, simply as a demonstration, evaluation of these models should be prepared after instruction and insight have had a chance to develop within the interpreting student. Models of interpretation may be evaluated either wholistically or at the proposition level. Wholistic evaluation includes evaluation such as overall view of message and naturalness of expression. At the propositional level each proposition which is interpreted will be evaluated for accuracy by comparing the source and target languages. Types of errors may be delineated. One way to categorize errors is to note omission of information, addition of information, or skewing of the message.

Once types of errors have been determined, reasons for errors may be inferred. Sample reasons would include: Inappropriate decalage, mis-comprehension of the source language, or lack of competence in the target language. Suggestions may be offered regarding the efficiency of processing in areas where errors occurred in the interpretation.

9. BACKUP INTERPRETING

This type of skill is commonly seen in sign language interpreting situations. Backup interpreting skills need to be taught for both Language A and Language B. Typically the "A" interpreter is the one who is delivering the interpretation and being assisted by the "B" interpreter. This scenario is often seen in ASL to English situations, and may also be seen in English to ASL interpretation settings, particularly in platform interpreting settings. In either case, specific instruction must be provided which deal with, at a minimum, the following ten factors:

9.1 Type of information to be provided by the backup

Information may be prepared in the form of a single lexical item, a phrase or a complete sentence. This information can be used by the A information to create a repair in the interpretation.

9.2 Format

The interpreter must determine whether it is appropriate to provide the back up information in a whisper, a low voice, or use normal volume. In the case of ASL to English interpretation, when the situation calls for backup, the format may be small, discrete signs, fingerspelling or larger signs, which are fed to the A interpreter who is typically on a stage or platform.

9.3 Methods of Backup Interpreting

It is generally the responsibility of the A interpreter to clarify how information is to be requested from the backup interpreter. In the case of ASL to English, typically the A interpreter provides some tactile indication that the backup information is requested. In situations where the A interpreter is working from English to ASL they may contact their backup through eye gaze or headnods as a means of requesting information. In any case, a variety of methods should be discussed and demonstrated with the caveat that fed information is not provided by the backup unless requested by interpreter A.

Interpreter B needs to be instructed in methods which relate to monitoring the source language while continuously monitoring the accuracy of the output of interpreter A. This practice will lead to greater accuracy in the event that backup information is required.

9.4 Amount of Information

The amount of information provided by the backup interpreter will vary in ASL to English interpretations. In general, more backup interpreting is used in ASL to English, than in English to ASL interpretation. Both of these types of information should alert the B interpreter to be especially aware and prepare to deliver information to their partner as needed.

In either ASL to English or English to ASL Interpreting, the A interpreter should have an awareness of how much information they would like to have their backup interpreter deliver. Methods can be devised in advance of an interpreting situation, so that the backup will know when to stop delivering information and the A interpreter can then proceed alone to deliver the remaining interpretation. In this context, it is most likely that assistance will be required for fingerspelled word recognition and digits. Backup is also likely to be required in English to ASL interpretations where the source contains numbers.

9.5 Timing of Information

Information must be delivered from the backup interpreter to the A or primary interpreter in a time efficient manner. This means that B interpreter must be ready to deliver the appropriate piece of information at a time which the A interpreter can effectively utilize it.

9.6 Control of the Process

This aspect of interpretation will come into play in either ASL to English or English to ASL interpreting. It is generally thought to be most effective if the A interpreter maintains control of the process. In situations where the B interpreter has not been able to fill in the requested or required information for A interpreter, it should be the A interpreter's responsibility to stop the speaker and request clarification.

9.7 Clarification

When it is necessary to stop the speaker, and the A interpreter has done so, specific methods of clarification must be used. In some situations the B interpreter is the one who may be most likely to know exactly what clarification should be requested. This is so because, the B interpreter is responsible for monitoring the source and the target and may be most quickly able to notice the discrepancy between the two. In any case, the B interpreter should provide suggestions for request of clarification to the A interpreter who should then ask the speaker for the clarification or repetition.

9.8 Repetition

In some situations the information was understood, but not enough time was allowed to process and deliver the interpretation. In such cases a repetition is needed rather than a clarification. Specific techniques for requesting repetition should be taught and evaluated. For example, it is rarely useful to ask the speaker to repeat the fingerspelled word. Effective uses of requests for repetition can be demonstrated for students.

9.9 Support from the B Interpreter to the A Interpreter

During ASL to English interpretation, support may take the form of slightly nodding the head so that the A interpreter can detect this non-verbal reassurance via peripheral vision. This reassurance may also take the form of subtle tactile pressure indicating that the A interpreter's interpretation is correct. In English to ASL interpretations, the B interpreter may nod and maintain eye contact within the sight line of the A interpreter.

9.10 Provision of Correction

In some situations the B interpreter will notice an error in the interpretation which the A interpreter did not notice. In this case, assistance has not been requested. In ASL to English interpretation, the B interpreter must find an effective place to insert the correction. The B interpreter must also use judgment and timing in order to determine if the correction is absolutely essential to the interpretation. Generally, these corrections must be inserted during a timeframe where the A interpreter is not also speaking. In the case of English to ASL interpretation the B interpreter is able to indicate that a mistake has been signed in the interpretation and prepare a clear visual correction in a useable format.

10. PROFESSIONAL TEAMING SKILLS

The area of Professional teaming skills includes and requires mastery of all the factors listed under backup interpreting. In addition to those, professional team members are also encouraged to demonstrate additional skills. The first is support for a team member while working, and during pre and post interpretation events that are directly relevant to the communication context. They are also expected to provide reasonable, useable feedback. Global feedback comments such as "nice work", or "good job", although welcome, will rarely provide enough information for the interpreter to make any changes or improvements to their interpretation. Rather, specific types of feedback in areas agreed upon by team members prior to beginning the interpreting assignment, can be meaningfully provided.

10.1 Scheduling

Professional Teaming Skills also include techniques related to scheduling. The team member responsible for scheduling interpreters must take into account the interpreter's abilities in interpersonal relations, linguistic content, topical content and endurance, as well as additional assignments which the interpreter may be faced with in the same timeframe.

10.2 Coordination

Coordination of interpreting events is an additional skill relating to team interpreting. The coordinator of interpreting service for a conference or other one-time event must take into account all of the factors just described and have a keen sense of awareness of the topical and cultural requirement of each text to be interpreted as well as information and insight into the needs of the audience. Also related to this aspect will be an awareness of the physical environment and lighting, sight lines, auditory constraints and other factors which will affect the interpreter's performance.

11. PROFESSIONAL SKILLS

There are at least nine factors which must be addressed within the area of professional skills.

11.1 Microphone Usage

Microphone usage information can range from techniques related to using a microphone effectively to stage presence and can best be taught through a series of exercises using a variety of microphones and amplification systems. These systems can range from large PA systems to FM, closed systems, or closed loop systems. Effective microphone techniques can only be taught through practical application.

11.2 Posture

How an interpreter stands during an interpreting assignment can subtly affect the message. Students of interpretation must be given information on non-verbal communication which will demonstrate the effect of posture on the interpretation. Posture will have at least two main effects on interpreting. One is the level of breath control that can be afforded by good posture. The second is the non-verbal aspects of demeanor that are conveyed by posture. For example, standing in a "careless" looking pose will affect, and possibly skew the message.

11.3 Movement

How to enter and exit a platform situation is important and rarely explicitly taught in interpreter education programs. Specific instruction and practice should be provided so that interpreters will know how to successfully and safely walk onto a stage, trade places with another interpreter, and safely exit the stage. There are a number of things which need to be considered and brought to the awareness of student interpreters. They include microphone wires, lighting, maneuvering steps and maneuvering in semi-lighted conditions.

11.4 Lighting

Working on a platform or other situation where additional lighting has been provided so that the interpreter maybe more clearly seen provides additional considerations. These situations often lead to diminished visibility for the interpreter. This is so because the lighting is often arranged in such a way as to shine directly on the interpreter. This will affect the delivery of the interpretation and the ability to see and use the backup interpreter.

11.5 Attire

Student Interpreters need to be carefully guided in terms of which type of professional attire is appropriate for various situations. The main goal of this unit will be to see that the interpreter's attire blends in with that of the other members of the communicative event. It is important that the interpreter not be over or under dressed for the situation. In appropriate attire can bring undue attention to the interpreter.

11.6 Safety

Issues related to safety include those related to occupational hazards. These include repetitive motion injuries, standing on raised platforms, entering and exiting semilighted situations, using electrical devices, such as microphones and lighting. Related to safety issues, are also issues which concern sexual harassment. Interpreters should be given information which relates to social boundaries which must be observed by interpreters and those who employ their services. For example, is it questions such as: "Is it ever alright for the consumer to touch the interpreter?". should be raised and addressed in safe environments.

11.7 Logistics

Discussions of logistics should include the ability to find the building where the interpreting assignment is held, the appropriate room, and the specific consumers who are requesting the service. Once in the appropriate environment, the interpreter may also be responsible for arranging seating, lighting and other environmental factors. Important in all of these are the interpersonal skills which the interpreter must use in order to accomplish seating arrangements and other interpersonal factors which relates directly to the arrangement of the communication environment.

11.8 Preparation

Professional interpreters need information and instruction on how to prepare for specific interpreting assignments. This may mean contacting people involved in the communicative event to obtain copies of the speeches which will be delivered. It may also include library work or other research work such as consulting on-line sources, which will prepare the interpreter with information to develop sufficient background knowledge to handle the topic of the interpreted event.

11.9 On site Management of Interpersonal Situations and Discretion

One aspect of professional skills relates directly to the application of the Registry of Interpreters for the Deaf Inc. Code of Ethics. All sign language interpreters who are members of the RID are obligated to adhere to all eight tenets of the Code of Ethics. One of these tenets is the use of discretion while on an interpreting assignment. Every interpreting situation requires that the interpreter function not only as a conveyer of linguistic information, but also as a mediator of cultural information. This requires high levels of effective interpersonal skills and strong discretionary skills. Of paramount importance are the issues of confidentiality and discretion in accepting assignments which are difficult or which develop into unacceptable situations.

In conclusion, a careful sequencing and delineation of the component tasks within the interpreter education program is the first step in developing a reasoned approach to student typology and materials development. When these three variables are considered, greater program cohesion and reliability results.

References

Cokely, D. (1992). Sign Language Interpreters and Interpreting. Burtonsville, MD. Linstok Press

Gile, D. (1995). Basic Concepts and Models for Intermediate and Translator Training. Philadelphia: Benjamins.

Lambert, S. (1992). Aptitude testing for simultaneous interpretation at the University of Ottawa. In L. Gran and J. Dodds eds. *The interpreter's newsletter*. University of Trieste: Trieste.

Longley, P. (1989). The Use of Aptitude Testing in the selection of students for conference interpretation training. In L. Gran, and J. Dodds, (Eds). *The Theoretical and practical aspects of teaching conference interpretation*. Campanotto Editore, Udine.

Roberts, R. (1993). Student Competencies in Interpreting: Defining, teaching and evaluation. In E.A. Winston (Ed.), Student competencies: Defining, teaching and evaluating. (pp. 3-16). Conference of Interpreter Trainers.

Schweda-Nicholson, N. (1989). Interpreter Evaluation: The whole does not equal the sum of its parts. In Wilcox, ed. *New Dimensions in Interpreter Education: Evaluation and critique. Proceedings of the Seventh National Conference of Interpreter Trainers Convention*.

Gallaudet University (1995). Master of Arts in Interpretation Curriculum, Department of ASL, Linguistics and Interpretation. Washington, DC.

GIST RECALL AS AN APTITUDE TEST IN INTERPRETER TRAINING

Jorma Tommola
University of Turku

Keywords: Interpreter Aptitude, Gist Recall, Summary Tests

Abstract: Selection tests for interpreter training often involve tests of recall for the central content of a prose passage. Motivation for these tests comes from the demands of both consecutive and simultaneous interpreting. In the first Finnish post-MA course on conference interpreting at the Centre for Translation and Interpreting, University of Turku, gist recall was employed as an eliminatory screening criterion. This paper describes the test given to candidates with English as their B language, and analyses the summaries against a propositional representation of the original. While numerical scoring of performances is desirable for reasons of objectivity, it is pointed out that the quantitative evaluation of summaries may contain problems that in fact make the measure less reliable than a criterion-referenced, intuitive judgement by a board of experienced examiners. It is also pointed out that it may not be obvious to candidates what is meant by a 'summary', and that one could speak of a 'recall and reproduction task', whose assessment could be based on a propositionalisation.

1. INTRODUCTION

1.1 Prose recall, CI and SI

Among the various characteristics that interpreter aptitude tests attempt to measure, a central one is the ability to analyse the source message into its semantic elements, to create a representation of the content, and to retain this representation in memory so that it can be rendered in the target language. Thus, selection tests for interpreters normally include tests of recall of prose passages.

Motivation for prose recall tests comes chiefly from consecutive interpreting (CI). In this mode, processing the message to the level of a hierarchical content representation is necessary because of the limited capacity of the working memory. The incoming information must be analysed and restructured into more abstract units of meaning which subsume the lower-level details; the logical relations between these units must be established, and the representation must be stored away in order for new material to be processed. Subsequent reconstruction in the target language relies on the network of these hierarchically organised higher-level elements. Thus, in the terms of van Dijk and Kintsch (Ref. 1), an essential prerequisite of successful consecutive interpretation is the analysis of the source text into a *text base* (set of basic propositions that the text contains), and the processing of this text base into a *macrostructure* (a representation of the hierarchical structure of content). An important aim of screening tests is to separate candidates who perform this kind of macroprocessing naturally from those who may have a habitual 'lazy macroprocessing strategy' (Ref. 2); i.e., adequate comprehension but absence of the effortful processing required for mastering larger entities of information.

In the simultaneous mode (SI), the interpreter follows temporally relatively close in the wake of the source text speaker, and the memory performance required is different from that of the consecutive interpreter. There is less emphasis on the memory for extensive segments of text, and more on high-speed target-text lexical access and syntactic formulation processes. However, the requirement to construct a meaning representation of the incoming message is in no way different from that of the consecutive situation. The language processing by a simultaneous interpreter cannot be any 'shallower' than that of the consecutive interpreter, and the simultaneous mode cannot in any sense be viewed as a simple 'transcoding' of the surface structure of the source text into the target text. This is essentially what Seleskovitch means when she speaks of the necessity to produce the 'sense' of the source message in SI (Ref. 3). When the simultaneous interpreter is working with typologically widely different languages, for example English and Finnish, extensive reorganisations of the semantic elements of the discourse may be necessary during the formulation of the target text. In this case, the interpreter must process the message to the level of macrostructural elements in order to be able to hold relatively long stretches of content in abeyance while waiting for the necessary source-text information. Furthermore, the ability to produce a simultaneous interpretation that involves a certain level of macrostructural condensing and summarising, instead of 'saying it all', is clearly a feature that distinguishes experts from novices (cf. Ref. 4). Thus, identifying the gist of the target discourse is important also in simultaneous interpreting. This is one reason why several interpreter training programmes begin with consecutive training and proceed to SI only later. For the same reason, recall tests seem to have some construct validity in the selection of trainees even from the point of view of simultaneous interpretation.

1.2 Previous studies

Recall for an orally presented discourse is mentioned as a desirable component in aptitude test batteries by several papers on the screening of interpreter trainees. Moser-Mercer employed oral recall for the main ideas in aurally presented passages on a monolingual course during which candidates were assessed for interpretation ability; a peculiarity of this exercise was that comprehension was accompanied by a subsidiary distractor task of counting numbers while listening (Ref. 5). Student performance in this and various other exercises reliably predicted success in final examinations on interpreting. Gerver *et al.* (Ref. 6) and Longley (Ref. 7) studied gist recall tests as a part of an extensive set of aptitude tests, and also found that one of their text memory tests, in which candidates wrote summaries of tape-recorded passages, differentiated between trainees who were later successful or unsuccessful in course-final interpreting exams. However, the finding of Gerver *et al.* seemed to be (p. 734) that the texts were not entirely suitable for the purpose, since they required the candidate to remember a large number of equally important points of content, rather than emphasising the ability to evaluate what is important and less important, and to extract gist information. Consequently, there was a floor effect on the scores. Bowen & Bowen employed a summarisation task of a written passage on which candidates had first performed a sight translation task (Ref. 8). They provide examples of summaries but do not discuss the efficiency of the summary task or the methods of assessing the performance of candidates. Gringiani (Ref. 9) and Dodds (Ref. 10) also mention a written summary task as a useful component of aptitude tests; however, according to Dodds, (p. 19), one problem with the summary test is that candidates may not know or may have widely different ideas about what a summary is. Lambert (Ref. 11) views the ability to grasp and convey the essential meaning of discourse as a quality that aptitude tests should look for, and describes an aptitude test battery that makes use of memory tests of the type developed by Wechsler (Ref. 12). The standardised texts, which were also used in translation to enable A to B and B to A testing, however emphasise total recall of the relatively limited amount of information contained in the test passages.

Prose recall and gist identification tests thus seem to have some theoretical construct validity, and seem to be a generally accepted type of test for screening interpreter trainees. In addition, they measure a prerequisite skill rather than, as some aptitude tests have tended to do, the actual skills of interpreting which should be the result of instruction. Interestingly, however, the papers dealing with these tests rarely touch upon the methods of assessment of the recalls or summaries produced, and seem to take it for granted that their scoring is unproblematic. With the exception of Dodds (Ref. 10), it is usually also taken more or less for granted that both the testers and the testees have a clear and uniform idea what is meant by an effective summary. Both of these points or problems became evident in the screening test for a recent Finnish post-MA-level course for conference interpreters. The following section describes the gist recall test employed and offers some comments on these topics.

2. GIST RECALL IN A SCREENING TEST FOR A FINNISH CONFERENCE INTERPRETER COURSE

The Centre for Translation and Interpreting at the University of Turku was chosen by the Finnish Ministry of Education to organise the first post-MA course for conference interpreters. The course, which started in March and will last until December 1995, is to a certain extent geared towards the needs of EU institutions, where there is at present an acute demand for interpreters between Finnish and the major EU languages. The candidates were required to have two foreign languages (from the set of English, French, and German) in addition to their A language (Finnish). In the selection tests that were organised for 35 candidates screened on the basis of written applications, performance on three gist recall tasks (B to A, A to B, and C to A) was used as the first eliminatory selection criterion before the candidates could proceed to interview tests. The observations on the performance by candidates and on the assessment of gist recall presented here are based on one of these tests, the B to A recall test for the 9 candidates who had English as their B language.

The text used as the basis for the test was a mixture of narrative, expository and descriptive elements. The topic was 'Unemployment in Britain before the Second World War'; the text discussed a government retraining scheme for unemployed youth which started in the thirties. The text had a relatively clear logical structure, with an introduction, a description of recruitment and conditions at the retraining camps, an explanation of their aims, and description of why the scheme failed and, how, in the end, the problem of unemployment was solved by the outbreak of war. In terms of propositional meaning elements (cf. Ref. 13), the text base consisted of 153 propositions divided into 24 sentences. The length of the text was 480 words.

Candidates were tested individually, and their performances were videotaped. The text was spoken (not read) to each candidate by a native speaker of English, and the instructions simply asked the candidate to summarise the text immediately into Finnish. The gist recall task thus also involved translation. The duration of the text was about 4 minutes. Notetaking was allowed. Thus, with regard to both the nature of the task and the text itself, the test bears some resemblance to the procedure employed by Gerver *et al.* (Ref. 6), and differs from the memory task emphasising verbatim recall of Wechsler-type short passages used, for example, by Arjona-Tseng (Ref. 14) or Lambert (Ref. 11). To some extent the test situation resembled the selection testing carried out by the European Commission for their interpreter trainees. A jury of 8 examiners[1] (professional interpreters with Finnish as their native language, two interpreters from EU institutions with some knowledge of Finnish, and members of staff from the Centre for Translation and Interpreting) assessed the summaries, and, on the basis of the three summary tasks, arrived at a consensus about whether or not the candidate was allowed to continue with the rest of the tests.

3. OBSERVATIONS ON THE TEST AND THE PERFORMANCE BY CANDIDATES

3.1 Types of summary

For the purposes of this paper, the 9 videotaped summaries produced by candidates with English as their B language were transcribed and examined in detail against the propositional elements and content structure of the source text. The aim was simply to find out, with a close analysis of the performances, how candidates coped with the recall and reproduction task, and what kinds of strategies were typical of good and less effective performances. The number of summaries is small, but even this limited material may allow some generalisations to be made.

Table 1 gives the first 7 sentences totalling 142 words (column 1 on the left), the 50 propositional text-base units (column 2) contained in these sentences, together with an intermediate macrostructure of the extract (column 3), expressed in key words. In principle,what the candidate in a recall task must first extract from the surface text is a representation similar to the base list in column 2. Although the propositional elements have here been given in English, this does not imply a hypothesis about the nature of the base list, which could in reality be represented in some language-independent format. It is clear that such a growing number of meaning elements cannot be handled by the working memory as a list; therefore, the candidate must immediately start imposing on the material an internal structure, a macrolevel representation, through the application of various strategies such as the deletion of information deemed secondary, or the construction and generalisation of new, higher-level propositions from the basic elements. The macrolevel representations can be constructed on various levels of abstraction, and it is these higher-level propositions that are likely to be remembered and produced in the recall task. They will also guide the reproduction of details subsumed under the higher-level units.

Columns 4, 5, and 6 of Table 1 contain instances of three distinct types of summary performance which could be identified in the material. The original Finnish summaries have here been back-translated as closely as possible. Unfilled hesitation pauses are marked with '...', and filled hesitations with 'xx'.

Summary 1 condenses the original information into high-level generalisations, from which the listener gets the barest essentials. All elaborations have been omitted. In spite of this, the impression one gets from this performance is that the summary clearly displays an effective capacity for 'on-line' identification of main points, and is thus a distinct indication of interpreting aptitude. Interestingly, the jury had initial problems with this summary, which was considered by some members to be too uninformative to indicate aptitude.

Summary 2 also seems indicative of efficient macroprocessing. The candidate has judged the 8 first propositions of sentence 1 and the adversative logical relation between its clauses to be irrelevant - a decision which reduces the informativity of the introduction but can nevertheless be defended. The typical feature of this summary is, in addition to its logical and even delivery, the inclusion, at relevant points, of material from further down the text (e.g. the notion of reintegrating the unemployed into society, which is a skillfully constructed macroproposition from information distributed over two sentences beyond this extract). The information from sentences 6 and 7, which mark a topic shift in the text, has been deleted, but appears later in the summary (again not shown here) as an efficient macro-level description of camp conditions. Summary 2 is no less effectively constructed than Summary 1, and may in fact be more informative.

Summary 3 is an example of a typical performance which produces a considerable number of the main ideas but which, however, seems to lack some of the coherence displayed by the first two examples. The strategy behind this performance could perhaps be characterised as a 'list' strategy, by which some detailed text-base propositions and macropropositions tend to follow each other linked with '*and*'. The order of information presented also seems to lack coherence (e.g. the inclusion of the information on camp conditions towards the very beginning of the summary). Yet, Summary 3, just as the other two, manages to include a well-constructed macroproposition after the introductory remark (how the British attempted to tackle the problem) - an organisational device which in fact is lacking from the original. Similarly, all of the three candidates decided that the information in sentence (4) was peripheral enough to be deleted.

If one had to decide which of these summaries indicated interpreter aptitude and trainability, the first choice would probably fall on 1 and 2. This is in fact the conclusion the panel of judges came to. A general observation from the 9 summaries analysed in detail for this presentation, and the remaining 26 performances as a whole, is that a large number of candidates - all of them MA-level university graduates - had problems in creating a gist representation adequate enough to enable them to render the main points of the passage and to link them together with the necessary indicators of logical relations. Thus, even though the text is not particularly difficult, and its internal organisation is relatively clear, the task is a difficult one. One is tempted to hypothesise that in several cases a large portion of the candidate's attentional capacity went into excessive and counterproductive note-taking, inspired perhaps by the pressure of the test situation. Another general observation from the summaries is that the clarity of delivery - an important factor in the screening situation - was correlated with the clarity of the gist representation the candidate has been able to extract. Efficient representations tended to be presented efficiently.

TABLE 1. Extract of the source text and three summaries

Text	Propositional text base	Key points in an intermediate macrostructure	Summary 1	Summary 2	Summary 3
			(backtranslations from Finnish)		
(1) We tend to think that unemployment is a problem that has never been so crucial in Europe before now, but in the run-up to the Second World War in Britain, it was a huge problem -	1 TEND-TO WE 2 2 THINK WE 3 3 ISA UNEMPLOYMENT PROBLEM 4 DEGREE-OF PROBLEM CRUCIAL 5 MOD MORE-THAN-EVER CRUCIAL 6 IN PROBLEM EUROPE 7 TIME 3 NOW 8 BUT 3 7 9 ISA UNEMPLOYMENT PROBLEM 10 MOD HUGE PROBLEM 11 IN PROBLEM BRITAIN 12 TIME 9 JUST-BEFORE WW2	**TOPIC:** UNEMPLOYMENT HUGE PROBLEM IN BRITAIN JUST BEFORE WW2	Unemployment is not merely today's problem. but people have suffered from it before. In Britain, before the 2nd Word War...	In the 1930's ... before the 2nd World War ... unemployment was ... a very big problem in Britain.	... xx Un-employment was a big problem in England at the end of the 30's ...
(2) a problem, one could say, that was solved by the outbreak of war.	1 BE-ABLE-TO $ 2 2 SAY $ 3 4 ISA UNEMPLOYMENT PROBLEM 5 SOLVE OUTBREAK PROBLEM 6 OF OUTBREAK WAR				and this un-employment... it did not get better until the outbreak of war
(3) In the late 1930s, unemployed young people were invited by labour exchanges to free viewings of a Ministry of Labour propaganda film, 'On the Way to Work'.	1 TIME 2 IN-THE-LATE-THIRTIES 2 INVITE LAB-EXCHANGE PEOPLE 3 MOD UNEMPLOYED PEOPLE 4 MOD YOUNG PEOPLE 5 IN-ORDER-TO 2 6 6 VIEW PEOPLE FILM1 7 MOD PROPAGANDA FILM1 9 LABEL FILM1 'ON-THE-WAY...'	**ATTEMPT TO SOLVE PROBLEM:** GOVERNMENT RETRAINING SCHEME (CAMPS) FOR UNEMPLOYED YOUTH LATE 30'S	... they... attempted to solve the problem by xx work-camps. Young men were invited ... xx with the help of propaganda films to work camps	Britain ... the Government of Britain xx tried to reduce un-employment by organising xx re-training camps to unemployed youngsters.	and ...the British xx fought xx against the un-employment ... so that they founded ... camps for the youngsters ... and... in these.. camps... youth c...

(4) It was one of the first official films to try to sell a government retraining scheme to unemployed youth.	1 ONE-OF FILM1 FILM2 2 MOD FIRST FILM2 3 NUMBER-OF FILM2 MANY 4 MOD OFFICIAL FILM2 5 TRY FILM2 6 6 SELL FILM2 RETR-SCHEME 7 MOD GOV'T RETR-SCHEME 8 FOR RETR-SCHEME YOUTH 9 MOD UNEMPLOYED YOUTH	ADVERTISED BY PROPAGANDA FILMS			these camps were military ... had a military discipline ... and ... they had to work ... and were paid xx only a few shillings... and a packet of cigarrettes ...
(5) The film showed youngsters plucked from dole queues enjoying themselves by day chopping down trees and damming streams, then relaxing by night playing darts and singing round a campfire in boy-scout style.	1 SHOW FILM 2 2 ENJOY YOUNGSTER 3 TIME 2 BY-DAY 4 PLUCK $ YOUNGSTER 5 FROM YOUNGSTER DOLEQUEUE 6 CHOP-DOWN YOUNGSTERTREE 7 DAM YOUNGSTER STREAM 8 AFTER 9 2 9 RELAX YOUNGSTER 10 TIME 8 BY-NIGHT 11 PLAY YOUNGSTER DARTS 12 SING YOUNGSTER 13 PLACE 12 CAMPFIRE 14 MANNER 12 BOY-SCOUT	FILMS SHOW DAILY WORK PLEASANT OUTDOOR JOBS AT NIGHT RELAX BOY-SCOUT STYLE	...the films displayed work xx as very attractive and a very xx beautiful picture was also painted of ... free time in the camps.	... xx Retraining camps were advertised ... with a film ... which showed how the unemployed have a nice time in the camp ... and get... and have a chance to reintegrate into society.	and youngsters were invited to these camps with the help of...propaganda films in which camps were shown as nice places to spend time in... and... where you could play darts and other games ...
(6) The reality was very different.	1 MOD DIFFERENT REALITY	**CAMP REALITY** DIFFERENT	The reality was, however, something else ...	These camps were very unpopular among the youth because of this ... hard discipline and ... xx concentration-camp ... conditions.	
(7) The camps were known to the inmates as 'slave camps'.	1 KNOW-AS INMATE RETR-CAMP 2 REF CAMP SLAVE-CAMP	'SLAVE CAMPS'			

3.2 Assessment of summaries

Judging the detailed propositions and the main points suggested in Table 1 against the three types of summary shown in Table 2, it is clear that it will be difficult to predict the strategic decisions people will make in a summarisation task. One tends to feel intuitively that the jury's decision was correct, but, as is pointed out in several papers on interpreter aptitude testing, it would be desirable to establish objective criteria for the adequacy of gist recall.

In the test described in this paper, an attempt to do this was made by constructing an assessment sheet where each of the three summarisation tasks was to be evaluated on three parameters: (1) adequacy of summarisation, including the presence of main points, and the coherence of causal and temporal relations (0 to 4 points); (2) language ability, including correctness of grammar, richness of vocabulary, pronunciation, articulation and intonation (0-4 points); and (3) public speaking skills (0-2 points). The total score was thus 10 points for each of the three tasks.

In assessing the adequacy of the summaries, some members of the jury did use the scoring sheet; others preferred to take notes of the summaries produced by the candidates and used them intuitively to evaluate the summary against the original. Some members of the panel felt that an attempt to quantify the summarisation performances with the scoring sheet was meaningless. In the end, the panel discussed their subjective views on the performances, and arrived at a general agreement about whether or not the total performance was above a certain intuitive standard. Thus, the test was used as a criterion-referenced measure of aptitude, and the candidates who reached what the panel thought was an adequate level, continued to further testing. The panel did establish individual scores for candidates, but this happened *post hoc*, possibly in anticipation of queries and appeals.

Since objectivity and quantitative assessment are often called for in aptitude testing (e.g. Ref. 5), one might hypothesise that a propositional representation of main points and detailed content could be used as the basis of numerical assessment of summarisation ability. The 'propositional points' mentioned by Gerver *et al.* (Ref. 6, p. 727), for example, suggest that methods of this kind have been successfully applied, but explicit accounts of how such scoring methods have functioned cannot usually be found in the papers where gist recall tasks have been discussed as part of interpreter aptitude testing.

The propositional analysis exemplified in column 2 of Table 1 above has been successfully applied in, for example, studies of summarisation and recall of fairly brief passages of technical prose (Ref. 13), and in experiments on simultaneous interpreting (Ref. 15). In a 'liberal' scoring of a memory test (i.e., scoring which does not presuppose verbatim recall), a recall protocol is scored by asking 'What propositions must the subject have had in mind in order to say what was in the protocol?' However, the gist recall task presents a problematic situation. Looking at the first sentence of Summary 1, for example, against the 12 propositions of the original sentence, one sees immediately that it is not possible in any objective sense to draw inferences about which details the subject had had in mind, whether the subject had actually forgotten to include some details, or whether the subject remembered everything but had decided to generalise extensively. Similarly, even if one wanted to assess the summary against the elements of meaning suggested in the intermediate macrostructure, there is no objective way of deciding which of those points are actually entailed by the effective generalisation produced by the candidate. Similarly, assessing the first sentence of Summary 2 against the text base and gist representations would mean penalising the subject for not including the first 8 propositions in the summary in some form, when this candidate may again simply have used a different selection strategy, which does not necessarily make the resulting summary any less effective.

Thus, the quantification of gist recall performances of extensive passages will remain problematic simply because the recalls are by definition constructive and generalising macropropositions, and the ways in which lower-level propositions are entailed in these higher-level statements may be difficult to count reliably and objectively. One solution might be to accept the intuitive, criterion-referenced pass-fail judgement by an experienced panel as reliable; this is actually what happened in the screening test described above. If the members of the panel have agreed on what the details and main points of the test passage are, and have established an agreement on the minimum of points of information that must be present, the danger of subjectivity in the pass-fail judgement does not seem particularly great. If adequate records are kept of the performance and the discussion, possible queries and appeals by candidates can be dealt with by going through their performances in detail afterwards.

Another problem in the test described above seemed to be that, just as Dodds points out (Ref. 10, p. 19), it is not clear whether all candidates, or assessors for that matter, understand the idea of a summary in a similar way. One way to get around this problem might be to give the candidates instructions that merely speak of a 'recall and reproduction task'. This instruction brings the task in the aptitude test slightly closer to the actual situation of consecutive interpretation, and will perhaps encourage the inclusion of whatever detail the candidate can reconstruct. When jurors have a clear agreement on the minimum amount of information that must be present for the candidate to reach an acceptable level, a numerical assessment of the performances, based on a scoring sheet developed from the propositional representation, might work.

4. CONCLUDING REMARKS

This paper has briefly reviewed studies which have dealt with the use of gist recall tasks in aptitude tests for interpreters, and has discussed some points of their construct validity. Summaries produced in a screening test for a Finnish conference interpreting course were examined in order to show that summarisation strategies vary extensively, and although it is intuitively quite possible to identify summaries which seem to be adequate representations of main ideas and which are produced efficiently in the test situation, the objectivisation of these intuitions is not an easy task. Consequently, one possibility would be to rely on the collective judgement of a panel consisting of members who have extensive experience in such analysis of texts and who have established among themselves common criteria for the assessment of summaries of a text. Another possibility would be to give task instructions that would more clearly indicate what is expected, and, in a sense, to expect from candidates a performance that comes closer to actual consecutive interpretation.

5. NOTES

[1]The writer was not a member of the panel.

6. REFERENCES

1 van Dijk, Teun A. & Kintsch, Walter. *Strategies of Discourse Comprehension.* New York: Academic Press, 1983.

2 Lorch, Robert F. Jr. & Lorch Pugzles, Elisabeth. "Effects of Organizational Signals on Free Recall of Expository Text," Manuscript, 1994.

3 Seleskovitch, Danica. "Interpretation, a Psychological Approach to Translating," in Brislin, D.W. ed. *Translation: Applications and Research.* New York: Gardner Press 1976, p. 92-16.

4 Sunnari, Marianna. "Processing Strategies in Simultaneous Interpreting: Experts vs. Novices," paper presented at the 36th Annual Conference of the American Translators Association, 1995.

5 Moser-Mercer, Barbara. "Aptitude Testing for Conference Interpreting: Why, When and How," in Lambert, Sylvie & Moser-Mercer, Barbara, eds. *Bridging the Gap: Empirical Research in Simultaneous Interpretation.* Amsterdam and Philadelphia: John Benjamins 1994, p. 57-68.

6 Gerver, David, Longley, Patricia E., Long, John & Lambert, Sylvie. "Selection Tests for Trainee Conference Interpreters." *Meta*, Vol. XXXIV, No 4, 724-735, 1989.

7 Longley, Patricia. "The Use of Aptitude Testing in the Selection of Students for Conference Interpretation Training," in Laura Gran & John Dodds, eds. *The Theoretical and Practical Aspects of Teaching Conference Interpretation.* Udine: Campanotto, 1989, p. 105-108.

8 Bowen, David & Bowen Margareta. "Aptitude for Interpreting," in Laura Gran & John Dodds, eds. *The Theoretical and Practical Aspects of Teaching Conference Interpretation.* Udine: Campanotto, 1989, p. 109-125.

9 Gringiani, Angela. "Reliability of Aptitude Testing: a Preliminary Study," in Laura Gran & Christopher Taylor, eds. *Aspects of Applied and Experimental Research on Conference Interpretation,* Udine: Campanotto, 1990, p. 42-53.

10 Dodds, John M. "On the Aptitude of Aptitude Testing." *The Interpreters' Newsletter* No. 3, 17-22, 1990.

11 Lambert, Sylvie. "Aptitude Testing for Simultaneous Interpretation at the University of Ottawa." *Meta* XXXVI, No. 4, 586-594, 1991.

12 Wechsler, D. "A standardised memory scale for clinical use." *Journal of Psychology* Vol. 19, No 1, 87-95, 1945.

13 Bovair, Susan & Kieras, David E. "A Guide to Propositional Analysis for Research on Technical Prose," in B.K. Britton & J.B. Black, eds. *Understanding Expository Text.* Hillsdale, N.J.: Lawrence Erlbaum 1985, p. 315-362.

14 Arjona-Tseng, Etilvia, "A Psychometric Approach to the Selection of Translation and Interpreting Students in Taiwan," in Lambert, Sylvie & Moser-Mercer, Barbara, eds. *Bridging the Gap. Empirical Research in Simultaneous Interpretation.* Amsterdam and Philadelphia: John Benjamins 1994, p. 69-86.

15 Tommola, Jorma. "Experiments on Interpreting: Which Dependent Variable?" Paper presented at the 1st International Congress on Translation and Interpreting, Universidad de Las Palmas de Gran Canaria, 1994.

PAST PRESIDENTS

1993-95	Edith F. Losa
1991-93	Leslie Willson
1989-91	Deanna L. Hammond
1987-89	Karl Kummer
1985-87	Patricia E. Newman
1983-85	Virginia Eva Berry
1981-83	Ben Teague
1979-81	Thomas R. Bauman
1977-79	Josephine Thornton
1975-77	Roy Tinsley
1973-75	William I. Bertsche
1971-73	Thomas Wilds
1969-71	William I. Bertsche (completed Moynihan's term)
1969	Daniel Moynihan (resigned after a few months)
1967-69	Boris Anzlowar
1965-67	Henry Fischbach
1963-65	Kurt Gingold
1960-63	Alexander Gode

CORPORATE MEMBERS

(as of 9/20/95)

A E Translations, Inc.

A L Madrid & Associates

A & M Logos International, Inc.

Academy of Legal and Technical Translations, Ltd.

Academy Translations

Accent Typography & Translation

Accento, The Language Company

Access Language Services

Accura International Translations

Accurapid Translation Services, Inc.

Accurate Translators of Florida

Accuword International, Inc./inlingua

ACE Translation Center

Ad Hoc Translation

Adams Technical Translations, ACS Inc.

Adaptive Language Resources, Inc.

AD-EX WORLDWIDE

Agnew Tech-II

Albors and Associates, Inc.

Alexandria Translations

Allen Translation Service

Alphatrad

Ambassador Translating, Inc.

America Translating Services

American Translation Company

Amway Corporation

ASIST Translation Services, Inc.

Atlanta Committee for the Olympic Games

AT&T Network Systems

Attache International, Inc.

Babel Translation Services

Benemann Translation Center - BTC

Bergen Language Institute, Inc.

Berkeley Scientific Translation Service, Inc.

Bilingual Services

Border Lingustic Services

Bowne Translation Services

BRADSON Corporation

BRAHLER ICS, Inc.

Bureau of Translation Services

Burg Translation Bureau

Business Benefit, Inc.

C. P. Language Institute

CACI Language Center

Caere Corporation

Calvin International Communications, Inc.

Cambridge Translation Resources

Canadian Union of Professional & Technical Employees

Carolina Polyglot, Inc.

Cato Research, Ltd.

Center for Professional Advancement

Cherry Valley Language Technology

Chicago Multi-Lingua® Graphics, Inc.

Cial-Lingua Service International

Ciba Corning Diagnostics Corporation

CinciLingua, Inc.

ComNet International

COMPUTAIN, Inc.

Contact International

Continental Communications Agency

Copper Translation Service

Corporate Language Services, Inc.

The Corporate Word, Inc.

Correo Argentino

Cosmopolitan Translation Bureau

Crimson Language Services
Cybertrans International, Inc.

Digital Publishing, Inc.
Diplomatic Language Services, Inc.
Direct Language Communications
Diversified Language Institute
Document Processing Center, Inc.
Documents International, Inc.
Dovali Translations, Inc.
DTS Language Services, Inc.
Dynamic Language Center, Ltd.

East-West Concepts, Inc.
EBON Research Systems
Egamco, Inc.
ELP LLC
Encompass Translations, Inc.
EOP Translation Service
Eriksen Translations, Inc.
Excelcom/Translex
Executive Linguist Agency
Executive Office for Immigration Review
Expert Translators
Federal News Services
Ferman & Company, Inc.
The Final Word, The Authority on Language Services, Inc.
FLS, Inc.
Foreign Language Center

G T Translation Services, Inc.
Galaxy Systems, Inc.
GARJAK International, Inc.
Geonexus Communications
Gerard Linguistic Services
The Global Institute of Languages and Culture, Inc.
Global Language Management, Inc.
Global Solutions, Inc.
Global Translation Services, Inc.
Global Translations & Interpreters Services, Inc.
GlobalDoc, Inc.
Globalink, Inc.
GLS International

Hartmann International Services
Harvard Translations
Hightech Passport Limited
Hispanic Service Center
Hitachi Micro Systems, Inc.
Honda R&D North America, Inc.
Housing Authority of the City of Los Angeles-Resident Relations

i. b. d., Ltd.
Iberia Language Services
IBM Software Manufacturing Solutions
ICN Language Services
IDC, Inc.
IDEM Translation
IDOC
Inlingua
Inlingua International Services
Institut für Fremdsprachen und Auslandskunde
INTERCONTACT - PERU
Intergraph Corporation
Interléxico
InterNation Communication, Inc.
International Business Communications Corporation
International Communication by Design, Inc.
International Communications, Inc.
International Contact, Inc.
International Effectiveness Centers

International Language Engineering Corporation
International Language Services, Inc.
International Technology Network, Ltd.
International Transition Services
International Translation and Publishing
International Word, Inc.
Interpreters International & Translations
Interpreting Services International, Inc.
Intertech Translations, Ltd.
IRU International Resources Unlimited
Iverson Language Associates, Inc.

Jackson Graphics, Inc.
Japan Pacific Publications, Inc.
Japanese Language Services, Inc.
JKW International, Inc.
JLS Language Corporation
John Benjamins Publishing Company
Josef Silny & Associates, Inc.

Kien snc
Kirchman Corporation
Koei Corporation

LangTech International
Langua Translations, Inc.
The Language Center, Inc.
Language Company Translations
The Language Connection
Language Intelligence, Ltd.
Language Interface, Ltd.
The Language Lab
Language Link Corporation
Language Matters
Language Plus
The Language Service, Inc.
Language Services Associates
The Language Solution, Inc.
Languages International
Languages Unlimited
Lawyers' & Merchants' Translation Bureau, Inc.
Lingua Communications Translation Services
Lingua Media, Inc.
Lingualink Incorporated
LinguaNet, Inc.
Linguistic Consulting Enterprises, Ltd.
Linguistic Services for Professionals, Inc.
Linguistic Systems, Inc.
Logos Corporation
LOGOS Group
Logos Publishing Company
LRA Interpreters, Inc.

M^2 Limited
Magnus International Trade Services Corporation
Master Translating Services, Inc.
MasterWord Services, Inc.
Gene Mayer Associates
Ralph McElroy Translation Company
MCS Language Connection
MediaMax
Mellon Bank
Mercury Marine
Merino Language Services, Inc.
Metropolitan Interperters
& Translators Worldwide, Inc.
Michael J. Sidor & Co., Inc.
Mitaka Limited
Morgan Guaranty Trust Company
MultiLing International, Inc.
Multilingual Communications Corporation
Multilingual Computing
Multilingual Translations, Inc.

NCS Enterprises, Inc.
NIS International Services

Occidental International
Exploration & Production Company
Omega International
Omni Interpreting & Translating Network
OmniLingua®, Inc.
Oriental Communication Services, Inc.
O'Sullivan Menu Corporation

P. H. Brink International
Pachuca New York, Inc.
Pacific Interpreters, Inc.
Pacific Quest Corporation
Pacifica Corporation
Peters Translation
Planning S.A.S.
Polyglot International
Premier Translation Services, Ltd.
Professional Translating Services
ProTrans, Inc.
PSC, Inc.

Quantum, Inc.
Quark, Inc.

Rapport International
Reliable Renditions
Rennert Bilingual Translations
Routledge, Inc.
Russian and East European
Partnerships, Inc.

Sally Low & Associates
SAP AG
SARJAM Communications, Ltd.
Schreiber Translations, Inc.
SH3, Inc.
Shoreline Translations, Inc.
SpaceLabs Medical
Spectrum Multilanguage Communications
Star Language Center LLC
STATISTICA, Inc.
Suzuki Myers & Associates, Ltd.
Sykes Enterprises, Inc.

Techlingua, Inc.
Techno-Graphics & Translations, Inc.
TechTrans International, Inc.
Terra Media International
Terra Pacific Writing Corporation
TEXTnology Corporation
TIICO-Translating Interpreting
International Company
The Toin Corporation
Toward, Inc.
Trade Typographers, Inc.
transACT
Transemantics, Inc.
Transglobal Translations & Immigration
Services, Inc.
Translation Aces, Inc.
Translation Company of America, Inc.
Translation News
Translation Solutions
TRANSLEX International
Translingua, Inc.
Trans-Lingual
Transperfect
Transperfect Translations International, Inc.
Transtek Associates, Inc.

United Parcel Service
Universe Interpreters and Translators
Corporation
University Language Center, Inc.
U.S. Translation Company

Verba Communications
VOCALESIS
Wang Laboratories, Inc.
Winter Wyman Contract Services, Inc.
WKI International Communications
The Wordmill
WORDNET, Inc.
World of Reading, Ltd.

World Trade Center Portland

Y A R Communications, Inc.

ZGS Communications, Inc.
ZML Software Systems, Inc.

INSTITUTIONAL MEMBERS

(as of 9/20/95)

Academy Interpreting and Translations International

Academy of Languages Translation & Interpretation Institute

Allied Languages Cooperative

American Bureau of Professional Translators

American Institute of Physics

AN-NAHDA Educational Office

Argo Translation and Interpretation

Asia Foreign Language Institute

Associação de Tradutores Publicos E Interpretes Comerciais Do Est. Do

Binghamton University (SUNY)

BYU Translation Research Group

Caribbean Institute of Languages

Center for Applied Linguistics

Christian Science Publishing Society

The Church of Jesus Christ of LDS

Colegio Colombiano De Traductores

Community Interpreter Services

Catholic Charities/Greater Boston

Community Management Staff

DBA Eureka - Foreign College Evaluators & Translators

The French Library & Cultural Center

Gallaudet University

Georgetown University

Georgia State University

Houston Independent School District

Instituto Superior de Interpretes y Traductores, S. C.

Inter-American Air Forces Academy

International Institute of Connecticut, Inc.

Iowa Council for International Understanding

J. M. H., Interpreter Service

Japan Translation Federation, Inc.

Jhpiego Corporation

Kent State University

The Language School

Laurentian University School of Translators and Interpreters

Marygrove College

Monterey Institute of International Studies

New York University

Northern Illinois University

Purdue University

Quba Institute of Arabic and Islamic Studies

Rose-Hulman Institute of Technology

San Diego City Schools

Seattle/King County Convention & Visitors Bureau

Southwest Washington Medical Center

Summer Institute of Linguistics

Thammasat University

Translation & Critical Languages Institute

Tru Lingua Language Systems, Inc.

Tucson Unified School District Title I Program

Universidad Autonóma de Cd. Juárez

The University of Findlay

University of Hawaii

University of Idaho

University of Miami

University of Nebraska at Kearney

University of Washington

University of Washington Medical Center

Western Michigan University

World Bank

CONFERENCE SITES

1995 .... Nashville, Tennessee
1994 .... Austin, Texas
1993 .... Philadelphia, Pennsylvania
1992 .... San Diego, California
1991 .... Salt Lake City, Utah
1990 .... New Orleans, Louisiana
1989 .... Arlington, Virginia
1988 .... Seattle, Washington
1987 .... Albuquerque, New Mexico
1986 .... Cleveland, Ohio
1985 .... Miami, Florida
1984 .... New York City, New York
1983 .... Atlanta, Georgia
1982 .... Arlington, Virginia
1981 .... Dallas, Texas
1980 .... San Francisco, California
1979 .... Kansas City, Missouri
1978 .... New York City, New York
1977 .... Pittsburgh, Pennsylvania
1976 .... Palo Alto, California
1975 .... Washington, DC
1974 .... Chicago, Illinois
1973 .... New York City, New York
1972 .... Washington, DC
1971 .... Boston, Massachusetts
1970 .... New York City, New York
1969 .... Washington, DC
1968 .... New York City, New York
1967 .... Philadelphia, Pennsylvania
1966 .... New York City, New York
1965 .... Washington, DC

1964	Philadelphia, Pennsylvania
1963	New York City, New York
1962	New York City, New York
1961	New York City, New York
1960	New York City, New York

FUTURE SITES

1996	Colorado Springs, Colorado
1997	San Francisco, California
1998	Hilton Head, South Carolina

AUTHOR INDEX